THE SEARCH FOR TERRESTRIAL
INTELLIGENCE

THE SEARCH FOR TERRESTRIAL INTELLIGENCE

BF
431
.T228
1998
West

M. TAUBE and K. LEENDERS
Neurology Dept., University of Zurich, Switzerland

World Scientific
Singapore • New Jersey • London • Hong Kong

Published by

World Scientific Publishing Co. Pte. Ltd.
P O Box 128, Farrer Road, Singapore 912805
USA office: Suite 1B, 1060 Main Street, River Edge, NJ 07661
UK office: 57 Shelton Street, Covent Garden, London WC2H 9HE

Library of Congress Cataloging-in-Publication Data
Taube, M. (Mieczyslaw), 1918–
   The search for terrestrial intelligence / M. Taube & K. Leenders.
     p.   cm.
   Includes bibliographical references.
   ISBN 9810232691
   1. Intellect.   I. Leenders, K. (Klaus), 1945–   II. Title.
  BF431.T228   1997
  153--dc21   97-41554
                                                                       CIP

British Library Cataloguing-in-Publication Data
A catalogue record for this book is available from the British Library.

Copyright © 1998 by World Scientific Publishing Co. Pte. Ltd.

All rights reserved. This book, or parts thereof, may not be reproduced in any form or by any means, electronic or mechanical, including photocopying, recording or any information storage and retrieval system now known or to be invented, without written permission from the Publisher.

For photocopying of material in this volume, please pay a copying fee through the Copyright Clearance Center, Inc., 222 Rosewood Drive, Danvers, MA 01923, USA. In this case permission to photocopy is not required from the publisher.

Printed in Singapore by Uto-Print

PREFACE

Maybe it is an important question for each intelligent being: What is the trick which makes us intelligent? Here is the answer of a scientist who can be called the father of artificial intelligence:

M. Minsky (1987): "The trick (of intelligence) is that there are no tricks."

Another approach is to question the origin of the human brain and its evolution. Perhaps here we can find the answer to the very nature of human intelligence. Here is the opinion of a Nobel laureate in the field of biochemistry:

A. Szent-Gyorgyi (1937): "The human brain is not an organ of thinking, only of survival, as the teeth or claws. It is constructed so that it forces us to perceive that truth which is for us convenient. Every person whose thoughts are entirely consistent must have an extraordinary, or even pathological constitution. From these types of people emerge martyrs, apostles or scientists. In most cases they finish up in a chair: academic or electric!"

But do we need a book about human intelligence? Why must we, intelligent beings, be involved in the search for the nature of intelligence, even if we are able to use the phenomenon in a responsible way? Could this quotation from a satirical English weekly magazine, from the middle of the 19th century not be a good answer?

Punch (1857): "What is matter? Never mind! What is mind? No matter!"

Let us assume that, in spite of these doubts, there exist people who are interested in deliberating human intelligence.

One of us (M.T.) during 1958-1968 had the chair of nuclear chemistry at the University of Warsaw and since 1969, was a lecturer at the Swiss Federal Institute of Technology (ETH), Zürich, and the University of Zürich on the topics of the cosmic evolution of matter and energy, and head of a laboratory at the Swiss Federal Institute of Energy Research, Würenlingen, Switzerland. He has published about hundred scientific papers and around a dozen books.

The other (K.L.) has as an academic neurologist since 1982 been concerned with normal and abnormal brain function through direct contact with patients. Since 1988 he has directed the biomedical research program related to positron emission tomography (PET) methodology and biochemical activity of the living human and primate brain at the Paul Scherrer Institute, Switzerland. He is also attached to the University of Zürich as professor of Neurology. He has published numerous scientific papers.

Perhaps this book can be considered as a further illustration of the opinion of M.F.Perutz (concerning the book of E.Schrödinger entitled "What is life"):

M.F. Perutz (1989): "All that was true in this book was not original, and most of what was original was still, at the time of writing, recognized as untrue."

Let us conclude with two final quotations:

F. Drake (1960): "Is there intelligent life on the Earth?"

J-P. Changeux (1985): "Readers of my book must judge for themselves whether the theories I am proposing can be applied to the brain of this author!"

Zürich August 1997   M.T. and K.L.

CONTENTS

1 INTELLIGENCE AND INFORMATION ... 2
1.1 INTELLIGENCE AND TOTALITY ... 2
1.1.1 The unity of object and method ... *2*
1.1.2 Why must we begin with the beginning? ... *4*
1.1.3 Is totality and intelligence comprehensible? ... *5*
1.2 INFORMATION AND MATTER ... 7
1.2.1 Information and matter ... *7*
1.2.2 Emergence of matter and of information ... *10*
1.2.3 Emergence of information: life ... *12*
1.2.4 What is signal and what is information? ... *12*
1.2.5 Information as a representation of the real world ... *15*
1.2.6 Quality and quantity of information ... *16*
1.3 PROCESSING OF INFORMATION ... 18
1.3.1 Overabundant production and selective elimination ... *18*
1.3.2 The bonus-malus system and semantics ... *19*
1.3.3 The bonus-malus system as a basis for information processing ... *21*
1.3.4 Semantics, definition ... *23*
1.3.5 Complexity, a definition? ... *25*
1.4 THE UNIVERSALITY AND 'TERRESTRIALITY' OF INTELLIGENCE ... 25
1.4.1 The universality of intelligence ... *25*
1.4.2 The 'terrestriality' of intelligence ... *27*
1.4.3 Intelligence—different meanings ... *28*
1.5 BIBLIOGRAPHY ... 30

2 UNIVERSE CREATES LIFE ... 32
2.1 UNIVERSE IS COMPREHENSIBLE ... 32
2.1.1 The Theory (of Everything) and the theory of intelligence ... *32*
2.1.2 Universality of the laws of Nature ... *33*
2.2 THE UNIVERSE AS TOTALITY ... 36
2.2.1 Matter and energy are superunified ... *36*
2.2.2 Four universal elementary forces ... *38*
2.2.3 Four eras in the evolution of the Universe ... *39*
2.3 STARS—SOURCES OF FREE ENERGY AND OF ELEMENTS ... 40
2.3.1 The birth, life, and death of stars ... *40*
2.3.2 Emergence of and changes in chemical elements ... *42*
2.3.3 The uniqueness of the Earth ... *45*
2.4 ELEMENTARY FORCES AND PARTICLES; BEARERS OF LIFE ... 46
2.4.1 Living beings—a universal definition ... *46*
2.4.2 Elementary forces as the carriers of life ... *48*
2.4.3 Elementary particles as the carriers of life ... *52*
2.4.4 Chemical forces as the carriers of life ... *52*
2.5 THE ANTHROPIC PRINCIPLE ... 53
2.6 BIBLIOGRAPHY ... 54

3 LIFE BREEDS INTELLIGENCE 56

3.1 LIFE IS RELATED TO INFORMATION PROCESSING 56
3.1.1 Intelligence is carried by life 56
3.1.2 Outline of a terrestrial living being 57
3.1.3 The emergence of life 59
3.1.4 Terrestriality of life 61
3.1.5 Chemical phenomena as carriers of life 62
3.1.6 Life and sources of matter and energy 63
3.2 ANIMALS AND INDIRECT INFORMATION PROCESSING 66
3.2.1 Material carriers of biotic indirect information 66
3.2.2 The emergence of the nervous system 67
3.2.3 The breakthrough: the emergence of the brain 69
3.3 BRAIN, BODY AND ENVIRONMENT 71
3.3.1 Inputs of information: internal and external-detectors 71
3.3.2 Brain size and body size 72
3.3.3 The evolution of mammals' brains 73
3.3.4 Brain and body connections 75
3.3.5 Brain's limits of growth; lateralization 77
3.4 BIOTIC INFORMATION PROCESSING 78
3.4.1 Reflex, the simplest genetically wired system 78
3.4.2 Instinct: genetically complex wired system 78
3.4.3 Discent: individual learning of new information 80
3.4.4 Intelligence: creator of new information 81
3.4.5 Hierarchical organisation of the four stages 81
3.5 GENUS HOMO EMERGES 83
3.5.1 Genus Homo 83
3.5.2 The occurrence of intelligence 83
3.5.3 The occurrence of genus Homo 84
3.5.4 Neoteny, the trick? 85
3.5.5 Homo habilis and Homo erectus 87
3.5.6 Homo sapiens neanderthalensis 89
3.5.7 Homo sapiens sapiens—modern man 93
3.6 BIBLIOGRAPHY 97

4 BRAIN CARRIES INTELLIGENCE 101

4.1 THE HUMAN BRAIN AND ITS MACROSTRUCTURE 101
4.1.1 Intelligence is carried by the human brain 101
4.1.2 The human brain must have a very complex structure 101
4.1.3 The brain's architecture 102
4.1.4 Brain organs: neocortex, thalamus 103
4.1.5 Other organs: amygdala, hippocampus 105
4.1.6 Cerebellum, etc. 105
4.1.7 Brain areas: sensory and motor 109
4.1.8 Speech processing: Broca's and Wernicke's areas 110
4.2 THE MICROSTRUCTURE OF THE BRAIN: NEURONS, SYNAPSES 111
4.2.1 Neurons and information processing 111
4.2.2 Neurons—'brainy' cells 113
4.2.3 Neuron categories: rather few 114
4.2.4 The glia cells 116
4.2.5 The synapse: the neuronal junction 117
4.2.6 Microtubules play a significant role 119

4.2.7 Neurons and synapses as electrical and chemical devices 119
4.2.8 Ion channels..121
4.3 THE BIRTH AND DEATH OF NEURONS AND SYNAPSES 122
4.3.1 The human brain: from embryo to adult 122
4.3.2 The birth and death of neurons 123
4.3.3 The production and elimination of synapses 127
4.3.4 Modular and laminar organization 128
4.4 PHYSICAL PHENOMENA IN THE HUMAN BRAIN..128
4.4.1 Do we need a special kind of physics? 128
4.4.2 Neuroscientists plan an 'atlas' of the brain..129
4.4.3 The energetics of brain 131
4.4.4 Electrical activity of the brain 135
4.5 CHEMICAL PHENOMENA IN THE HUMAN BRAIN 137
4.5.1 Chemical composition seems at first sight simple 137
4.5.2 Neurotransmitters: 'brainy' molecules...139
4.5.3 The biochemical basis of the bonus-malus system 141
4.5.4 The nervous and endocrine systems 142
4.5.5 Blood-brain barrier..144
4.6 THE BRAIN AS THE PRODUCT OF GENETIC INFOR-MATION 145
4.6.1 Genetics and the brain..145
4.6.2 Genetics and epigenetics 147
4.6.3 Genes and brain normality...149
4.6.4 Brain plasticity 150
4.6.5 Genes and brain abnormality...150
4.7 BIBLIOGRAPHY...151

5 BRAIN PROCESSES INTELLIGENCE 154

5.1 THE NETWORK OF NEURONS AND SYNAPSES 154
5.1.1 Brain with a complex structure and function: carrier of intelligence..................154
5.1.2 Dimensions of the information processes 155
5.1.3 Time perception..155
5.1.4 Spatial pictures or linear structures...156
5.1.5 The brain: a permanently changing system..159
5.2 MEMORY; INFORMATION STORAGE..161
5.2.1 Memory, different types, etc...161
5.2.2 Memory overabundance and selective forgetfulness 164
5.2.3 Chunks, the units of memory..165
5.2.4 Where is memory stored? 167
5.3 EMOTIONS, DRUGS AND THE BONUS-MALUS SYSTEM..169
5.3.1 Neural functions; own body and the world outside 169
5.3.2 The bonus-malus system and emotions 170
5.3.3 Psychoactive agents, pain and the bonus-malus system 173
5.3.4 Sleep, one third of life 177
5.4 COGNITIVE OPERATIONS AND THEIR PHYSICAL BASIS..178
5.4.1 Cognitive operations could be physically measured..178
5.4.2 Visual information processing; the best known system.......................................178
5.4.3 Visual illusions as gate to cognitive activity..179
5.4.4 Complex information is initially decomposed 180
5.4.5 Synchronization of neural networks 182
5.4.6 The processing of just one word is measurable...185
5.4.7 Attention! Attention is very important 186

    5.4.8 Planning and execution *188*
  5.5 THE HUMAN BRAIN AND ITS FUNCTIONS ARE ASYMMETRIC 189
    5.5.1 Macro- and micro-asymmetry *189*
    5.5.2 Emergence of the asymmetric human brain.. *189*
    5.5.3 Lateralization—significant or not? *191*
    5.5.4 Female and male peculiarities of the brain *194*
    5.5.5 Handedness... *198*
  5.6 BIBLIOGRAPHY 199

6 TOOLS OF INTELLIGENCE 205
  6.1 DEFINITION OF HUMAN INTELLIGENCE.. 205
    6.1.1 Common sense, intellect, reason, wisdom *205*
    6.1.2 Does a theory of human intelligence exist? *205*
    6.1.3 Terrestrial or universal definition *207*
    6.1.4 Intelligence and different branches of science *209*
  6.2 INTELLIGENCE AS THE HIGHEST STAGE OF INFORMATION PROCESSING..............210
    6.2.1 From reflex to intelligence; evolution *210*
    6.2.2 Intelligence influences reflex, instinct and discent *212*
  6.3 EMOTIONS, IMPORTANT COMPONENTS OF INTELLIGENCE 212
    6.3.1 Emotions, the human predominance *212*
    6.3.2 Animal intelligence: yes or no? *215*
    6.3.3 Lost stage of evolution: Sub-intelligence or superdiscent? *217*
  6.4 LEARNING ABILITY AND UNDERSTANDING 217
    6.4.1 Learning ability *217*
    6.4.2 Semantics; what do we understand as 'understanding'?.............................. *219*
    6.4.3 Learning and the bonus-malus system *220*
    6.4.4 Highest complexity of the bonus-malus system.. *221*
  6.5 SPEECH, THE MOST IMPORTANT MEDIUM OF INTELLIGENCE 222
    6.5.1 Four stages in the evolution of communication *222*
    6.5.2 Animal communication; some remarks *225*
    6.5.3 Levels of speech development *226*
    6.5.4 Universal grammar: some doubts *232*
    6.5.5 How a child learns to speak *234*
    6.5.6 Continuity of human speech.. *235*
  6.6 SPEECH AND ABSTRACTS 237
    6.6.1 Percepts, concepts, etc.. *237*
    6.6.2 Heuristic and algorithmic problem solutions.. *239*
    6.6.3 Qualia.. *240*
    6.6.4 Mutual basis of speech: world knowledge *240*
    6.6.5 How important are questions? *241*
  6.7 SPEECH, SOME ABNORMALITIES 243
    6.7.1 Brain damage and speech *243*
    6.7.2 Deaf-and-dumb language *244*
  6.8 WRITING AND READING: INTELLECTUAL ABILITIES 246
    6.8.1 Writing and reading; human uniqueness *246*
    6.8.2 Pictorial representations *248*
  6.9 BIBLIOGRAPHY 250

… Contents …

7 EVERYDAY INTELLIGENCE **254**

7.1 COMMON SENSE, HUMANITY'S GREATEST ATTRIBUTE 254
- *7.1.1* Common sense: what is it? *254*
- *7.1.2* Common sense and heuristic problem solution *255*
- *7.1.3* The quality of biotic information: the bonus-malus system *256*
- *7.1.4* Wisdom *258*
- *7.1.5* Science and common sense *260*

7.2 DIFFERENTIATION BETWEEN INDIVIDUALS 263
- *7.2.1* The 'IQ' question—is human intelligence measurable? *263*
- *7.2.2* Lateralization as one of the causes of personal differentiation *265*
- *7.2.3* Creative thinking, fantasy *266*
- *7.2.4* Genius, the summit of intelligence? *267*
- *7.2.5* Idiot savants, remarkable individuals *268*
- *7.2.6* Artistic skills *270*
- *7.2.7* Mental disorders and brain diseases *271*

7.3 PERSONALITY: THE PRODUCT OF GENES OR OF CULTURE 274
- *7.3.1* How does a personality emerge? *274*
- *7.3.2* Intelligence: nature or nurture *278*
- *7.3.3* School and intelligence *280*
- *7.3.4* Taboos, sexuality, placebos, hypnosis and psychosomatics *280*

7.4 THINKING: THE QUINTESSENCE OF INTELLIGENCE 281
- *7.4.1* What do we think about thinking *281*
- *7.4.2* Paradoxes and the principle of incompleteness *282*
- *7.4.3* Myths and rationality *286*
- *7.4.4* Deception, fraudulence, cheating *287*
- *7.4.5* Humour is unique *289*

7.5 CULTURE: IDEAL ARTIFACT 289
- *7.5.1* Culture: product and source of intelligence *289*
- *7.5.2* Psychoactive agents and culture *290*
- *7.5.3* Mental game *295*
- *7.5.4* Aesthetics, some remarks *296*

7.6 BIBLIOGRAPHY 299

8 INTELLIGENCE INSIDE **302**

8.1 SELFCONSCIOUSNESS, OBJECTIVE OR MYSTERIOUS? 302
- *8.1.1* Self as the supreme concept *302*
- *8.1.2* Selfconsciousness must be defined *303*
- *8.1.3* In asymmetric brain the consciousness could also be asymmetric *307*
- *8.1.4* Thinking and intuition *308*
- *8.1.5* Selfconsciousness: origin and evolution *311*
- *8.1.6* Are animals conscious? *312*

8.2 CONSCIOUSNESS FROM DIFFERENT POINTS OF VIEW 314
- *8.2.1* Unconsciousness and the natural sciences *314*
- *8.2.2* Para-psychology etc. *315*
- *8.2.3* Vagueness of information; illusions *317*

8.3 FREE WILL, THE REAL MYSTERY OF HUMANS 319
- *8.3.1* Necessity, chance and free will *319*
- *8.3.2* Hypothesis of free will mechanisms *320*
- *8.3.3* Free will and the natural sciences *323*
- *8.3.4* Free will and psychic force? *325*

8.4 INTELLIGENCE AND MORALITY 326

8.4.1 Intelligence and morality *326*
8.4.2 Genocide, homicide and crimes *329*
8.4.3 Consciousness about own death and altruism *331*
8.5 BIBLIOGRAPHY 333

9 ARTIFICIAL INTELLIGENCE 336

9.1 INTELLIGENCE AND ITS ARTIFACTS 336
9.1.1 Artifacts; what are they? *336*
9.1.2 The brain and the computer: similarity? *338*
9.1.3 Stages of computer evolution *341*
9.2 COMPUTERS AND INTELLIGENCE; FAULTLESSNESS AND COMPLETENESS 342
9.2.1 Computers faultlessness? *342*
9.2.2 Fuzzy, connectial, and other computers *343*
9.2.3 The real world and jigsaw puzzles *344*
9.2.4 Virtual reality *345*
9.2.5 Semantics and information processing *345*
9.3 ARTIFICIAL REFLEX, INSTINCT, AND DISCENT 346
9.3.1 Artificial reflex; robots *346*
9.3.2 Artificial instinct; robots. *347*
9.3.3 Artificial instinct; chess player *348*
9.3.4 Is communication with animals possible? *349*
9.3.5 Artificial discent; experts *350*
9.3.6 Artificial discent; learning robots *351*
9.3.7 Artificial language translators *351*
9.4 ARTIFICIAL INTELLIGENCE, COMPETITOR OR SLAVE 353
9.4.1 Computational brain *353*
9.4.2 Artificial intelligence; chance and reality *355*
9.4.3 Artificial consciousness? *358*
9.5 BIBLIOGRAPHY 360

10 EXTRATERRESTRIAL INTELLIGENCE 362

10.1 EXTRATERRESTRIAL INTELLIGENCE? 362
10.1.1 Extraterrestrial intelligence, some assumptions *362*
10.1.2 Extraterrestrial intelligence could be different and much older *364*
10.2 EXTRATERRESTRIALS COMMUNICATE WITH US ? 367
10.2.1 Extraterrestrial communication? *367*
10.2.2 How to communicate with aliens? *367*
10.2.3 Extraterrestrial intelligence; friend or foe? *369*
10.2.4 Extraterrestrial artificial intelligence *370*
10.3 BIBLIOGRAPHY 370

11 FUTURE OF INTELLIGENCE 372

11.1 THE SURVIVAL OF TERRESTRIAL INTELLIGENCE 372
11.1.1 Can terrestrial intelligence survive? *372*
11.1.2 Why has terrestrial intelligence emerged only once? *373*
11.1.3 Human brain; further evolution? *373*
11.2 HOW STABLE IS OUR COSMIC ENVIRONMENT? 376
11.2.1 Totality is always growing *376*

    11.2.2 The distant future of the Universe and the Milky Way.................................376
    11.2.3 Our Sun will remain unchanged for 5 gigayears...378
11.3 HOW STABLE IS OUR PLANETARY ENVIRONMENT?..............................380
    11.3.1 Earth, distant future...380
    11.3.2 A catalogue of natural catastrophes..381
    11.3.3 Natural biotic catastrophes 384
    11.3.4 Biosphere, distant future...384
11.4 HOW STABLE IS OUR CULTURAL ENVIRONMENT?..................................385
    11.4.1 Man-made catastrophes..385
    11.4.2 New catastrophes: a superdrug and quasi-reality....................................386
11.5 SUPERINTELLIGENCE, THE PROBABILITY OF ITS EMERGENCE..............387
    11.5.1 A thousand future generations around one table....................................387
    11.5.2 Are there limits to the growth of intelligence?.......................................389
    11.5.3 Superintelligence, the future highest level?..390
    11.5.4 Doubts concerning the 'brave future'; cloning 392
    11.5.5 Can an artificial thinker be selfconscious?...395
11.6 THE VERY DISTANT FUTURE OF INTELLIGENCE 396
    11.6.1 Will intelligence exist for ever?..396
    11.6.2 Is science for ever?...397
    11.6.3 Mind and brain—an excellent unification theory 399
11.7 FUTURE OF THE INTELLIGENCE; SOME DOUBTS..401
11.8 BIBLIOGRAPHY...402

12 GLOSSARY..**404**

FIGURES

Figure 1-1 What is information ? 8
Figure 1-2 Emergence of information during evolution 11
Figure 1-3 Vagueness of signals and information 22
Figure 1-4 From atoms to ideas about atoms...24
Figure 1-5 Evolution of biotic information processing...27
Figure 1-6 Evolution of complexity...28

Figure 2-1 Superunified scheme of evolution..34
Figure 2-2 Why space must be three-dimensional?...37
Figure 2-3 Four eras in the evolution of the Universe 44
Figure 2-4 Stars and four elementary forces..46
Figure 2-5 Abundance of elements..47
Figure 2-6 Chemical bonds, carrier of complexity 50
Figure 2-7 Universal scheme of living being...51

Figure 3-1 Scheme of terrestrial living being 57
Figure 3-2 Evolution of species..64
Figure 3-3 Human detectors and effectors...72
Figure 3-4 Evolution of nervous system..74
Figure 3-5 Brain and number of legs..76

Figure 3-6 Human and animal brains ... 79
Figure 3-7 Evolution of biotic information processing ... 82
Figure 3-8 Four stages of biotic information processing ... 84
Figure 3-9 Genealogy of the modern man ... 86
Figure 3-10 Natural, sexual and mental selection ... 92

Figure 4-1 Scheme of human nervous system ... 104
Figure 4-2 Brain and computer ... 106
Figure 4-3 Scheme of human brain ... 108
Figure 4-4 Human brain: motor and sensor, primary, secondary, tertiary ... 110
Figure 4-5 Number of neurons in the central nervous system ... 112
Figure 4-6 Interneurons are very numerous ... 115
Figure 4-7 Scheme of neuron and synapse ... 118
Figure 4-8 Microtubules: significant component of neurons ... 120
Figure 4-9 Ion channels ... 122
Figure 4-10 From embryo to adult-brain's mass ... 124
Figure 4-11 The birth and death of neurons and synapses ... 130
Figure 4-12 How to make a map of human brain ... 132
Figure 4-13 Energy flow in a child's brain ... 134
Figure 4-14 Energetics of brain and intellectual activity ... 136
Figure 4-15 Chemical composition of human brain ... 138
Figure 4-16 Number of genes for human brain ... 148

Figure 5-1 Maps: local and general ... 158
Figure 5-2 Hierarchy and functions of maps ... 160
Figure 5-3 Different levels of memory ... 163
Figure 5-4 Mechanism of memory ... 168
Figure 5-5 Complex information processing ... 170
Figure 5-6 Drugs ways in human brain ... 173
Figure 5-7 Complex information is initially decomposed ... 181
Figure 5-8 Synchronization of neuronal signals ... 183
Figure 5-9 The space scale of brain structure ... 185
Figure 5-10 Evolution of lateralization of brain ... 191
Figure 5-11 Asymmetry of human brain ... 195

Figure 6-1 Intelligence—a scheme ... 206
Figure 6-2 Four stages of intelligence ... 211
Figure 6-3 Tennis player ... 213
Figure 6-4 Levels of learning ... 216
Figure 6-5 Reading known and unknown words ... 218
Figure 6-6 Windows of opportunity ... 223
Figure 6-7 Evolution of biotic communication ... 227
Figure 6-8 Stages of human communication ... 230
Figure 6-9 Human speech—evolution ... 234
Figure 6-10 A thousand past generations at one table ... 236
Figure 6-11 Material objects and true and false concepts ... 238
Figure 6-12 Speaker-listener communication ... 245

Figure 7-1 Common sense ... 257
Figure 7-2 Common sense and bonus-malus system ... 259
Figure 7-3 Intelligence quotient ... 264
Figure 7-4 Development of personality ... 276
Figure 7-5 Nature via nurture ... 281

Figure 7-6 Taboos, placebo, celibacy, pornography, etc. 283
Figure 7-7 Gödel´s incompleteness 285

Figure 8-1 Is selfconsciousness explainable? 304
Figure 8-2 Thinking is conscious or unconscious? 308
Figure 8-3 Evolution of consciousness 310
Figure 8-4 Emergence of soul? 313
Figure 8-5 Illusions 318
Figure 8-6 Mind and vagueness 321
Figure 8-7 Emotions and morals 328
Figure 8-8 Evolution of altruism 330

Figure 9-1 Plasticity of brain and computer 354
Figure 9-2 Price of the intellectual energy 357

Figure 11-1 Stages of evolution 374
Figure 11-2 The spectrum of natural catastrophes 383
Figure 11-3 A thousand future generations at one table 388
Figure 11-4 Human intelligence and artificial impact 396
Figure 11-5 Excellent unification: mind-matter 400

TABLES

Table 1-1 Properties of a 'good' book about intelligence 3
Table 1-2 Intelligence as part of totality 7
Table 1-3 Matter and information 13
Table 1-4 Three kinds of information 15
Table 1-5 Multiple dimensions of information 20
Table 1-6 Four stages of biotic information processing 26

Table 2-1 Elementary forces 39
Table 2-2 Particles, atoms and molecules 41
Table 2-3 Stability of elementary particles 43
Table 2-4 Universal definition of living being 49

Table 3-1 Types of nutrition 65
Table 3-2 Velocity of the information processing 69
Table 3-3 Mass explosions (good genes) and mass extinction (good luck) 70
Table 3-4 Homo erectus, habilis, sapiens 90
Table 3-5 Sexual selection activities in all four stages of biotic information processing 94

Table 4-1 Neurotransmitters, brainy molecules 143
Table 4-2 Genetics and brain 146
Table 4-3 DNA size and number of neurons 147

Table 5-1 Short, working and long memory 165
Table 5-2 Procedural and declarative memory 166
Table 5-3 Drugs, psychoactive agents 174

Table 5-4: Left and right hemisphere differences 193
Table 5-5 Differences in the brain structure between gender 196

Table 6-1 Stages of thinking machine 208
Table 6-2 Semantics-learning about the real world 220
Table 6-3 Four stages of language 224
Table 6-4 Components of human speech 229
Table 6-5 Development of children's language 233
Table 6-6 Volumes of average individual knowledge 242
Table 6-7 Written and drawn representations; quality 247

Table 7-1 Laterality and personality 266
Table 7-2 Human inventiveness 269
Table 7-3 Hallucinations and others 273
Table 7-4 Five main determinants of personality 277
Table 7-5 Brain's dictionary 284
Table 7-6 Psychoactive agents, drugs 292
Table 7-7 Similarity of some social action and sexual selection 294

Table 8-1 Consciousness versus unconsciousness 314
Table 8-2 Sleep and dreams 316

Table 9-1 Brain and body relation 339
Table 9-2 Human brain and computer/robot 340
Table 9-3 Realm of puzzles and real world 344
Table 9-4 Artificial information processing 352

Table 10-1 Planets bearing intelligence 363
Table 10-2 Is there intelligent life on the Earth 366
Table 10-3 Alien artificial intelligence 368

Table 11-1 Past and distant future of the Sun 379
Table 11-2 Future of the Earth 381
Table 11-3 Points of no return 385
Table 11-4 Human brain: future evolution 392
Table 11-1 Somatic and embryonic cloning and the human 394

CHAPTER 1
INTELLIGENCE AND INFORMATION

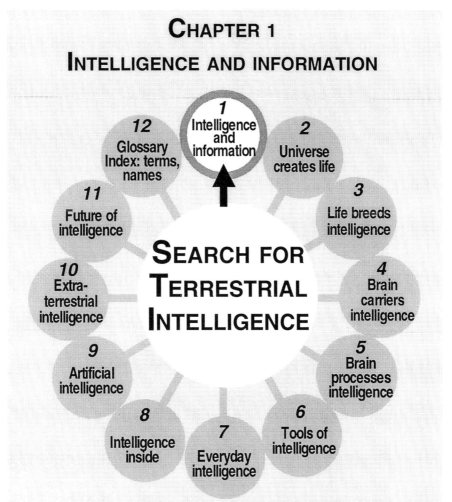

Intelligence is able to investigate the Totality of all things existing in the real world, including the phenomenon of intelligence itself. What is signal and what information? Quantity and quality of information. Processing of information: overabundant production and selective elimination. The bonus-malus system as basis for information processing. Four stages of biotic information processing: reflex, instinct, discent and intelligence. The universality and the terrestriality of the phenomenon of intelligence. Different meanings of term 'intelligence'.

1 INTELLIGENCE AND INFORMATION 2
1.1 INTELLIGENCE AND TOTALITY 2
1.1.1 The unity of object and method............ 2
1.1.2 Why must we begin with the beginning?............ 4
1.1.3 Is totality and intelligence comprehensible?............ 5
1.2 INFORMATION AND MATTER 7
1.2.1 Information and matter............ 7
1.2.2 Emergence of matter and of information 10
1.2.3 Emergence of information: life............ 12
1.2.4 What is signal and what is information?............ 12
1.2.5 Information as a representation of the real world............ 15
1.2.6 Quality and quantity of information 16
1.3 PROCESSING OF INFORMATION 18
1.3.1 Overabundant production and selective elimination 18
1.3.2 The bonus-malus system and semantics............ 19
1.3.3 The bonus-malus system as a basis for information processing............ 21
1.3.4 Semantics, definition............ 23
1.3.5 Complexity, a definition?............ 25
1.4 THE UNIVERSALITY AND 'TERRESTRIALITY' OF INTELLIGENCE............ 25
1.4.1 The universality of intelligence 25
1.4.2 The 'terrestriality' of intelligence 27
1.4.3 Intelligence—different meanings 28
1.5 BIBLIOGRAPHY   30

1 INTELLIGENCE AND INFORMATION

1.1 INTELLIGENCE AND TOTALITY

1.1.1 *The unity of object and method*

Our task is the investigation of human intelligence.

Our tool is human intelligence itself.

The unity of goal and method is unique and astonishing. Is this unity of goal and method, of object and tool, a positive, self-amplifying process, or, on the contrary, does it make our task more difficult, or even impossible?

The simplest and most natural way to solve our task would be to begin with a description of human intelligence, but this could be exactly the last thing to begin with.

How to write a book about intelligence?

Before saying anything about the nature of human intelligence, it would seem to be our duty to say how we propose to write a book about it. Let us arbitrarily formulate some criteria for judging what a 'good' book about human intelligence should be like. The book must fulfil the criteria presented on Table 1-1. It must treat intelligence as a cosmic phenomenon, and not as a specific terrestrial event. As far as possible, it should be sufficiently general, disregarding local terrestrial peculiarities. It must be clear to the author and reader that fulfilling this criterion is a difficult task, and can be realised only in a limited manner. This property we will call 'universality'. The properties which are obviously of a local character, we will call 'terrestriality', that is, specific to the Earth, 'Terra'. The term 'terrestriality' is a neologism, but is intended to be helpful in our study. We must confess that elsewhere in this book other neologisms will also be necessary.

Table 1-1 Properties of a 'good' book about intelligence

'Physicality'	Based on established physical theories (in a broad sense, including chemistry, biology, etc.). Avoids postulating new, specially created, phenomena and entities, but some neologism are necessary. Also the physical uncertainty of information, 'vagueness' of information processing must be taken into account.
'Evolvability'	Explains the origin of intelligence exclusively on the basis of biotic evolution.
'Socialability"	Considers human speech as the highest level of communication between intelligent individuals in a given human section of society, and the role of the emergence of a collective intelligence in the form of culture. The vagueness of semantic and weakness of syntactic nature of words and sentences, including the decisive impact of the 'reward-punishment' phenomena (bonus-malus system).
'Selfcognizability'	Explains the unique and peculiar properties of human intelligence, especially the ability for self-cognition (including the ability for formulating a definition of intelligence), but also self-deception.
'Comprehensibility'	Comprehensible not only for a narrow circle of scientists but for the average educated person.

A. Einstein (1950): "The fundamental ideas of science are essentially simple and may as a rule be expressed in language comprehensible to everyone."

On the other hand we must taken into account that the information and the information-processing are from the very nature rather unclear, uncertain, vague and ambiguous. Intelligence, as the highest stage of biotic information-processing, cannot be, and is not, free of all these 'negative' properties. Finally, the definition of the term intelligence will be uncertain, vague and ambiguous. (Remark: in this book the term 'vague' is used as defined by Webster's Third Dictionary:—not firm, lacking in clarity or definition, and 'vagueness' the quality or state of being vague).

Unfortunately this threat cannot be eliminated in writing a book concerning the intelligence. To minimize vagueness and ambiguity we will use the form of tables and schemes, which seems to be more exact than a prosaic text.

1.1.2 *Why must we begin with the beginning?*

After thousands of years of the evolution of human intelligence (here is not the best place to give an estimate of how many thousands of years, this will be done later), the situation from our point of view can be briefly characterised as follows:

The human being, has created a science which approaches the 'Theory of Everything', a unified self-consistent theory dealing with the Universe (the largest object) and its elementary particles (the smallest objects) and elementary forces. It is believed that this theory, will be able to explain all phenomena from the very beginning of the Universe to our time, and even some 'simplest' future events, up to the very end of the Universe. The 'Theory of Everything' claims to be a theory covering all phenomena, living and inanimate structures, all material and all incorporeal entities (providing that such things exist). Doubtless the 'Theory of Everything' will firstly consider the most simple entities. The Universe seems to be rather simple. The most complex things are, and probably will remain so in the future, the human brain and its product—intelligence. Is there a 'Theory of Everything' capable of helping us in our task?

Human beings have investigated the phenomenon which we call 'mind', 'consciousness' or 'intelligence' for thousands of years. It must be said that, in spite of enormous successes in this field, our knowledge about our own intelligence is less developed than our knowledge in other branches of sciences. This is the current opinion both in scientific and lay circles. Why is there such a discrepancy between the state of knowledge of Everything, and our knowledge about our own intelligence?

G.M.Edelman (1989): "To construct a global brain theory that is compatible with physics and evolutionary theory.. cannot begin with consciousness and proceed to physics and evolutionary theory, although this is obviously the order in which these subjects arose in the history of human interests. We must travel in a loop in order to comprehend the physical means that gave rise to consciousness... The theory assumes that world is structured as described by quantum field theory, relativity theory and statistical mechanics. In addition to these physical constraints, the theory also assumes that all aspects of the brain, structural and functional, may be attributed ultimately to evolution as described by the theory of natural selection. In constructing the brain theory, no additional attributes are assumed for matter or its interactions."

We must accept the consequences of this situation and try to do the best we can with it. Instead of beginning this book by immediately considering the phenomenon of intelligence, we must begin with the 'Totality of Everything'. What is the 'Totality of Everything'? Totality includes all material (without asking what this means) and incorporeal (if this exists), assuming that a third type of objects does not exist. This assumption gives us at least one advantage—we can include intelligence (terrestrial and extraterrestrial) as part of totality.

The question now arises of whether human intelligence, being itself a part of totality is able to discuss, investigate and understand (what is 'understanding'?) the term 'Totality'. One attribute of human intelligence we have postulated: that intelligence is able to create the term 'totality' and at the same time claiming to be a part of totality. Totality contains the whole of space, the here and the elsewhere. The last term means that all regions of space always lie within totality.

Totality contains the whole of time—past, present and future (and in the future all events will belong to totality), even if the latter is a suspect claim. Totality embraces all material things, from the smallest elementary particles to the largest structures such as galaxies.

Totality comprises all so-called 'immaterial', or 'spiritual' things, such as thoughts (good and bad, permitted and prohibited, valuable and valueless), mind, consciousness, and intelligence itself (human, that is terrestrial, extraterrestrial, artificial and all other kinds).

F.Hoyle (1957): "The Universe is everything: both living and inanimate things, both atoms and galaxies. And if the spiritual exists as well as the material, of spiritual things also; and if there is a Heaven and a Hell, Heaven and Hell too; for by its very nature the Universe is the totality of all things."

We must add that the concept of 'Heaven' and 'Hell' exists in the mind of humans and therefore exists in the Universe. Totality encloses all entities, including its own limits, and bounds—beginning and end. Totality does not allow the existence of anything which is outside itself; there is no outside. The number of different types of material and immaterial things, of inanimate objects and living beings, increase continuously with time. At present, Totality includes intelligence and, therefore, also the inherent ability to self-define the term 'Totality'.

1.1.3 *Is totality and intelligence comprehensible?*

All our discussions are based on the assumption that totality is related in such a way to intelligence that the conscious creature has the feeling that it is able to understand totality. The way to a better understanding of the world around us does not seem to be straightforward, and is to a large extent intricate and rough. However, it will probably lead to the desired direction—towards an understanding of the world and the nature of human intelligence.

The last sentence depends upon our understanding of the term 'understanding' itself. This paradoxically sounding statement will be repeated in this book in different guises as variants of the principal question: Can intelligence explain itself?

We are a long way from the extreme point of view of the 'reductionists', who try to explain even the most complicated phenomena by means of very simple mathematical descriptions or equations, or by means of clearly formulated physical principles.

In spite of the fact that all phenomena are based on the same elementary particles and forces, the explanation of complex events cannot be explained by describing their elementary components. The interaction of large numbers of particles and different forces with very different boundary conditions and histories, with new events which emerge during these histories, cannot be explained by the simplest particles and forces alone. It calls for a special kind of investigation and a special type of treatment.

Physicality is the evidence that totality is self-comprehensible. Someone has said that a physicist is the means by which atoms investigate and explain themselves.

What is the most astonishing property of intelligence? Is it the ability to understand and explain totality? (Table 1-2). Or the ability to investigate and talk about intelligence itself? The most amazing feature of intelligence is the emergence of the feelings of selfconsciousness and 'free will'. However, it is not obvious if this is a specifically human property or a general, universal property of intelligence. In this book the term 'selfconsciousness' is used as 'the quality or state of being selfconscious'; conscious of one's own acts or states as belonging to or originating in oneself: aware of oneself as an individual that experiences, desires, acts, and thinks and knows himself as thinking" (Webster's Third New International Dictionary, 1976). Finally, the really 'mysterious' problem is why human intelligence is constructed in such a way that it is able to describe the totality (from quark to Universe, from crystal to living and intelligent being) by means of man-made language of mathematics and physics and by means of experiments and practical actions such that its finds coherence of 'reality' with the man-made abstract pictures of totality and of intelligence itself.

A.Einstein (1930): "The essential mystery of the world is its comprehensibility..... The hardest thing to understand is why we can understand at all."

Table 1-2 Intelligence as part of totality

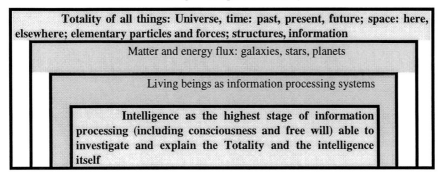

Within the relatively recent evolution of the Universe a new type of structure, a new intelligent creature has emerged. Man not restricted to explaining things; he can, consciously, attempt to change the Universe, thereby producing something which did not exist before. He thus enriches Totality with new, unique and specific things, thoughts, concepts and ideas. The intelligent being is not only an observer, it is also, or even primarily, the constructor of new entities. He is an active co-author, or joint constructor, of totality. If there is a phenomenon in Totality which requires a new force, or another agent, as the carrier responsible for the emergence and existence of intelligence, then such a phenomenon could be intelligence itself.

One of the principles of this book is that the intelligence has an inherent strength and an inherent weakness. The strength is in the knowledge of the own weakness, and its weakness is in trying to neglect this knowledge.

1.2 INFORMATION AND MATTER

1.2.1 *Information and matter*

Totality consists of matter, in the broad sense of this term. It is difficult to discuss totality without systematic differentiation. The first, perhaps principal, division of Totality is into two classes: matter and forces. However, such a classification is far from satisfactory. Matter, in the narrow sense, consists of particles, and the definition of particles is almost clear intuitively. Forces act between particles. Particles and forces are mutually and indivisibly connected. Particles are directed and influenced by forces. Forces influence particles and are able to destroy or to create them. In the next chapter this division will be discussed in more detail.

Figure 1-1 What is information ?

F. Wilczek (1992): "Information is a fuzzy concept. It's very important to be careful to specify what you mean."

K. Devlin (1991): "Imagine yourself suddenly transported back in time to Iron Age. You meet a local ironsmith and you ask him 'What is iron?'. What kind of answer are you likely to get? He is an acknowledged expert on ironship, his product sells well, and he knows a good piece of iron when he sees it. To anyone trying to understand the nature of information in today's Information Age, the situation must surely seem not unlike that facing your Iron Age man. But what exactly is information? The difficulty in trying to find an answer to this question lies in absence of an agreed, underlying theory upon which to base an acceptable definition. Like the Iron Age man and his stock in trade, Information Age Man can recognize and manipulate 'information', but is unable to give a precise definition as to what exactly is being recognized and manipulated." (Fig. 1-1).

Matter has some secondary properties, such as 'complexity' and 'information', which from our viewpoint are of importance. The formulation of a suitable definition of complexity is not easy. A definition of information itself and information processing systems will be formulated below, in spite of all the difficulties. Among others there are some 'information-processing structures' such a living being, either animal or human. Another member of the same subclass is the man-made information-processing machine, the computer. The latter can be described by the laws of nature, that is the laws of physics.

R.Penrose (1989): "Are minds subject to the laws of physics? What, indeed, are the laws of physics?"

A critical remark: The laws of physics are concepts (products of human intelligence) which in more-or-less abstract form represent the physical phenomena. Our concepts are based on rather 'vague' signals, on vague man-made detectors, on inherent vague transformation of signals into concept, on vague information processing the concepts and on vague action or experiments to confirm our vague predictions.

Before we are going further the two kinds of information should be defined: direct acting and indirect acting. Direct acting information carrier acts on the information processing system in a specific way, determined by chemical and physical properties of its material carrier. The most significant example of direct acting information carrier is deoxyribonucleic acid DNA acting on the information processing system, living cell. Direct acting information is not influenced in a specific way by external signals, and therefore could be considered as long-term stored information.

Biotic indirect acting information carrier acts on the intrasomatic information processing system in a non-specific way, mostly in form of a 'standardized' electrical signal or 'standardized' chemical molecule or on the extrasomatic system, individuals of the same species, by means of acoustical signals or chemical signals (pheromones) or so-called body language. The information processing system in this case is mostly an assembly of highly specialized cells: muscles, glands, nervous ganglia, substructures of brain. This kind of information is influenced in a very short time by external or internal signals, but could also be stored during relative long-time in the intrasomatic memory.

Information processing is the search for compression, the search for pattern, the relation between events. On the highest stage of information processing, that is intelligence, the compression leads to the algorithm of the observed events (outcomes) which allows to make hypotheses about the 'forces' and 'particles' or about species and environment, or between numbers and figures.

1.2.2   *Emergence of matter and of information*

Figure 1-2 summarises the history of the Universe, from the viewpoint of the emergence and evolution of information. The evolution of the Universe can be broken down into the following periods: 1) Creation of the Universe; no information exists. 2) Emergence of life; with a directly acting carrier of information (genetic information). 3) Animals emerge; a biotic indirectly acting information processing system, the nervous system, comes into existence. 4) Intelligent beings emerge and evolve a system of artificial (extrasomatic) carriers of information. 5) Intelligent beings develop technical information processing machines.

It must be clear that, in this book, all kinds of information are treated as belonging to the category of matter, in the broad sense of this term. It is of importance that we postulate that information includes such phenomena as thoughts, ideas, mathematical theories, human speech, writing, painting, and also animal communication. It must also be clear that a more exact definition of information will be derived in the course of this book.

One characteristic peculiarity of the present time is the insufficient amount of information about information, as a scientifically well-defined entity. To where this may lead us can be illustrated by the discussion started by T.Stonier (Fig. 1- 3).

T.Stonier (1990): "1) All organized structures contain information. No organized structure can exist without containing some form of information. 2) The addition of information to a system manifests itself by causing a system to become more organized, or reorganized. 3) An organized system has the capacity to release or convey information."

Other authors have written in the same manner:

D.Layzer (1990): "Because the word 'information' refers explicitly (rather than implicitly) to a theoretical description and partly because the word has subjective connotations: information is something we know (or don't know) about something out there. In fact, information, as here defined and used, is as objective as quantities like mass and energy. It differs from mass and energy in being a property of the Universe as a whole, rather than of individual subsystems."

K.Haefner (1992): "From my point of view information processing systems are always material structures. And vice versa, all material structures are information processing systems!"

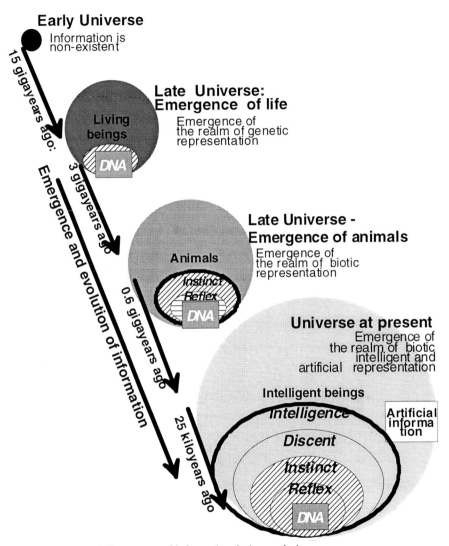

Figure 1-2 Emergence of information during evolution

In this book, we defend the point of view that, in the very early stages of the evolution of the Universe, matter and energy (forces) were void of any information. The reason is that there did not exist any structure able to process information.

G.M.Edelman (1989): "Information is linked to order. Information is one type of order which represents the real world. Disorder cannot contain information, but all order carries information. The contrary to order is disorder—in some cases also called 'chaos'. "

D.J. Chalmers (1997): "The basic principle (in the Theory of Consciousness) that I suggest centrally involves the notion of information. I understand information in more or less the sense of Shannon (1948). ... This leads to a natural hypothesis: that information (or at least some information) has two basic aspects, a physical aspect and a phenomenal aspect....If so, then information is a natural candidate to also play a role in a fundamental theory of consciousness."

Nowadays the term 'chaos' is used to characterize aperiodic activity arising in a dynamic system, or in a set of equations describing the system's temporal evolution as a result of a deterministic mechanism that has sensitive dependence on initial conditions. Chaos is a system that responds disproportionally to stimuli.

In the 1960s, E.N.Lorenz, a meteorologist at the Massachusetts Institute of Technology, reasoned that the mere flapping of a butterfly´s wings could spawn hurricanes, because a 'chaotic' system is very sensitive to initial conditions. A very complex, non-linear, so-called 'chaotic' system depends on even tiny changes in the initial conditions and can easily snowball in an unexpected and unpredictable direction. We mention this phenomenon because it will play a significant role in our discussion on intelligence and free will (Section 8.3).

1.2.3 *Emergence of information: life*

If chaos (in the broad sense) corresponds to disorder, then life corresponds to order. Life is the most complex order that we know. The tendency to self-organization is inherent in atomic matter. At the moment when the conditions on a cosmic object are appropriate (i.e. the temperature is below a thousand degrees Kelvin; the pressure is not too high or too low; the concentration of the light elements, such as hydrogen, oxygen, carbon and nitrogen, is significant; and a constant source of free energy, such as the proximity of a long-living star, exists), then chemical forces lead to the emergence of self-catalytic self-evolving systems—living structures.

The carrier of the information on how to construct a living being from the structural material existing in the environment, using the flow of free energy, is a macromolecule—deoxyribonucleic acid, DNA, encapsulated in a cell. (The properties of life will be examined in Section 2.4). The emergence of DNA on the Earth is equivalent to the emergence of information processing systems in this part of the Galaxy. Before this time, cosmic matter was devoid of information.

1.2.4 *What is signal and what is information?*

The terms 'signal' and 'information' are mutually related, but in this book they are strictly differentiated. What is a signal? For example, solar radiation is a signal. It acts on stone and increases its temperature; or on the leaves of a plant and causes photochemical reactions, resulting in the synthesis of energy-carrying substances. The signal acts directly upon the material structures, changing their properties.

Between signal and information there is the detector (sensor) which transforms the signal into information. Each signal/information transformation is heavily accompanied by losses of quantity and quality (Section 1.2.6).

Table 1-3 Matter and information

	Matter	**Nonmaterial things**
In this book	Only small part of matter always with very complex structure, such as primary central animal nervous system or artificial processing machine, carries information. The rest of matter carries and exchanges signals. Signals can be transformed in more or less exact or vague information.	Does not exist
According to T. Stonier, 1993	Each part of matter carries information, no differentiation between signals and information.	Does not exist
Idealistic hypothesis	Matter is separated from information.	Information is non-material

What is information? The same solar light received by the eye of a bee, and enabling it to find the direction to the desired source of food, must be considered as a signal carrying information. The transformation from signal into information began in the bee's eye and ended as a nervous signal controlling the bee's wings. Solar radiation acts on a spectrometer, can be interpreted by a man-made computer and then a physicist can deduce the proportions of chemical elements in the Sun. Information does not exist independently of an observer who is able to process the incoming signals as internal information. The very information is the thought in the mind of a scientist, its written record, or code in a computer.

From this point of view, external signals (such as solar radiation) belong to the realm of 'signals', and the nervous signal and written record to the realm of 'information'. Because of a loss of quality and quantity during the transformation of signal into information, we can say that the incoming external signal is at least partially vague, indistinct, in short: vague.

A 'message', as defined by C.E.Shannon, the father of information theory (more exactly theory of the communication of information), is a highly abstract structure, susceptible to an endless variety of concrete interpretations. Information theory considers signals or messages to be general information, which do not have to be meaningful in any ordinary sense.

According to the Encyclopaedia Britannica, 'meaning' is quite irrelevant to the problem of transmitting information. Information theory quantifies information and the capacity of various systems to transmit, store, and otherwise process, information.

J.Horgan (1995): "Created by C.E.Shannon in 1948, the (information) theory provided a way to quantify the information content in a message. Efforts to apply information theory to other fields, ranging from physics and biology to psychology and even the arts, have generally failed—in large part because theory cannot address the issue of meaning."

The emergence of information is linked to the emergence of living beings—the first information processing structures. There are three different kinds of information, all intimately related, but in spite of this not to be confused (Table 1-4):

1) Genetic information; carried by the deoxyribonucleic acid macromolecule DNA (or, in some simpler forms, RNA) acting directly on the molecular environment inside a living cell.

2) Biotic information; carried by nervous signals (unfortunately the term 'signals' is here confusing, but must be used for historical reasons), and processed inside a multicellular animal, in its 'nervous system'.

3) Artificial information; often thought of as information itself, in the narrow sense. This is carried by structures mostly existing outside living organisms, and belongs to the class of artifacts (Chapter 9). In the case of an intelligent being, this kind of information corresponds to writing, painting, books or tapes, created or interpreted by artificial processing devices such as computers.

In this book, information is considered to be the representation of things or events (processes), carried by different types of matter, and independent of the nature of the carrier.

In information theory, the term 'IGUS' is used for an 'information gathering and using system'. It can be a Maxwell's demon-like entity, capable of performing measurements and of modifying its strategies; for example, for extracting useful energy (W.H. Zurek, 1990).

Table 1-4 Three kinds of information

	Direct acting information included in an executive system (DNA in cell)	First kind of information Genetic information carried by DNA	Change in codon contents leads to significant even fatal consequences
Information in the broad sense of the term	Indirect, (non-specific) acting information transported towards executive mechanisms (nervous signal towards muscle)	Second kind of information Biotic nervous signals system.	Change in nervous signal leads to significant but not 'fatal' consequences
		Third kind of information Speech processing and processing of artificial information	Deliberate or random changes in sentences of transmitter often cannot be 'discovered' by receiver

M.Gell-Mann (1990): "The IGUSes, including human beings, occupy no special place and play no preferred role in the laws of physics."

The processing information in animals and in humans finishes with an output. The specialized organs for transforming the information output into action of own body or action toward the world is here called 'effector' or 'motor'. Often or maybe always the action of the body is not very exact described, that is includes different information, therefore is also 'vague'.

1.2.5 Information as a representation of the real world

The above comments about signals and information are based upon the assumption that information represents some properties of an external signal, in such a way that the most significant properties have the correct relationship, corresponding to the 'true' or 'real' world. Of course the correspondence information /real world is weakened by losses during the transformation in the effectors (motors). The correspondence between information and the real world belongs to the branch of science called semantics (in the broad sense) (Section 1.3.4). In the narrow sense of the term, semantics is the study of the correspondence between words, sentences, symbols, etc. in human speech, and their correspondence to the real world.

There are many reasons why information theory, as one of the bases of modern computer science, must ignore the semantical meaning of information processed by the computer. The information theory was developed by Shannon and Weaver for a very specific purpose: to calculate information transmission across a channel in a relatively well defined and simple physical system, such as a telephone network.

C.E. Shannon (1949): "The semantic aspects of communication are irrelevant to the engineering aspects."

V.B. Mountcastle (1990): "I regard as one of the central dogmata of neuroscience: that representations of material reality are instantiated in patterns of neural activity within the brain, and that these representations are essential components of the mechanisms of mind. Only rarely are our representations of the external world at central neural levels isomorphic with physical reality."

We must really not forget that computerized information has nothing to do with semantics; that is, with correspondence to the real world. This is not so in the case of animals, for whom the significance of the information processed depends absolutely upon the semantical meaning of information, upon its correspondence to the real world.

K. Haefner (1992): "Information per se doesn't make sense at all! The 'traditional' concept of information must be replaced by completely new approaches to the definition of information which, however, are not at hand at present. If we would derive an understanding of crucial principles, this might help to give rise to a 'Grand Unified Theory of Information Processing which must be developed, but which is not at hand yet."

P. Grice (1989): "False information is not a inferior kind of information; it just is not information."

G. t'Hooft (1997): "The elusive mystery of quantum mechanics gave rise to a great deal of controversy, and the amount of nonsense that has been claimed is so voluminous that a sober physicist does not even know where to start to refute it all. Some claim 'that 'life on Earth started with a quantum jump', that 'free will' and 'consciousness' are due to quantum mechanics. Even paranormal phenomena have been ascribed to quantum mechanical effects...(W)e can argue that no brain cell or chemical reaction exists that can bypass the quantum mechanical uncertainty relations. Perhaps life on Earth originated as a result of an extremely improbable coincidence of events, but this has nothing to do with quantum mechanics. ....Probabilities and statistics are mistreated a great deal, even by physicists. Some have uttered the theory that all possibilities for certain events are being realized in 'parallel worlds', with their given probabilities. This is called the 'many worlds' interpretation of quantum mechanics. This is how crazy it becomes if one tries to 'quantize the universe'. To my sober mind, all this is nonsense.....Anyway, for me, the 'hidden variables' hypothesis is still the best way to ease my conscience about quantum mechanics."

1.2.6 *Quality and quantity of information*

Current 'information theory' is concerned only with the quantitative side of information flow—the amount of information, measured in bites, the redundance of information, the effectiveness and the rate of information transfer and transformation, the amount of stored information, etc. At present, theory neglects not only correspondence to the real world but also does not try to understand a very important property of each piece of information—its quality. This principal weakness of modern information theory has been fully accepted by its originators, amongst others, C.E. Shannon, A. Weaver, and L.N. Brillouin. The following quotations need no comment:

W.Weaver (1949): "In this theory the information is used in a special sense that must not be confused with its ordinary usage. In particular, 'information' must not be confused with meaning. In fact, two messages, one of which is heavily loaded with meaning and the other of which is pure nonsense, can be exactly equivalent from the present viewpoint as regards information."

C.E. Shannon, W. Weaver (1964): "The concept of information developed in this theory at first seems disappointing and bizarre—disappointing because it has nothing to do with meaning, and bizarre because the words 'information' and 'uncertainty' find themselves to be partners."

L.N. Brillouin (1956): "Returning to information theory, we must start with a precise definition of the word 'information'. .But we are in no position to investigate the process of thought, and we cannot, for the moment, introduce into our theory any element involving the human value of the information. The elimination of the human element is a very serious limitation, but this is the price we have, so far, to pay for being able to set up this body of scientific knowledge. The restrictions that we have introduced enable us to give a quantitative definition of information and to treat information as a physically measurable quantity. This definition cannot distinguish between information of great importance and a piece of news of no great value for the person who receives it... The 'value' of the information, on the other hand, is obviously a subjective element, relative to the observer. The information contained in a sentence may be very important to me and completely irrelevant for my neighbour. All these elements of human value are ignored by the present theory. This does not mean that they will have to be ignored forever, but, for the moment, they have not yet been carefully investigated and classified. These problems will probably be next on the program of scientific investigation, and it is to be hoped that they can be discussed along scientific lines."

Biotic information is always labelled as 'good' or 'bad'; causing pain or pleasure; connected with reward or punishment; carrying signs of bonus or malus; and, therefore, directly influenced by the biotic 'hedonic system' (in this book called 'bonus-malus' system).

It seems that, in modern literature, the quality of information is neglected. S. Lloyd, in 1990, considered briefly the different cost of information, claiming that cost is a measure of the amount of information required by a process. However, cost does not really capture the distinctions between different types of quality of information.

A. Einstein (1927): "As far as the mathematical theorems refer to reality, they are not sure, and as far as they are sure, they do not refer to reality..... The axioms are voluntary creations of human mind.... To this interpretation of geometry I attach great importance for should I not have been aquatinted with it, I would never have been able to develop the theory of relativity."

Consequently a rigorous definition of information is probably impossible. We must be satisfied with an approximately 'reasonable' definition (Table 1-5).

How is it possible that atoms in our brain 'think' that they 'exist really', that the world is real, that reality exists independent from the process of thinking? What evolutionary gain emerges from this kind of biotic information processing?

1.3 PROCESSING OF INFORMATION

1.3.1 *Overabundant production and selective elimination*

Biotic information processing relies on the strategy of overabundant production and selective elimination. The amount of perceived and processed information is very large. Subsequent confrontation with the real world leads to the elimination of all processed and stored information which does not adequately represent reality. Some of the information which represents reality more or less adequately, are strengthened and 'survive'.

Phidias, an Athenian and one of the most outstanding of all sculptors, lived in the 5th century BC, and was, amongst other things, the creator of the marble sculptures of the Parthenon. It is said of Phidias that he alone had seen the exact image of the gods. An anecdote (perhaps about another sculptor) says that, when Phidias began to prepare his sculptures from one large block of marble, he was asked how he was going to do this. His answer was, 'It is very simple. I am taking marble block which must be much larger than the final product, and I remove all those pieces of marble which are superfluous'. Did he know from the beginning what the final product was to be? How large should the marble block be for a sculpture of a goddess or a horse?

Perhaps Phidias only had an indistinct mental picture of the end product, and the final piece was the result of deliberate and accidental changes made during the processing of the marble block. But one thing is certain, the materials used and Phidias´ technique, limited the features of the final product.

Nature very often, but probably not always, uses the same strategy of overabundant production coupled with selective elimination. It is a simple, but at first glance very inefficient, method. However, because it achieves the desired final result, it must be considered to be efficient enough. Nature does not ask how high the price is. The only measure of success is that the probability of genetic reproduction is high enough.

1.3.2 The bonus-malus system and semantics

This realm of information includes some properties which have been completely neglected in information theory. These are:

–Semantic correspondence between information content and the 'real world'.
–The quality of information.
–The evolutionary character of information processing systems.

In the realm of animals, neither the truth or falseness of information, nor its completeness or incompleteness, are of significance. Also, neither precise semantics nor the correspondence of information to the real world is of value. The only essential property of information is whether it helps an animal, or its offspring, to survive, or not.

All information coming from the environment, from other individuals of the same species, or from the owners body or memory must be initially evaluated in a simple way to see if it is good or bad. Every animal must have its own scale of 'good or bad'. In this book, we will speak about the 'bonus-malus system'. All remembered information, old or new, has a quality rating which is measured on a continuous 'bonus- malus scale'. It is impossible to overestimate the significance of the bonus-malus system in biotic information processing, especially at the highest stage—that of human intelligence. We will often return to this value scale. In short, we can say that, in a broad sense, the following systems are more or less equivalent: 1) bonus-malus 2) pain-pleasure 3) reward-punishment 4) appetite-aversion 5)hedonic system (anhedony? = loss of pleasure, delight).

These involve the whole range of feelings, from pleasant to unpleasant. The term 'hedonic system' is commonly used in food testing, to indicate the extent of like or dislike of particular foods. 'Hedonism', in the narrow sense of the word, is a doctrine of conduct that considers pleasure, of one kind or another, to be the ultimate criterion of what is good. The term 'bonus-malus', therefor, seems to be more abstract and will be used in this book (Fig. 7-2, Fig. 8-7).

The absolute value of bonus and malus increases with the evolution from lower to higher level of animals.

–Four stages in the development of biotic information processing—reflex, instinct, discent and intelligence—cover specific parts of the bonus-malus scale, and these stages will be mentioned in detail later on (Table 1-6, Fig. 6-1).

–Different behaviour corresponds to a different value being given to different information.

Table 1-5 Multiple dimensions of information

Dimension of information	Content		Reference
Quantity of information	'Theory of information' allows exact quantitative 'management' of information, without giving 'semantic' definition of information, and without biotic basis.		C.E. Shannon (1948)
Quality of information	There does not exist a theory of quality (semantic, biotic, social etc.) of information. A 'bonus-malus' system including individual experience is described.		This book.
Qualia in information	In the last decade emerges a discussion about the so-called 'qualia', the subjective, 'private' experience of some phenomena, such as 'redness'.		F. Crick (1994)
Uncertainty of information and ambiguity in physical theories	Exact, certain formulation of uncertainty in atomic and subatomic phenomena. The problem is known in 'quantum mechanics'. Certainly we are uncertain if we at all are certain or uncertain in all our thoughts.		Heisenberg (1928) uncertainty principle
Vagueness of information processing	The inherent vagueness and deception of human information processing (Fig. 1-3).		'Fuzzy': not firm, lacking in clarity
Complexity of information processing	Depending on the world knowledge of communicating persons the complexity changes significantly. There does not exist an acknowledged theory of complexity, in spite (or maybe because) of so wide use of this concept.		M.Gell-Mann (1994); J.Horgan (1995)
Conservation of information	Quantum theory requires that information about particles must be conserved.	The general relativity theory claims that a black hole destroys infalling particles and information.	S.Hawking and R.Penrose (1996)
Information as **representation** of the world	Positivist; does not know if real world exists: N.Bohr.	Realist, 'Platonist'; the real world exists, our theories represent the real world: A.Einstein.	S.Hawking and R.Penrose (1996)
Example: **concept** of prime numbers	Prime numbers are a construct of human brain and exist as 'mental object'.	A. Connes: prime numbers exist in real world and the intelligent being discovers it.	JP Changeux and A. Connes (1992)
Information at the level of **intelligence**	The way in which the atoms (in brain) investigate the existence and activity of atoms (in real world) by means of theoretical considerations and experiments.		This book

–A bonus-malus value changes for each individual with time, and is continuously influenced by experience.

The significance of bonus-malus labelling in the processing of biotic information is often underestimated. General principles are as follows:

–A bonus-malus label is given to each piece of information processed by an animal.

–Bonus has a positive value, reaching a maximum of +1 in the human mind—an arbitrarily chosen value.

–Malus has a negative value, reaching a maximum of -1 in the human mind, again arbitrarily set.

At the highest level the information processing, which posses the most sophisticated tool, namely speech, the processing is more complex. The reason is the easiness to influence the semantic content of the message, by means of a lie, deception, cheat, cunning. For an individual's bonus-malus system the alternative 'true-lie' opens a new chance for optimalization. Deception and/or lie can often 'improve' the bonus level. From this viewpoint intelligence (including the speech-ability), as the highest level of biotic information processing includes a very specific tool which distinguishes Homo sapiens from all other animals. The intimate bonds between true and intentionally false information is a unique and most human property.

C.Shannon (1990): "Somehow people think it (the information theory) can tell you things about meaning, but it can't and wasn't intended to."

1.3.3 *The bonus-malus system as a basis for information processing*

The central question about the nature of the encoding of the real world by the nervous system is: what is the organism's strategy for balancing the need for extensive knowledge with the need for specific action?

A correct correspondence between the real world and its internal representation must be reinforced, and the lack of correspondence must be punished before it causes the death of the individual.

B.F. Skinner (1984): "(i)What is good for the species is whatever promotes the survival of its members until offspring have been born and, possibly, cared for. Good features are said to have survival value. Among them are susceptibilities to reinforcement by many of the things we say taste good, feel good, and so on. (ii)The behaviour of a person is good if it is effective under prevailing contingencies of reinforcement. We value such behaviour and, indeed, reinforce it by saying 'Good!' Behaviour towards others is good if it is good for the others in these senses. (iii)What is good for a culture is whatever promotes its ultimate survival, such as holding a group together or transmitting its practices. There are not, of course, traditional definitions; they do not recognize a world of value distinct from a world of fact and, for other reasons to be noted shortly, they are challenged."

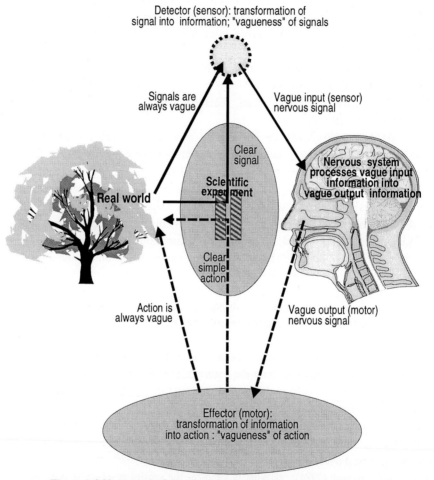

Figure 1-3 Vagueness of signals and information

L.L. Cavalli-Sforza (1995): "We know one thing for sure: our motivations appear to be controlled by nerve centres with a known position in the brain, and these determine whether a given sensation is pleasant or unpleasant. These centres undoubtedly influence our behaviour in complex fashion, but the nature of their action is entirely unknown. We know of certain substances that probably play a part in determining pleasure, such as endorphins (short for endogenous morphine). Through the complex network of nerve fibres in the brain, nearly every sensation and action, as well as every memory, takes on an emotional colour that may be positive or negative and is used to orient the behaviour. The exact way this happens is still shrouded in mystery, and is one of the greatest physiological challenges of our time."

The bonus-malus system is the base for the emotional experience, achieving the highest stage in humans. Intelligence is intimately connected with emotionality (Section 6.3.1). The existence of a bonus-malus system in each biotic information processing influences the representation of the real world significantly. Biotic information processing is not aimed to finding the 'real', the 'true' representation, but only such a representation which is most effective in survival and in the reproduction of species. This has influenced strongly our thinking, including the scientific considerations. The definition of intelligence, the research in this field, are heavily influenced by individual and social bonus-malus labels. This is one of the causes that in most books and papers concerning the problem of intelligence the 'successes' and 'achievements' of intelligence are strongly accentuated and the dark sides, such as lies, crimes, genocide etc. are mostly understated. The weakness of intelligence is the existence of the 'dark sides' and its strength is the knowledge of these weaknesses.

1.3.4 *Semantics, definition*

Semantics, according to Webster's dictionary, is the scientific study of the relations between signs or symbols, and what they mean, or denote, and of behaviour in its psychological and sociological aspects as it is influenced by signs. In a wider context, semantics means the indispensable correspondence between the real world and the internal nervous signals which enable animals to exist—living beings with the ability to rapidly process information. Semantics refers to the general concept of meaning. As such, it encompasses both linguistic and non-linguistic knowledge. At the primitive stage of biotic evolution, semantics emerges as a result of the elimination of individuals with inappropriate correspondence between the real world and their internal nervous signals. At the higher stage, the learning processes must also lead to semantically effective responses.

If the afferent nerve impulses in a flatworm, caused by the sensor reacting to some organic substances, and the efferent nerve impulses, steering the muscles, do not correspond to the real situation in the environment, then the flatworm dies. The correspondence of the stimuli, their processing, and the response in the real world are the principal properties of effective information processing. Evolving from reflex to instinct and discent, the internal representation of 'own body' and of 'world outside' is more complex, and more strongly influenced by the quality label added by the hedonic system; also called the 'bonus-malus system'.

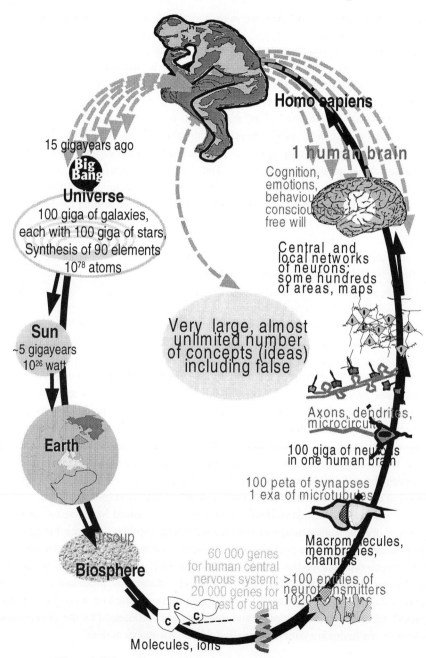

Figure 1-4 From atoms to ideas about atoms

The representation of 'own body' changes to the concept 'own body and myself', and the representation of 'world outside' develops into the concept of 'world outside and my cultural environment'.

R.B.Barlow Jr. (1990): "The brain and its sensory organs are not merely passive recipients of information from the outside world. Instead, the brain actively controls those organs to optimize the information it receives. People and birds are complicated, however, no one knows exactly how they see, much less how their brains modulate that vision. Ultimately, a series of ... more complex studies, may explain how the incomplete and unstable picture that sensory organs provide, modulated both by brain and the environment, gives rise to such direct and incontrovertible impressions as the image of a sunset, the smell of a rose or the sound of a Bach fugue."

1.3.5 *Complexity, a definition?*

Rather often in information processing, especially in verbal and written speech, the term 'complex' and 'complexity' is used. There are systematic efforts to investigate and to define the concepts of complexity (M.Gell-Mann, 1994, J.Horgan, 1995). Because in this book we will use this term it seems to be reasonable to give a short description (a quasi definition) of it. Fig. 1-6 is probably self-explaining.

There exist 45 different definitions of complexity (J.Horgan, 1996).

1.4   THE UNIVERSALITY AND 'TERRESTRIALITY' OF INTELLIGENCE

1.4.1 *The universality of intelligence*

One of the few postulates in this book is that terrestrial intelligence, which is the only example of intelligence we, at present, know from our own experience and self-examination, is the product of cosmic evolution; and there is nothing extraordinary about it.

The consequences of such an opinion are far-reaching. In Section 2.5 we will consider the so-called 'anthropic principle', which is one consequence; even if it apparently seems to be, paradoxically, the converse.

On the other hand terrestrial intelligence is obviously nothing super-mundane. It must be the product, not only of the physical boundary conditions existing on this planet (the criterion of physicality), but also of the specific terrestrial history of evolution, especially of biotic evolution. Being produced by this evolution, it has been influenced by random, accidental events.

Table 1-6 Four stages of biotic information processing

Stage	Reflex	Instinct	Discent	Intelligence
Time of emergence	800 megayears Precambrian era	500 megayears Ordovician era	250 megayears Permian era	0.13 megayears Pliocene epoch
Order of stage	First stage, the lowest	Second stage	Third stage	Fourth stage, the highest
Principles of information processing	Very simple network, genetically designed	Complex network, almost only genetically designed	Very complex network designed by epigenetic	Highest stage of complexity, mostly designed by epigenetic
Examples of animals	Corals, sponges, worms	Spiders, insects, fish	Birds, reptiles, hominoids	Homo sapiens sapiens
Nervous system	Nervous strings	Ganglia with nervous strings	Small to large brain	Large cortex, lateralization
Number of neurons	Some kilo	Up to mega	Hundreds of mega	Hundreds of giga
Art of communication	Pheromone (chemical molecular messengers)	Pheromones acoustic communication	Body expression acoustic communication	Complex body expression, speech, creation of information
Bonus malus system	Primitive	Lower stage	Highly developed including social position	Highest stage: emotions, belief, controlling reflex, instinct
Self-manipulation			Some very limited use of drugs?	Use of drugs, alcohol, hallucinogens, tobacco
Learning abilities	Habituation only	Small	High	Almost unlimited?
Repertoire of performance	Very small	Larger but limited	Very large	Almost unlimited
Social structure	Developed	Extreme highly developed	Highly developed	The highest stage of social structures
Material artifacts	Coral; own exo-skeletons	Termite's hill; spiders cobweb; excreta	Bird's nest, beaver's dam:	Global civilization, machine information processing
Ideal artifacts	None	None	None	Global science culture, fantasy

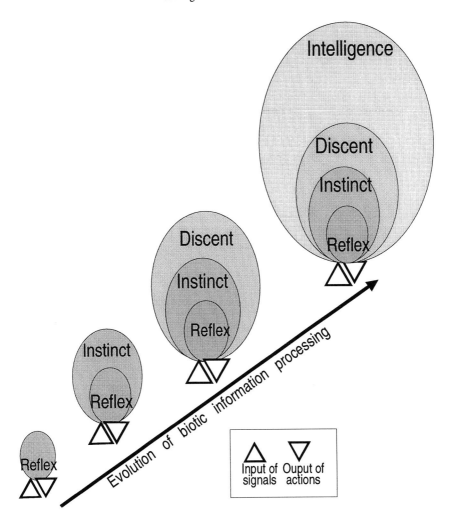

Figure 1-5 Evolution of biotic information processing

1.4.2 The 'terrestriality' of intelligence

May be that the shortest definition of intelligence is that this is an ability to produce new, before not existing, information, a kind of 'mutation' of information.

Normal visual perception always represents the effort to impose a single interpretation on ambiguous visual input. As a slightly strained example, consider the full stop at the end of this sentence. It could be a speck of dirt. It could be a small portal into a black void beyond the page. You see it as a full stop because your visual system renders the verdict that a full stop is the most plausible interpretation of that black point in the image.(J.M. Wolfe, 1996) (see also the 'vagueness of information' Fig. 1.3) (N.K. Logothetis et al, 1996).

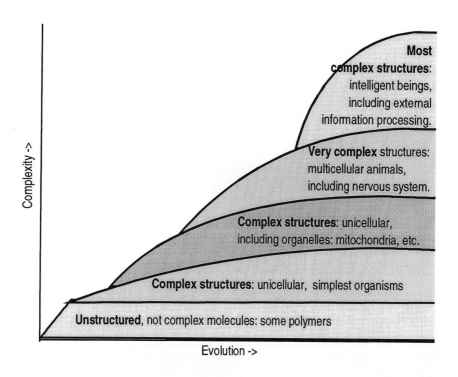

Figure 1-6 Evolution of complexity

1.4.3 *Intelligence—different meanings*

Some remarks: firstly there is no general accepted definition of intelligence, as biotic information processing. Secondly, definition of intelligence is formulated by intelligence itself. Third, the influence of judeo-christianity is of significance, which claims that human being is the highest stage of 'creation' (anthropomorphism in science).

Encyclopaedia Britannica and Webster's Third International Dictionary (1976): "Intelligence has two meanings: 1) the cognitive disposition distinct from the emotional or motivational or the faculty of understanding; capacity to know or apprehend (example: IQ, Intelligence Quotient); 2) the acquisition of information about a rival or enemy; interchange and evaluation of information concerning an enemy (example: CIA, Central Intelligence Agency in USA)."

In this book the term 'intelligence' is used exclusively in the first meaning, that is as cognitive disposition, as faculty of understanding, capacity to know or apprehend, including the emotional aspects.

1.5 BIBLIOGRAPHY

Brillouin L (1956) *Science and information theory.* Academic Press, N.York
Chalmers DJ (1997) The puzzle of conscious experience. *Sci Am Spec.Iss.* Vol.7 30
Changeux J-P, Connes A (1989) *Matiere a` pense'e.* Ed.Odile Jacob, Paris.
Devlin K (1991) *Logic and information.* Cambridge Univ Press, Cambridge, N.York
Gell-Mann M (1994) *The quark and the jaguar.* Freeman Co, N.York
Grice P (1989) *Studies in the way of words.* Harvard Univ Press, Cambridge, Mass.
Haefner K (1992) Evolution of information processing—basic concepts, in *'Evolution of information processing systems'* (K.Haefner edit), Springer, Berlin
Haken H (1992) The concept of information seen from the point of view of physics and energetics, in *'Evolution of information processing systems'* (K.Haefner edit), Springer, Berlin
't Hooft G (1997) *In search of the ultimate building blocks.* Cambridge Univ Press, Cambridge UK
Horgan J (1995) From complexity to perplexity. *Sci Am* **272** 6 72
Lloyd S (1990) Valuable information in complexity, entropy, and the physics of information. (WH Zurek edit). Addison-Wesley, N.York
Logothetis NK et al (1996) What is rivalling during binocular rivalry? *Nature* **380** 621
Popper K (1983) *Realism and the aim of science.* Hutchinson, London
Shannon CE, Weaver W (1949) *The mathematical theory of communication.* Univ Illinois Press, Urbana, Ill.
Stonier T (1990) *Information and the internal structure of the Universe.* Springer, London
Taube Mortimer (1961) *Computer and common sense—the myth of thinking machines.* Columbia Univ. Press, N.Y

Chapter 2
Universe creates life

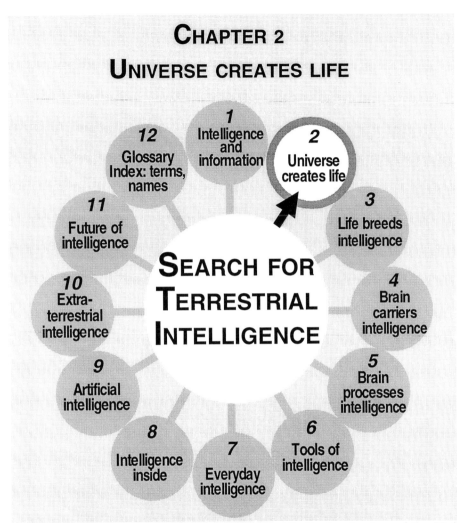

The Universe and the Natural Laws. In the beginning of the Universe are matter and energy unified. Emergence of four elementary forces: gravitational, electro-magnetic, strong and weak nuclear forces.

Universe evolves: four stages of the evolution. The birth, life and death of stars. The synthesis and evolution of chemical elements.

Emergence of life: universal definition of living being. Elementary forces able to carry life. Some light elements are able to bear life. Chemical forces as carrier of life. Anthropic principle?

2 UNIVERSE CREATES LIFE	**32**
2.1 UNIVERSE IS COMPREHENSIBLE	32
2.1.1 The Theory (of Everything) and the theory of intelligence	*32*
2.1.2 Universality of the laws of Nature	*33*
2.2 THE UNIVERSE AS TOTALITY	36
2.2.1 Matter and energy are superunified	*36*
2.2.2 Four universal elementary forces	*38*
2.2.3 Four eras in the evolution of the Universe	*39*
2.3 STARS—SOURCES OF FREE ENERGY AND OF ELEMENTS	40
2.3.1 The birth, life, and death of stars	*40*
2.3.2 Emergence of and changes in chemical elements	*42*
2.3.3 The uniqueness of the Earth	*45*
2.4 ELEMENTARY FORCES AND PARTICLES; BEARERS OF LIFE	46
2.4.1 Living beings—a universal definition	*46*
2.4.2 Elementary forces as the carriers of life	*48*
2.4.3 Elementary particles as the carriers of life	*52*
2.4.4 Chemical forces as the carriers of life	*52*
2.5 THE ANTHROPIC PRINCIPLE	*53*
2.6 BIBLIOGRAPHY	*54*

2   UNIVERSE CREATES LIFE

2.1   UNIVERSE IS COMPREHENSIBLE

2.1.1   *The Theory (of Everything) and the theory of intelligence*

We have mentioned that one of the numerous peculiarities of human intelligence is the unexpectedly fast development of the physical theory concerning the totality of all phenomena, of all types of elementary particles, of all elementary forces, of the beginning of space-time, of the early evolution and fate of the Universe, of most if not all of the laws of Nature, and of the so-called natural constants, etc.

All is changing, is in evolution; from elementary forces and particles, from Big Bang to human beings, only the Laws of Nature are immutable, unchangeable for ever.

All these theories and all this information have had to go through the sieve of confrontation with experiment and with observations, according to our principles of semantics and verification, or, what is more important, of falsification.

On the other hand the direct attack on the problems of human intelligence made by psychologists, psychiatrists, neurologists, information theory experts, cognitive scientists, and other students has allowed the creation of an astonishingly large and multispatial scientific picture of human intelligence, though this is far from complete. We are still a long way from having a full theory of human intelligence or of the human mind.

There are a number of possible reasons for this situation:

– The quality and quantity of proved physical terrestrial experiments and astronomical observations is good enough to allow mankind to formulate a universal, all-embracing theory.

–A human being is better prepared for the investigation of its environment than of his own mind.

–The best scientists are more interested in investigation of the field of physics, which promises relatively fast scientific, social and financial success. This is not the case with the study of intelligence.

–There exist more taboos—moral, religious, etc., in the field of investigation of the human mind than in other sciences.

–And, last but not least, maybe the human brain the mind, is much more complex and difficult to investigate, and much more effort, more time, more experimental devices, more financial means, and simply more geniuses are needed to research them.

We must try to use the results of the physical and 'cognitive' sciences to describe the present state of our knowledge of human intelligence. Nothing which is not accepted by the physical sciences can be used in the investigation of the human mind. Also, nothing especially invented for this task can be used, as required by Ockham's razor.

W. Ockham (1285-1349): *"Non sunt multiplicanda entia."*

A careful reading of Ockham's statement: 'Plurality is not to be assumed without necessity' and 'What can be done with fewer assumptions is done in vain with more'. This means that Ockham's principle is independent of the nature of the world.

2.1.2 *Universality of the laws of Nature*

The human intelligence experiences Nature as being in an amazing way comprehensibly structured. There has been formulated a number of rules, the so-called Laws of Nature, which are valid in all our everyday practical activities, in experiments, in observations, and, in one way or another, in our scientific methods and theories. For an intelligent being, the following quotation from Albert Einstein is of deep significance:

A.Einstein (1952): "The human mind is not capable of grasping the Universe. We are like a little child entering a huge library. The walls are covered to the ceiling with books in many different tongues. The child knows that someone must have written these books. It does not understand the languages in which they are written. But the child notes a definite plan in the arrangement of the books—a mysterious order which it does not comprehend, but only dimly suspects."

Of course, in practice, for scientists, there is no place for doubts about the real nature of the laws and principles of Nature.

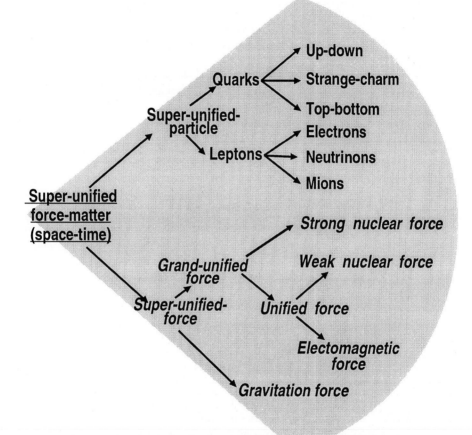

Figure 2-1 Superunified scheme of evolution

A. Salam (1970): "Nature is not economical of structures—only of principles."

S.Hawking (1987): "The way the Universe began would be determined by the laws of science. I would succeed in my ambition to discover how the Universe began. But I shall not know why it began."

S.Weinberg (1987): "There is one clue in today's elementary particle physics that we are not only at the deepest level we can get to right now, but we are at a level which is in fact in absolute terms, quite deep—perhaps close to the final source. There is reason to believe that, in elementary particle physics, we are learning something about the logical structure of the Universe at a very deep level."

P.J.E. Peebles et al (1994): "We do not know why there was a Big Bang or what may have existed before... We do not understand why the fundamental constants of Nature (e.g. light velocity) have the values they have."

If there is something very peculiar in our description of Nature, it is the lack of peculiarity of the Laws of Nature; the overwhelming principle of the universality of these laws. The whole construction of our science and logic is based on this absolute principle.

S.Weinberg, in 'Dreams of final theory' (1992) predicts that within a generation or two physicists may be able to explain the origin and nature of mass and energy in terms of a single, unified theory.

Weinberg even nominates a candidate: superstring theory, which traces all known forces and particles to infinitesimal loops of energy. He notes that many higher-order phenomena 'form turbulence to thought', would remain to be investigated. But he argues that the theory would bring an end to 'the ancient search for those principles that cannot be explained in terms of deeper principles'. Some 15 years ago Weinberg notes that "it would be wonderful to find in the laws of nature a plan by a concerned creator in which humans played some special role"... an adds: "I find sadness in doubting that we will".

The future superunified theory of the Universe will be based on the two models: the quantum mechanical world and the relativistic world. The world is quantum mechanical! Such a statement seems unprecise to us. The appropriate formulation is: The world shows an inherent property described as quantum fluctuations. Our description (representation) of the world corresponds to quantum mechanics theory. It is also wrong to claim that the Universe is relativistic. Universe could be 'better' understood using the language of general relativity theory. These two greatest achievements of the 20th century provoked significant differences in interpretation of natural phenomena (S.Hawking and R.Penrose, 1996) (G. t' Hooft, 1997).

In the last five years of our century a new jump in the elaboration of the 'Theory Of Everything' (TOE) may arise. It may be better to cal TOE just 'The Theory', because claims of finding the TOE met with so much ridicule in the 1980s that now the physicists are allergic to that sobriquet. The present state of the development of the theory of superunified forces and particles is based on a 'dual symmetry'. Note that the word 'super' was overused in theoretical physics. This theory, which is sometimes called 'M-theory' (according to taste, M stands for magic, mystery, marvel or membrane) is seen by many as a likely candidate for a complete description of Nature (E. Witten, 1996).

2.2 THE UNIVERSE AS TOTALITY

2.2.1 Matter and energy are superunified

It is probably the property of totality to be initially as simple as possible. Or, perhaps, it is the most important property of the human mind to think that the beginning must be as simple as possible. It cannot be excluded that both statements are correct, because the human brain ultimately has the ability to investigate totality in an adequate way; that is, to find the truth, or almost the complete truth, about the environment, including the cosmic.

At the beginning of time and space, and of the Universe, the only existing force was perhaps the so-called superunified force (superforce), acting on only one type of particles—the superunified particle (superparticle). It is also possible that at that moment, the differentiation of the phenomena into the two classes forces and particles is meaningless (Fig. 2-1).

At the beginning of the Universe, the realm of elementary material particles, the 'bricks' of construction and the associated quasi-particles, the carriers of the elementary forces which play the role of 'mortar', was rather simple and contained a very limited range of types of particle; perhaps only one type, which we can call a superparticle. Some modern theories claim that these superparticles did not have point-like (that is zero-dimensional) or spherical (three-dimensional) geometry, but were rather in an exotic form of strings (one-dimensional, but linearly curved, in three-dimensional space).

These hypothetical superparticles have the name 'strings'. It will not be a simple and easy task for theoretical and experimental physicists to justify the existence of these postulated 'superstrings', but, at present, there exists no other hypothesis which is able to present a more elegant and deeper picture of the Big Bang.

The question arises whether modern science has evolved enough to be able to discuss such phenomena as the beginning of totality. Nevertheless, the idea of a super- unified force and superunified particles at least partially explains many phenomena and can help us to investigate our primary aim—Nature and the origin of human intelligence. What is the answer? At the present stage of science, the most common opinion is that all known phenomena, and even the yet unknown, can be explained on the basis of the superunified force and its derivative forces; that is, the four elementary forces which we will discuss below.

There exists no need, at least at the present time, to assume that biotic information processes, including human intelligence, the mind and phenomena of the soul, are based on another elementary force, such as a psychical force, also called a 'psi' force, or anything else.

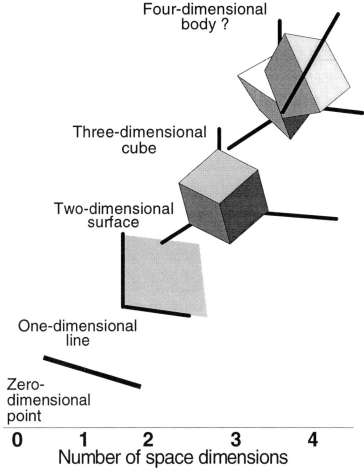

Figure 2-2 Why space must be three-dimensional?

Evidently life exists in three-dimensional space (Fig 2-2). Of course, it is also far-fetched to believe that, at some time in the future, physicists might not discover a further, let us say ultra-weak, elementary force, and try to explain the phenomena of human psychology by means of that force. This is not excluded or forbidden, a priori, but does seem to be very improbable and, from the point of view of Ockham's razor, even prohibited.

2.2.2 *Four universal elementary forces*

At the present state of physics, it is claimed that the primordial super-unified force evolved and spontaneously divided, firstly into two forces (the gravitational and grand-unified force), then into three forces (the grand-unified force decays into unified and weak nuclear forces), and, finally, into the following four types of elementary force (from the weakest to the strongest) (Table 2-1):

1) Gravitational Force, as the weakest and the most universal force, acting on all components in the Universe, without exception. This force acts at very great distance, to the greatest possible distance—the cosmic horizon. This is the force responsible for the emergence and existence of galaxies, stars, and planetary systems.

2) Weak Nuclear Force acts over a very short distance, inside the atomic nucleus. Its most important characteristic is the transformation of one type of quark in another type, by simultaneous emission of leptons: neutrino and electron. This process is called 'beta-decay' and plays an enormous role in, for example, the nuclear processes in our Sun. It is responsible for the very quiet and slow 'nuclear burning' of the Sun. The emergence and development of life and intelligence on the third planet in the Solar System is directly influenced by the properties of the Weak Nuclear Force.

3) The Electromagnetic Force. This force is responsible for most of the processes in our environment, such as solar light, electricity, magnetism, heat, mechanical properties, the gaseous, liquid and solid states, chemical structures and processes, biochemical phenomena, and, last but not least, most of the techniques of information processing. From the point of view of this book, the most important force which directly influences most of the phenomena which occur in the brain, and is responsible for those which we call intelligence, is the electromagnetic force. It must also be stressed here that, in spite of the richness and complexity of the realm of electromagnetic phenomena, the theory of electromagnetism—more exactly, the quantum theory combined with the theory of special relativity—achieves the highest level of scientific beauty and exactness.

4) The Strong Nuclear Force; The influence of the nuclear forces on the phenomenon of intelligence is only of an indirect nature, for example the activity and longevity of the Sun are controlled by Strong and Weak Nuclear Forces (L.M. Lederman and D.N. Schramm, 1989).

The evolution of elementary forces runs in parallel with the evolution of elementary particles (quarks, leptons) and later with atoms and molecules (Table 2-2 and 2-3).

Table 2-1 Elementary forces
CoupConst = Coupling Constant; a value characteristic for the strength of force

	Super-unified force (only during Big Bang)		
Gravitational Force in high and low temperature *CoupConst*: 10^{-38}	**Grand-unified Force** (prompt after Big Bang, only in very high temperature)		
	Strong NuclearForce in high and low temperature *CoupConst: 1*	**Unified Force** ('Elektroweak Force') in relative high temperature	
		Weak Nuclear Force low temperature *CoupConst*: 10^{-12}	**Electromagnetic Force** low temperature *CoupConst*: 10^{-2}
Attraction only	Attraction only	Transformation (radioactive decay)	Attraction and repulsion
Acts on all particles: quarks, leptons, and force carriers: gravitons, photons **Attraction decreases** with distance	Acts only on the heavy elementary particles: quarks **Attraction increases** with distance	Acts as transformer between both kinds of particles: quarks and leptons **Attraction and repulsion** is very weak	Acts only on electrical and magnetic charged particles and photons. **Attraction** and **repulsion decrease** with distance
Carrier: **Graviton.** Very long distances up to the edge of the Universe	Carrier: **Gluon.** Short distance, of a femtometre, radius of **atomic nucleus**	Carriers: **W,Z-Boson.** Short distances, only inside **elementary particles**	Carrier: **Photon.** Very long distances up to the edge of the Universe

2.2.3 *Four eras in the evolution of the Universe*

The Universe must have at least the following two properties: extreme simplicity in the beginning, and an inherent ability to evolve in the direction of decreasing simplicity, that is, of a tremendous complexity.

For the sake of greater comprehensibility, we can distinguish, more or less arbitrarily, four eras in the evolution of the Universe—from the Big Bang to the present moment and far into the future. For our needs Figure 2-3 gives the crucial data concerning the evolution of the Universe. Its future will be discussed in Section 11.2.

It sounds rather ridiculous that science claims to know more about the very beginning of the Universe than about the much later phenomena of the emergence of the largest material structures in the Universe—clusters of galaxies and the galaxies themselves. One hypothesis claims that neutrinos, very light but not massless neutral particles, from the very early stages of the evolution of the Universe, were not distributed homogeneously, but tended to agglomerate into large clusters. The cause for this inhomogeneity is unclear. If it is true that nine-tenths of mass of cosmic matter is composed of neutrinos, then the neutrino clusters were the right place for the emergence of galaxies.

The Universe contains approximately a hundred giga of galaxies. The age of these galaxies is not much shorter than the age of the Universe, but it is unclear how long they will continue to exist and if they will die, in a Universe which becomes colder as it evolves. Of course, most of cosmic space lies beyond the galaxies—even beyond the clusters of galaxies. Intergalactic space, the 'voids' between the clusters, is much larger than the volume inside the clusters. 'Empty' space is also the scene of different processes, but it seems that the extended existence of intelligent beings is not possible here. The existence of intelligence and the galaxies is indivisible.

Galaxies contain tens to millions of giga stars, plus a great deal of matter between the stars. Perhaps even 90 per cent of cosmic matter exists in the form of invisible intergalactic matter between the visible stars. However, the stars and their environment are the right place for the emergence and continuing existence of life (M. Fukugita, 1996).

2.3 STARS—SOURCES OF FREE ENERGY AND OF ELEMENTS

2.3.1 *The birth, life, and death of stars*

The birth of galaxies corresponds to the birth of the first generation of stars. The primordial matter of very young galaxies accumulated by mutual gravitational attraction into spheres with masses of approximately a thousand giga giga giga kilograms; that is, between ten times smaller, or hundreds of times larger, than the mass of our central star—the Sun.

If the mass of a protostar is somewhat small, say half the mass of the Sun, its evolution proceeds rather slowly and the star lives for tens of giga of years. If its mass is tens of times larger then the solar mass, its evolution progresses over tens of millions of years and ends with an enormous explosion, in a time interval of some seconds. This process is called a 'supernova'.

Table 2-2 Particles, atoms and molecules

Quarks include 3 families of elementary particles each of two kinds of particles. **Heavy particles** with electrical charge of 1/3 and 2/3: components of proton and neutron, (bonding force: strong nuclear force). Three quarks result in one proton or neutron.		**Leptons** include 3 families of elementary particles each of two kinds of particles: **Light particles** with electrical charge equal **1 or 0**: electron components of atom (bonding force: electromagnetic force).	
up-quark	down-quark	electron	electron-neutrino
charm-quark	strange-quark	mion	mion-neutrino
bottom-quark	top-quark	tau	tau-neutrino
Protons and neutrons as components of atomic nuclei, having quantized energy levels		**Electrons** as components of atoms, moving on quantized orbitals	
Atoms, conglomerate of atomic nuclei and electrons (bonding force: **electromagnetic force**) with generic properties called '**chemical forces**', which are due to quantum fluctuation properties, include strictly directional 'chemical' bonds.			
Molecules, conglomerates of atoms (bonding force: **electromagnetic force**). The orientated bonds lead to spatial complex molecules, being the appropriate liquid environment for more complex structures.			
Macromolecules, conglomerates of molecules (bonding force: **electromagnetic force**). The basis for more complex structure including living being and intelligence, the highest biotic information processing.			

The motor of a star´s evolution is, on the one hand, the Gravitational force, and, on the other hand, the nuclear reaction between atomic nuclei; that is, the Strong and the Weak nuclear forces. The transportation of generated energy from the centre of a star to its surface, and then into cold cosmic space, is controlled by the Electromagnetic force (Fig. 2-4).The small, long-lived stars of the first generation have probably continued to exist up to the present time.

A stellar explosion transports a large part of its matter into interstellar space in the form of dust and gas clouds. The primordial elements (hydrogen, helium) and the products of the nuclear synthesis inside a star of the first generation (mostly light elements: carbon, nitrogen, oxygen) are mixed with the gas and dust particles from other stars. After some time, the gravitational force accumulates interstellar matter once more into a protostar of the second generation.

Some stars of the second generation achieve large masses and, therefore, evolve very rapidly, producing new heavier elements (from iron up to uranium) and then ending in a stellar explosion—a so-called supernova. The possibility for the emergence of a star of the third generation then exists. Part of the matter remaining after a supernova explosion is in the form of a very dense, very small, rapidly rotating star—a so-called 'neutron star'. The matter in such a star is in the form of free neutrons, bound together by gravitation. More massive stars form 'black holes'.

2.3.2 *Emergence of and changes in chemical elements*

The history of the Universe, of galaxies and of stars is, at the same time, the history of the evolution of the matter; that is, the origin and transformation of chemical elements, from the simplest and the lightest (hydrogen) to the heaviest and most complex (uranium). In Nature, altogether, around a hundred elements exist, some of which are radioactively unstable. Atoms contain an atomic nucleus and electrons in atomic shells. Atomic nuclei are built of only two types of particle: protons and neutrons, which can even be regarded as two states of the same particle, also called 'nucleon' (D.Arnett, 1996).

How many different atomic nuclei can be constructed? Using from one to around hundred (why no more?) protons and from zero to a hundred and fifty neutrons (why no more?), some thousands. Under the influence of the Weak Nuclear Force and the Strong Nuclear Force, only a small proportion of all these potentially constructable atomic nuclei are stable or at least, stable enough to exist in the Universe over giga of years. These stable atomic nuclei number only 272 isotopes, belonging to 80 stable elements. All other isotopes are radioactively unstable, and are extremely rare in cosmic matter. Here can also be observed, even in modified form, the strategy of the overabundant generation of atomic nuclei and the selective elimination of the unstable ones, leaving only a small number of stable isotopes.

Our central star, the Sun, is probably a star of the third generation. This is the reason why all possible stable and quasi-stable elements (long-lived radioactive elements) are present in the Sun. The presence of all these elements is perhaps also one of the necessary conditions for the emergence of life-carrying planets.

From the point of view of cosmic abundance, cosmic matter contains three classes of elements:

1) The most abundant elements; These are hydrogen, with about 93 per cent of all atoms in Universe, and helium, the lightest noble gas, with about 6 per cent of atoms. Hydrogen is the lightest, simplest, and oldest element from

approximately ninety stable and quasi-stable elements. Hydrogen and helium are the products of the first minutes of existence of the Universe.

Table 2-3 Stability of elementary particles

Half-life of particles	Stable and quasi-stable particles	Unstable (radioactive) particles
Extreme long (>10^{33} years)	**Proton and bounded neutrons, also electron**	
Gigayears (10^9 years)		
1 year		
1000 seconds	**Free, unbounded neutron**	
Nanosecond (10^{-9} s)		**All other elementary particles (mesons, almost all hadrons)**
Attosecond (10^{-18} s)		**are extreme unstable**

2) Elements of average abundance, given in decreasing cosmic abundance; These are oxygen, carbon, nitrogen, neon (a noble gas), iron, calcium, sodium and sulphur. All of these together have an abundance of less than a ten-thousandth of that of hydrogen. These elements are products of the evolution of stars such as Red Giants.

3) All other elements, totalling about eighty, have a cosmic abundance significantly smaller than that of hydrogen. Included here are not only the light metals, with atomic mass between helium and carbon (lithium, beryllium and boron), but also the very heavy elements such as thorium and uranium, the latter being the products of supernova explosions.

In spite of the great importance of chemical matter in our later considerations, we cannot ignore the fact that the larger proportion of cosmic matter is not in the form of chemical elements, that is, not in the form of protons and neutrons, bonded in atomic nuclei, and in the form of electrons, together with atomic nuclei, resulting in atoms. It is possible that in the Universe, tenfold times more matter exists in other than the chemical form. There are different opinions about the properties of this 'dark matter'. The most probable is that it is in the form of neutrinos—stable, very light, neutral particles—which act with other types of matter through the Weak Nuclear Force.

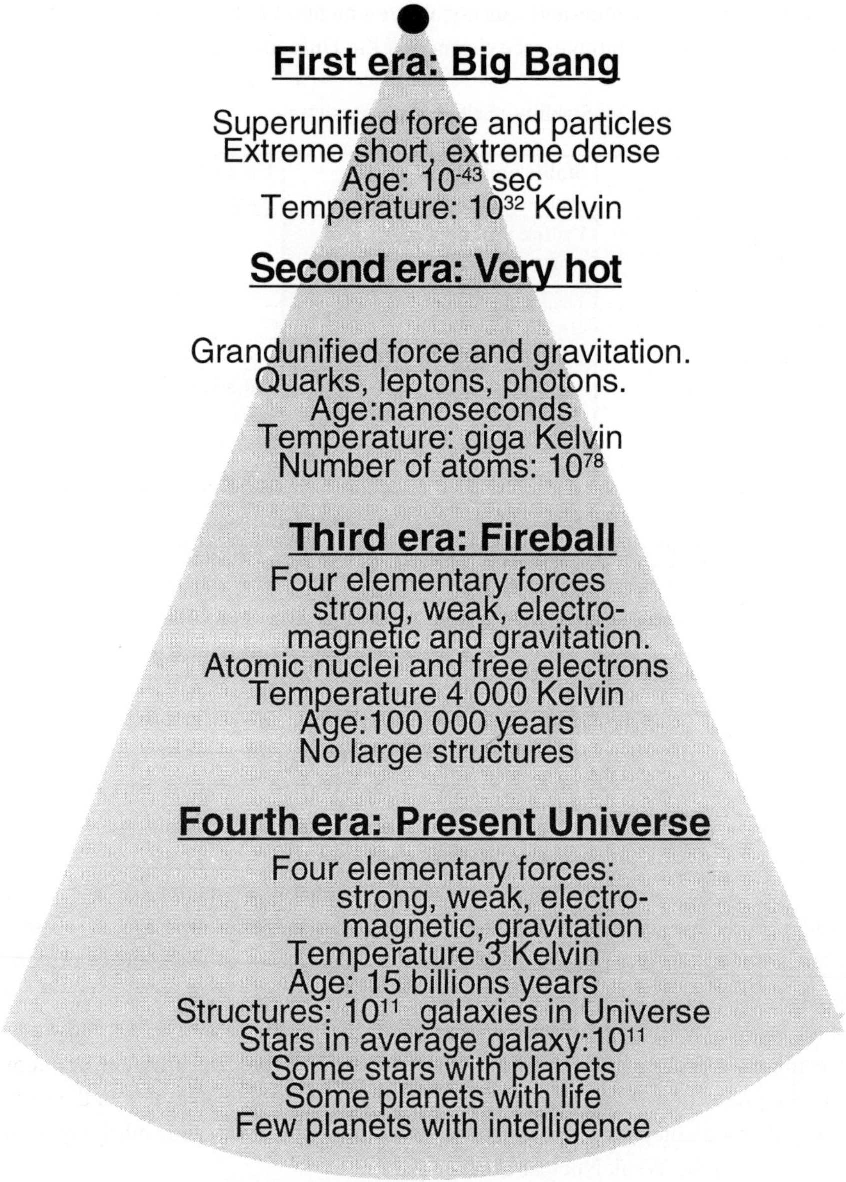

First era: Big Bang

Superunified force and particles
Extreme short, extreme dense
Age: 10^{-43} sec
Temperature: 10^{32} Kelvin

Second era: Very hot

Grandunified force and gravitation.
Quarks, leptons, photons.
Age: nanoseconds
Temperature: giga Kelvin
Number of atoms: 10^{78}

Third era: Fireball

Four elementary forces
strong, weak, electro-
magnetic and gravitation.
Atomic nuclei and free electrons
Temperature 4 000 Kelvin
Age: 100 000 years
No large structures

Fourth era: Present Universe

Four elementary forces:
strong, weak, electro-
magnetic, gravitation
Temperature 3 Kelvin
Age: 15 billions years
Structures: 10^{11} galaxies in Universe
Stars in average galaxy: 10^{11}
Some stars with planets
Some planets with life
Few planets with intelligence

Figure 2-3 Four eras in the evolution of the Universe

From our previous considerations, it must be clear that it is impossible to make a stable complex construction, not to say an intelligent structure, using neutrinos. Another type of matter, whose existence is all but certain, is 'neutronic matter', contained in 'neutron stars' also called 'pulsars'. Our previous considerations allow us to claim that, from neutrons alone, mutually bonded by the Strong Nuclear Force and Gravitation, both acting only by means of attraction, the only possible construction is a simple homogeneous sphere, of very large mass: a neutron star. An intelligent structure cannot be created from neutronic matter.

2.3.3 The uniqueness of the Earth

Around our central star exist many planets and many more secondary cold smaller objects—moons. Also present are thousands of very small and cold objects (asteroids), giga of comets, and many giga of giga of meteorites and meteors. On the other hand, the information about the existence of planetary system around other stars is very vague. Is the planetary system the exception or the rule? Is it very rare or very common? All these questions have an indirect influence on our discussion about terrestrial human intelligence and the possibility of extraterrestrial intelligence.

In one way or another, our planet contains all the stable and quasi-stable long-lived radioactive elements, such as thorium and uranium. (Fig 2-5). There is no doubt that many phenomena such as internal heat in the interior of the Earth, continental drift, volcanoes, degasification of rocks and the emergence of atmosphere and part of the hydrosphere, and, indirectly, global climate, all result from the radioactive decay of thorium and uranium. Therefore, the influence of radioactivity, even if indirect, upon the emergence of life and intelligence is of significance.

Only Venus and Earth are carrier of all three phases, gaseous, liquid, solid. Mercury and Mars include only solid phase. The large planets, Jupiter, Saturn and Neptune have a large dense atmosphere and a small inner partially liquid and solid core. Jupiter and Saturn, having an extreme large mass (roughly 300 Earth masses), act as scavengers of comets in space around the Earth; a condition sine qua non for long-term existence and evolution of terrestrial life. Without this comets scavenger process (for example the collision of Shoemaker-Levy comet with Jupiter in summer 1992) catastrophic events such as Cretaceous/Tertiary 'Great Extinction' would occur every 100 thousand years on the Earth instead of about once in 100 megayears.

The obliquity of a planet orbit (the angle that the spin axis makes with the perpendicular to its orbital plane) determines seasonal variations, and the evolution of long-term climatic phenomena. The Earth has been saved from chaotic variations of the obliquity by the stabilizing influence of the Moon—the only large satellite in the inner Solar System. The modest Earth obliquity of 23.5 degrees has small variations of approximately 1.3 degree, which result in triggering ice ages. For an obliquity greater than 54 degree the equator annually receives less radiation than the poles. The forecast for a 'Moon-less Earth' would have been bleak.

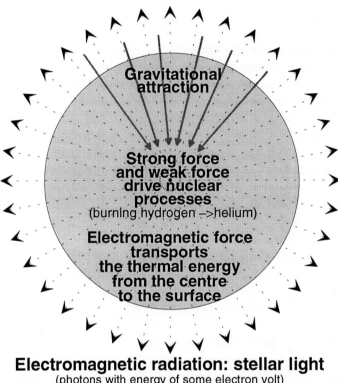

Figure 2-4 Stars and four elementary forces

2.4 ELEMENTARY FORCES AND PARTICLES; BEARERS OF LIFE

2.4.1 *Living beings—a universal definition*

We begin with a general definition, in which the phenomenon of life is independent, not only of the properties of this or another planet, but, if possible, also of one or other property of matter. We must limit ourselves to the general properties of matter, without knowledge of its detailed structure. Physics has a general method of describing material structures, independent of details, such as the properties of particles and the nature of forces. This is the language of thermodynamics, operating with the terms: order and disorder (randomness), free and bounded energy, entropy, spontaneous and forced transformations, equilibrium and steady-state. Our universal definition of life, which seems to be valid not only here and now on the Earth, but elsewhere in the Universe, contains five statements:

1) A living being has in a limited space and during a limited time, a relatively high order, much higher than the order of its direct environment. The living being and its environment (direct and distant, including the cosmic neighbourhood) create an indivisible totality. To summarize: Life is a highly-ordered system limited in space and time.

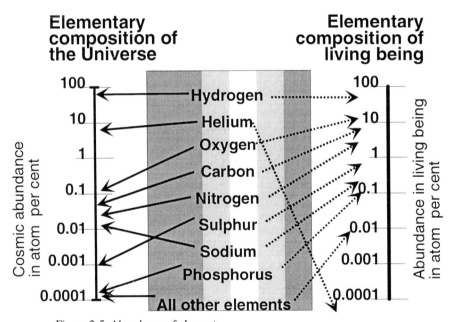

Figure 2-5 Abundance of elements

2) A living being contains in its interior a specific medium which is the carrier of the specific order in steady-state; that is, in continuous decay and reconstruction, and with the continuous inflow and outflow of matter, from the interior towards the environment, and vice versa. The internal medium cannot be in a gaseous state, because of the high disorder of movement of gaseous particles. Also, the solid, crystalline state, with very high order but without any particle movement (material transport) at all, seems to be excluded from the role of bearer

of life. The only appropriate state for the bearing of life seems to be the liquid or quasi-liquid state of matter; for example, a quasi- crystalline liquid. To summarize: Life requires an internally ordered liquid medium.

3) Each ordered system spontaneously decays and transforms into disorder. The only way to reconstruct the order is to use externally supplied free energy. The intake of external free energy, and its storage and use for removing the disorder, is the third property of each living being. The source of the free energy must be present in the direct environment of the living being, over a very long period; so long that the spontaneous emergence of life and its continuous evolution is permitted over giga of years. The reasons why gigayears, and not thousand of days or a giga of gigayears, will be discussed later. To summarize: Life needs the continuous inflow of free energy and fresh matter, and the more-or-less continuous removal of bonded energy and 'used' matter.

4) A living being has the ability to control the rate of its intake of free energy and the rate of removal of bounded energy. It must control the rate of reconstruction of its specific internal order, and must be able to co-operate with, and react to changes in the environment, especially to changes in the source of free energy. The control of the rate of all these processes also needs a kind of catalyst, a substance which increases or slows down selected processes. One important problem is the construction and reconstruction of the wall, the barrier which divides the interior of the living being from its environment and from other living beings. To summarize: Life is a self-controlling and self-constructing system.

5) The existence of a living being is limited. Sooner or later, it will be destroyed by external or internal processes. The ability for self-reconstruction is the most characteristic property of each living being. The plans for self-reconstruction allow continuous restoration of its own structure, and the control of all processes during life, including information processing, must be prepared by the living being itself, before its own decay. These plans, so-called 'genetic information', must be carried by as small a material carrier as possible, in order to keep the energetic and material cost as low as possible. This means that genetic information depends very strongly on its past, and on the past results of numerous accidental events in its history. To summarize: Life is a self-reproducing system, resulting from its own history, and based on overabundant production and selective elimination (Table 2-4).

2.4.2 *Elementary forces as the carriers of life*

Our definition of life contains five components, which can be formulated in the following hierarchical way. Life is: 1)an ordered system, 2)with an internal medium, 3)with free energy inflow, 4)with self-control and self-structuring, and

5) self-reproduction. In a more-or-less arbitrary way, we will assume that for each of these five properties there is a corresponding carrier—'brick' and 'mortar'. The problem is: which types of elementary particles and forces are able to carry the phenomenon of life?

Table 2-4 Universal definition of living being

Property	Characteristic
High order system in time-space	A living being has in a limited space and during a limited time, a relatively high order, much higher than the order of its direct environment. The living being and its environment (direct and distant, including the cosmic neighbourhood) create an indivisible totality.
Internal, ordered medium	A living being contains in its interior a specific medium which is the carrier of the specific order in steady-state; that is, in continuous decay and reconstruction, and with the continuous inflow and outflow of particles, from the interior towards the environment, and vice versa. The internal medium cannot be in a gaseous state, or solid, crystalline, state. The appropriate state seems to be a quasi-crystalline liquid.
Flow of matter and energy	Each ordered system spontaneously decays and transforms into disorder. The only way to reconstruct the order is to use externally supplied free energy. The intake of external free energy, and its storage and use for removing the disorder, is the third property of each living being. The source of the free energy must be present over a very long period; over giga of years.
Self-control and self-construction	A living being has the ability to control the rate of its intake of free energy and the rate of removal of bounded energy. It must control the rate of reconstruction of its specific internal order, and react to changes in the environment. The control of the rate of all these processes needs a kind of catalyst, and the construction of the wall, the barrier which divides the interior of the from its environment.
Self-reproduction of mutation	The existence of a living being is limited. The ability for self-reconstruction is the most characteristic property of each living being. The plans for self-reconstruction allow restoration of its own structure, and the control of all processes during life. These plans, so-called 'genetic information', must be carried by as small a material carrier as possible, in order to keep the energetic and material cost as low as possible. Life is a self-reproducing system, resulting from its own history, and based on overabundant production and selective elimination.

Let us begin with the 'mortar', which is able to hold together the 'ordered system' of a living being, taking into account the four elementary forces (in order of decreasing 'strength'): Strong Nuclear Force, Electromagnetic Force, Weak Nuclear Force and Gravitational Force. If life is a highly ordered system, it is obvious that it cannot be a simple homogeneous sphere, because the order of such a sphere is very simple and trivial and, therefore, very low. For making a sphere, it is enough to use a 'mortar' which is only adhesive; that is, to use a force which acts only attractively. For a more complex structure, of nontrivial high order, the use of only an attractive force cannot be sufficient—a repulsive force must also be present.

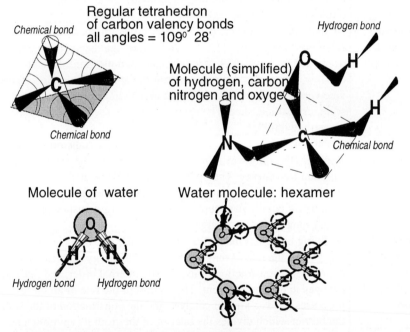

Figure 2-6 Chemical bonds, carrier of complexity

Two actions, attraction and repulsion, are able to hold two or more bodies at a given distance, without allowing them to adhere into one homogeneous sphere. All four elementary forces have attractive properties, but only one has both attraction and repulsion. This is the Electromagnetic Force. This is the reason for claiming that living beings throughout the whole Universe are constructed on the basis of the Electromagnetic Force.

Also, the strength of this elementary force is rather large, to allow a relatively condensed system to be built, even if the number of bound elementary particles is relatively small. Of course, the radius of action of the corresponding elementary force cannot be too small. Both nuclear forces, the Strong and the Weak, act only over a very small distance, resulting always in very condensed, but also very small, structures.

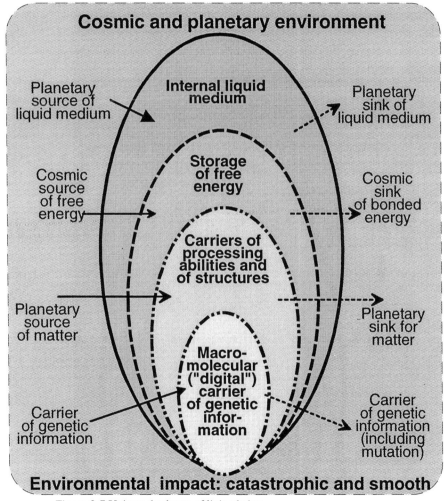

Figure 2-7 Universal scheme of living being

For example, the Strong Nuclear Force, being about 140 times stronger than the Electromagnetic Force, results in the emergence of small condensed spheres: atomic nuclei. Atomic nuclei are almost spherical, even if they contain just two or tens of nuclear particles (neutrons and protons), because the strong nuclear force acts only attractively. The Gravitational Force is a giga giga giga giga times weaker than the Electromagnetic Force. For making a condensed body, it needs a giga giga giga giga atoms. The result is a more-or-less spherical body, a planet or star.

Only the Electromagnetic Force, with large radius of action, with relatively high strength, and with both actions—attraction and repulsion—can be the universal carrier of life, as a cosmic phenomena. If this is a correct conclusion, then we must claim that human intelligence, being a product of the evolution of life on this planet, is also carried by the Electromagnetic Force.

If life is carried by the Electromagnetic Force, we have thereby assumed that only those elementary particles, elementary 'bricks' which are charged electrically or magnetically, can be involved in the structure of living beings. However, before we begin to search for the corresponding elementary particles, we must take into account a further very important element: time. On the basis of our definition of life, it is not clear enough how long life must exist in order to fulfil all the criteria of the definition of life. We must return to this question later, but for now we will postulate more-or-less arbitrarily that life, as a phenomenon, especially if at one moment in its evolution it produces an intelligent being, must endure for giga of years; that is, of the same order of magnitude as the existence of the Universe itself.

2.4.3   *Elementary particles as the carriers of life*

To summarize: Life must be built of elementary particles, which are electrically charged and stable enough to enable life to exist for giga of years. In such a case, our choice is unexpectedly small. There do exist some tens, if not hundreds, of elementary particles which are electro-magnetic charged, but almost all of them are unstable, with a lifetime of less than one-millionth of a second or even many factors of ten less than this (Fig. 2-5). Only very few elementary particles fulfil our criteria of longevity: these are the proton (and its nuclear partner, the neutron, electrically uncharged, but magnetically charged) and the electron. The proton is electrically positively charged and the electron negatively charged.

2.4.4   *Chemical forces as the carriers of life*

The carriers of life are clearly defined; these are the stable structures of the stable elementary particles: nucleons (the common name of proton and neutron) and electrons, that is atoms of chemical elements. There exist 80 stable elements and, in addition, around 10 naturally occurring radioactive elements. All of these exist on the Earth.

The elementary force acting between atoms is the Electromagnetic Force. Because of the very complex system of the mutual interactions of atoms inside the molecules and macromolecules, the acting forces have been called 'chemical forces' (Fig. 2-6 and 2-7).

2.5 THE ANTHROPIC PRINCIPLE

The Universe, the totality of all things, is designed in such a way that the emergence of an intelligent being, an intelligent observer and co-designer of this Totality, is not only possible, but even indispensable. Who knows? Perhaps the most exciting achievement of the natural sciences in the twentieth century, next to the Theory of Everything, will be the refinement of the Anthropic Principle. This book is not the correct place for a broad discussion of this principle. Its most important aspect, from our point of view, can be formulated in the following way:

A.Einstein (1935): "What I'm really interested in is whether God could have made the world in a different way; that is, whether the necessity of logical simplicity leaves any freedom at all."

Of course, the form of the Anthropic Principle given above is more a caricature than an idea which can help us better understand totality. One way or another, there seems to be a strong relationship between the laws of Nature, the rules of the Universe, elementary particles, elementary forces, the Big Bang and the fate of the Universe, and the emergence of life and of intelligent selfconscious beings.

From this, it follows that all these phenomena (the Totality of Everything including consciousness) are of the same nature, are ruled by the same universal laws: by the same forces, and are made of the same material components. Perhaps, from this, it follows that there is nothing else, nothing strange, and nothing unique in the human mind.

J.D. Barrow and F.J. Tipler (1986): "Intelligent life, or rather consciousness, is essential to bring the entire Cosmos into existence,... consciousness must continue to exist as long as the Universe does."

P.C.W. Davies (1984): "It is sometimes suggested that, if only we understood the workings of Nature in sufficient detail, we should discover that only one sort of Universe is logically consistent. If that were so, God would indeed have had an easy task of it, because the Universe we observe would be, in some sense, inevitable."

S. Weinberg (1996): "The so-called anthropic principle, which holds—in logic that many physicists find circular—that fundamental constants have the values they do because only those values allow human beings to exists and measure them.....The only kind of theory that is today respectable in which you can understand the cosmological constant problem is theories based on some kind of anthropic reasoning,"

At least now, it must be clear that there exists a paradoxical link between the Anthropic Principle and the criterion of universality (Section 1.4.1). The Anthropic Principle claiming that the *Anthropus* ('man') is the measure of the Universe, corresponds very well to the human bonus-malus system. In spite of this, it has not been declared as a general principle of natural sciences. Homo sapiens remains 'sufficiently' self-censorious.

2.6 BIBLIOGRAPHY

Alle`gre CJ, Schneider SH (1994) The evolution of the Earth. *Sci Am* **271**,4,44
Arnett D (1996) *Supernova and nucleosynthesis.* Princeton Univ Press, Princeton, N.J.
Barrow JD, Tipler FJ (1983) *The anthropic cosmological principle.* Oxford Univ Press, Oxford
Dyson FJ (1979) Time without end. Physics and biology in an open Universe. *Rev Mod Phys* **51**,3,447
Guth AH (1997) *The inflationary Universe* Helix Books Addison-Wesley, Reading, Mass
Hawking SW (1980) *Is the end in sight for theoretical physics?* Cambridge Univ Press, Cambridge UK
Hawking SW (1988) *A brief history of time.* Bantam Books, N.York
Hawking SW (1991) The beginning of the Universe. *Annals N.York Acad Sci* **647**,315
Hawking SW, Penrose (1996) The nature of space and time, *Sci Am* **275**,44
Hooft 't G (1997) *In search of the ultimate building blocks.* Cambridge Univ Pr, Cambridge UK
Layzer D (1990) *Cosmogenesis.* Oxford Univ Press, Oxford
Lederman LM, Schramm DN (1989) *From quark to cosmos.* Sci Am Library, N.York
Leslie J (1996) *The end of the world.* Routledge, London
Peebles PJE et al (1994) The evolution of the Universe. *Sci Am* **271**,4,28
Penrose R (1997) *The large, the small and the human mind.* Cambridge Univ Pr, Cambridge UK
Schwartz JH (1987) Superstrings. *Phys Today* **11**,33
Taube M (1982) An empirical formula for the coupling constants of the four elementary interactions. *AtomKernenergie* **40**, 2,128
Taube M (1982) A formula for the calculations of the ratios of the masses of some elementary particles. *AtomKernenergie* **40**, 3, 208
Taube M (1992) Evolution of matter and energy, in O.Hutzinger (edit) *The handbook of environmental chemistry.* Springer, Berlin.
Yuan C, You J (edit) (1995) *Molecular clouds in star formation* World Scientific Pub. Singapore
Weinberg S (1993) *Dreams of a final theory.* Hutchison, London
Weinberg S (1994) Life in the Universe. *Sci Am* **271**,4, 22
Wheeler JA (1974) The Universe as home for man. *Am Sci* **62**,683
Witten E (1996) The holes are defined by the string. *Nature* **383**,215

CHAPTER 3
LIFE BREEDS INTELLIGENCE

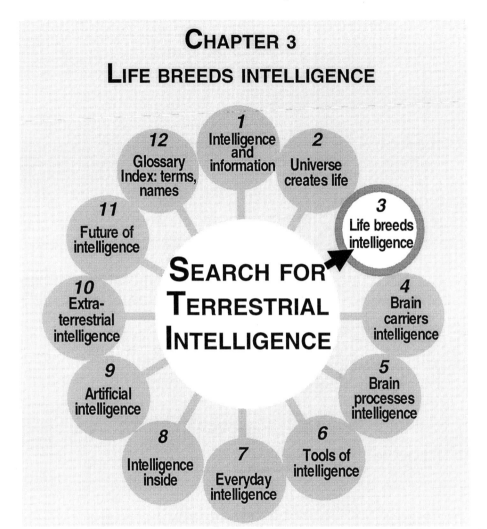

Outline of a terrestrial life. The emergence of life. Chemical processes, sources of matter and energy for living beings. Carriers of biotic direct information: genetic information.

Emergence of nervous system as carrier and processing system for indirect information. The breakthrough: the emergence of brain. The evolution of mammals' brain; limits of growth and lateralization. Lower stages of biotic information processing: reflex, instinct, and discent.

Homo habilis and Homo erectus. Homo sapiens neanderthalensis and Homo sapiens sapiens: modern man. The occurrence of intelligence.

3 LIFE BREEDS INTELLIGENCE 56
3.1 LIFE IS RELATED TO INFORMATION PROCESSING 56
3.1.1 Intelligence is carried by life *56*
3.1.2 Outline of a terrestrial living being *57*
3.1.3 The emergence of life *59*
3.1.4 Terrestriality of life *61*
3.1.5 Chemical phenomena as carriers of life *62*
3.1.6 Life and sources of matter and energy *63*
3.2 ANIMALS AND INDIRECT INFORMATION PROCES-SING 66
3.2.1 Material carriers of biotic indirect information *66*
3.2.2 The emergence of the nervous system *67*
3.2.3 The breakthrough: the emergence of the brain *69*
3.3 BRAIN, BODY AND ENVIRONMENT 71
3.3.1 Inputs of information: internal and external-detectors *71*
3.3.2 Brain size and body size *72*
3.3.3 The evolution of mammals' brains *73*
3.3.4 Brain and body connections *75*
3.3.5 Brain's limits of growth; lateralization *77*
3.4 BIOTIC INFORMATION PROCESSING 78
3.4.1 Reflex, the simplest genetically wired system *78*
3.4.2 Instinct: genetically complex wired system *78*
3.4.3 Discent: individual learning of new information *80*
3.4.4 Intelligence: creator of new information *81*
3.4.5 Hierarchical organisation of the four stages *81*
3.5 GENUS HOMO EMERGES 83
3.5.1 Genus Homo *83*
3.5.2 The occurrence of intelligence *83*
3.5.3 The occurrence of genus Homo *84*
3.5.4 Neoteny, the trick? *85*
3.5.5 Homo habilis and Homo erectus *87*
3.5.6 Homo sapiens neanderthalensis *89*
3.5.7 Homo sapiens sapiens—modern man *93*
3.6 BIBLIOGRAPHY 97

3 LIFE BREEDS INTELLIGENCE

3.1 LIFE IS RELATED TO INFORMATION PROCESSING

3.1.1 *Intelligence is carried by life*

The only known example of intelligence, at least of the phenomenon which in this book we are calling 'intelligence', is carried by living beings—by humans. It is obvious that the definition of the term 'intelligence', at first glance, is intimately related to the term 'life'. The spontaneous evolution of life on this planet led to the emergence of information processing creatures and, ultimately, to the intelligence carried by living beings.

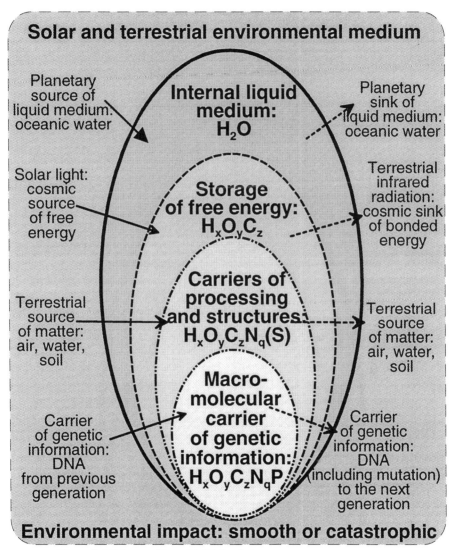

Figure 3-1 Scheme of terrestrial living being

3.1.2 *Outline of a terrestrial living being*

What is life? What is terrestrial life? Or life anywhere in the Universe? The proposed universal definition of life (M.Taube, 1962), which because of its independence of the kind and property of the involved particles and forces, seems to be valid throughout the whole Universe, contains five components, which are ordered hierarchically and are as follows:

Living being has following properties: 1) is a complex ordered system, assuming that ordered systems emerge spontaneously form a disordered environment, 2) includes an internal medium, 3) with inflow of free energy and outflow of bonded (degraded) energy, 4) with self-control and self-structuring abilities, 5) self-reproducing complex subsystem including ability to more or less 'exact' reproduction and being impacted by more or less catastrophic events in their environment. The environment includes the sources and the sinks for energy and for structural matter, and also other living beings of the same and of other species (Fig. 3-1).

In a more-or-less arbitrary way, we will assume that, for each of the five properties there is a corresponding carrier—'bricks' and 'mortar' (M.Taube, 1962, 1985). We know that the only appropriate carriers of life, the bricks, are the chemical elements (Section 2.4.4) and that the only 'mortar' is the electromagnetic force, in the form of a specific short-distance force—the 'chemical force'. Now we must search for the most suitable elements for carrying life throughout the whole Universe, at least on the Earth.

1) Because the chance for the spontaneous emergence of an ordered structure is, according to the Laws of Nature (formulated in thermodynamics), principally very low, then the structural material must be sufficiently abundant. When a very rare structural material is considered and the probability of constructing the structure is very low, then the likelihood of the emergence of the final product is negligible. The obviously highest probability occurs when using structural material with the highest cosmic abundance—hydrogen (Fig. 2-5).

2) The internal medium of a living being must be liquid (allowing relative fast material transport—diffusion etc.) and, in addition, must be well ordered, e.g. quasi-crystalline. The best candidate is a chemical compound containing hydrogen and oxygen, the next most cosmic abundant element (if helium is excluded because of its chemical inertness). The result is as simple as it is trivial: the best carrier of the ordered internal medium is hydrogen oxide in liquid form—water.

3) A living being depends upon a continuous inflow of free energy. Free energy must be stored by means of a more-or-less stable energy carrier. Passing on from a molecule of water, the best candidate for this role is a molecule containing not only hydrogen and oxygen, but also a third element, combined with these two atoms. The next most abundant element in the whole Universe is carbon. Molecules containing hydrogen, oxygen and carbon are very numerous, and a good number of them can carry free energy, even in an aqueous medium. Carbon has the unique

property of being able to form long-chain molecules. Some examples of these are sugar-like and fat-like substances. The biotic carriers of long-term stored free-energy are carbon-containing molecules, in an appropriate chemical environment.

4) A living being must be separated from its less-ordered environment and must have walls, membranes, etc. It must have catalysts for controlling the chemical reactions of synthesis and decomposition. These and other properties can be realized by means of macromolecules, built of molecules containing, beside hydrogen, oxygen and carbon, also a fourth element—nitrogen. Molecules containing all these elements cover a very large spectrum. Among others, the most important, at least from our point of view, are the amino acids. Nitrogen is the component of highly organized macromolecules (polymerized amino acids: peptides): catalysts and membranes.

5) The final property of a living being is the ability to carry, the information for self-reproduction (genetic information). To do this, in addition to the above-mentioned four elements (hydrogen, oxygen, carbon, and nitrogen) a fifth element allowing construction of very long, linear and relatively stable macromolecule must also be considered. This role plays phosphorus, in form of phosphate, as it is contained in ribonucleotide or deoxyribonucleotide molecules, polymerised in very long chains called RNA (ribonucleic acid) or DNA (deoxyribonucleic acid). These are the molecules which carry genetic information.

C. Sagan (1994): "Biochemical definition-for example, defining life in terms of nucleic acids, proteins and other molecules-are clearly chauvinistic. The definition that I like best—life is any system capable of reproduction, mutation and reproduction of its mutations."

But even Sagan (1994) claims that "life seems to need liquid water. Water-rich worlds something like ours, each provided with a generous complement of complex organic (carbon including) molecules".

3.1.3 *The emergence of life*

The elements are products of a long and complex evolution beginning with Big Bang primordial synthesis from elementary particles, through the later transformation in the interior of stars and explosions of supernovae. The elements of our planet are cooked twice in stellar interiors. The long evolution of elements results in the following mix. The most abundant elements in the Universe are: hydrogen (around 93 per cent of all atoms), helium (around 6 per cent), oxygen (less than 0.1 per cent), carbon (less than 0.1 per cent), and nitrogen (around 0.01 per cent). All the other, approximately 80, elements together make up 0.1 per cent. From our point of view, however, the amounts of sulphur, phosphorus, sodium, magnesium, calcium, iron and chlorine present in cosmic matter are also significant.

It can be expected that, anywhere where the temperature decreases to some hundreds Kelvin, the most abundant elements will react and produce simple chemical substances—mostly hydrogen-containing molecules, such as hydrogen oxide (water), carbon dioxide, the simplest hydrocarbon (methane) and nitrogen hydride (ammonia). Oxides must also be abundant; such as carbon dioxide and carbon monoxide. In addition, carbon-nitrogen and carbon-oxygen-nitrogen compounds must be expected.

The complexity of liquid water is due to the combination of the small size and distinct polar charge distribution of the water molecule. The charge distribution of the electrical field allows each water molecule to participate in strong polar (electrostatic charge-dipole or hydrogen-bonding) interactions with a high degree of spatial directionality. The strong hydrogen-bonding water-water interaction results in large cohesive energy or latent heat, a high boiling point, a high surface tension, and a reluctance to dissolve inert (nonpolar and hydrophobic) solutes with which it cannot interact through similarly strong polar forces (J. Israelshwili, H. Wennerstrom, 1996).

The emerging planets (the candidate for life bearer) contained in their atmospheres the above-mentioned chemical compounds, which are capable of further spontaneous chemical reactions. Among the more complex and well-ordered products of these spontaneous reactions are sugar-like, such as ribose, and fat-like molecules and numerous amino acids. Different sulphur-containing molecules also play an important role. Some other chemical compounds are relatively abundant, including those based on phosphorus, in the form of phosphates, so-called nucleo-bases, and their appropriate derivatives. For around 4 gigayears, on the surface of Earth, microscopic crystals of clay (a product of the chemical action of hot liquid water on certain minerals, mostly compounds of oxides of silicon, aluminium, magnesium, sodium, and iron) began to reproduce themselves by the simple process of crystal growth in aqueous solution. These crystals are of a relatively high order, and, if a crystal should fracture, each piece may inherit a copy of the pattern, sometimes with a slight change. Such defects can have a dramatic effect on a clay's physical and chemical properties. Clays are powerful chemical catalysts.

The probable way of evolution toward life on this planet include following steps: (Fig. 3-2)

–Formation and growth of RNA precursors PNA (Peptide Nucleic Acid), so called 'Pre-RNA-world' .

–RNA-world, in which RNA (RiboNucleic Acid), since around 3.7

gigayears plays both the role of carrier of genetic information and the role of primitive catalyst, promoting synthesis of proteins from amino acids. The presence of mineral surfaces (montmorillonite for nucleotides, illite and apatite for amino acids) induces the abiotic formation of rather long chains up to 55 monomers long, which suffices for emergence of genetic carriers (RNA) and peptides (Ferris JP, 1996).

Recent experiments show that the non-enzymatic replication is conceivable in a wide range of synthetic chemical systems. The challenge now facing these studies is how to develop information-coding systems from simple prebiotic precursors (L.E. Orgel, 1994). The oldest fossils of cellularly preserved filamentous microbes approx. 1.7 gigayears old have been recently discovered. But doubtless life exists on the Earth since approx. 3 gigayears in form of simplest unicellular prokaryotes.

3.1.4 *Terrestriality of life*

The properties of small molecules, such as amino acids, fatty acids, sugars, etc., are clearly defined. They have the same properties throughout the whole Universe. This is not so in the case of large polymeric molecules, such as proteins. The same, or almost the same, properties, (for example, catalytic properties (enzymes) or spatial structures (membrane, organelles)) can be exhibited by different macromolecules.

Different macromolecules, with different evolutionary pasts, and, therefore with different genetic information, function in similar ways. These functions have some universal properties, such as controlling the photolysis of water and the synthesis of sugars. The past of the macromolecules is obviously connected with the past of the direct and indirect environment, which is itself of local character. Macromolecules and their genetic information are of non-universal, local (that is, terrestrial) character.

It currently seems to be more justifiable to assume that the process of spontaneous self-polymerization began with ribonucleic acid (RNA), and not with deoxyribonucleic acid (DNA). The latter is carried by most unicellular organisms and all multicellular organisms, while the former is only found in simple organisms and viruses. However, RNA has some unique properties which allow it to be considered as the primordial carrier of life. This macromolecule is not only a carrier of genetic information but also has some self-catalytic properties. When building an ordered system, such as a cell, very complex chemical reactions must take place—among others, the polymerisation of amino acids and also of ribonucleic acid itself. Catalysts can be very helpful here. Polymerized amino acids, so-called peptides, are excellent catalysts.

Particular note must be taken of the fact that RNA, as well as DNA, represents genetic information, ordered in a line, in the form of string, which is a one-dimensional structure. There must be good reasons for not using two-dimensional (surface-like) or three-dimensional (space-extended) structures. As we will see later, the linearizing of information organization allows much more exact reproducibility, and also easier readability, of part of the information chain. The next steps on the long, and very difficult, pathway to the emergence of very primitive unicellular living beings are at present not very clear, but will become clearer in the future (Ch. de Duve, 1996).

The real story of life is meaningful only if all the 'unexpected' events, which played a decisive role in the history of the Earth, including all extraterrestrial impacts, have been taken into account. The terrestrial life is only one example of universal life.

Elsewhere in Universe there surely may exist other 'mutants' of planetary life.

3.1.5 Chemical phenomena as carriers of life

Hydrogen is not only the most important carrier of life, it is also the overwhelming component, out of every one hundred atoms in a terrestrial living creature around 60 are hydrogen, some 21 are oxygen, 17 are carbon, 1 is nitrogen or sulphur, and 1 is phosphorus. All the other dozens of elements constitute the rest (Fig. 2-5) Using the five elements of hydrogen, oxygen, carbon, nitrogen and sulphur, some hundreds or even thousands of amino acids can be built. From this enormous overabundance of amino acids, only twenty have been selected to construct a terrestrial living being.

Just these twenty amino acids are able to produce millions of different proteins; that is, macromolecules. Of the vast number of possible proteins, only a vanishingly small proportion exist. Human body includes around 100,000 proteins (proteins of the immunological apparatus are here neglected) with different twists and bumps utterly unique to the owners. It is of interest to relate the number of the amino acids to the number of letters and the number of proteins to the number of words, in a modern language (Section 6.5).

Ch. Darwin (1858): "It is often said that all conditions for the first production of a living organism are present, which would have ever been present. But if (and oh, what a big if!) we could conceive in some warm little pond, with all sorts of ammonia and phosphoric salts, light, heat, electricity etc. present, that a protein compound was chemically formed ready to undergo still more complex changes, at the present day such matter would be instantly devoured or absorbed, which would not have been the case before living creature were formed."

The most important elementary carriers of life—hydrogen, oxygen, carbon, nitrogen and phosphorus—do not exhaust the list of elements which from our point of view, are of significance. The criteria in the search for other elements are the following. Assuming water to be the internal medium of a living being, one must consider the fact that water is an excellent medium for dissolving inorganic salts, especially those having an ionic crystalline structure. The aqueous solutions of such salts, being chemical compounds of so-called metallic and non-metallic elements, have excellent electrical conductivity which opens a broad field of operation for electrical phenomena in the processes of living beings. It is obvious, and will be discussed later in detail, that there is a mutual dependence between information transport and processing, and electrical signal transport (nervous signals) and processing.

The next criterion for appropriate ionic salts is, of course, the cosmic abundance of their elementary components. It must be remembered that the lightest metallic elements—lithium and beryllium—and the lightest non-metallic element—fluorine—are present in extremely small amounts, resulting from the properties of their atomic nuclei, and from the history of the nuclear origin of elements during the evolution of the Universe. Taking all these things into account, we find that the most abundant metallic elements are sodium, potassium, magnesium calcium and iron and that the most abundant non-metallic elements chlorine and sulphur. These elements form the ionic components of the electrically conducting aqueous solution within each terrestrial living being.

3.1.6 *Life and sources of matter and energy*

S.J.Gould (1994): "The history of multicellular animal life may be more a story of great reduction of initial possibilities, with stabilization of lucky survivors, than a conventional tale of steady ecological expansion and morphological progress in complexity."

In spite of this opinion the evolution of the living being on this planet shows a general tendency in the kind of the energy intake and energy transformation. Because a living being has a very highly ordered structure, it is always threatened by the spontaneous decay of this order. Only the inflow of free energy, which is able to restore order, enables the long-term existence of life. This free energy can be carried in the form of photons, the massless carrier of electromagnetic energy, or in the form of chemical molecules, which are able to transform into other molecules, giving up stored free energy.

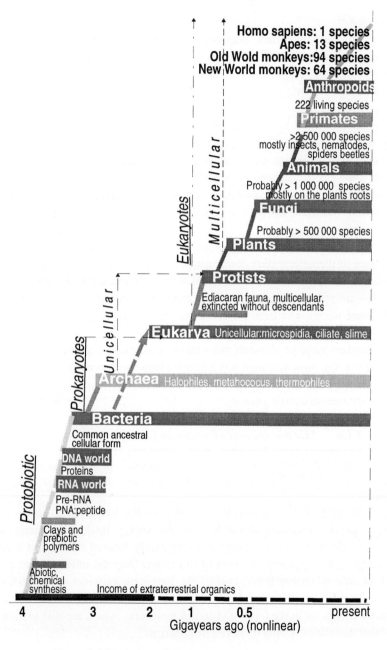

Figure 3-2 Evolution of species

S. Boltzmann (1904): "The general struggle for survival of the living creatures is:–not a struggle over primary materials,–also not a struggle over energy (which in form of heat is abundantly available in everybody, but unfortunately not convertible to other energy forms),–but a struggle over entropy which is made available in the transfer of energy from the hot Sun to the cold Earth."

Despite the importance of molecular self-replication of live-bearing peptide the first experimental has been realized only recently on some selected 32-residue a-helical peptide which can act auto-catalytic in templating its own synthesis in dilute aqueous solutions (D.H. Lee et al, 1996; G. Kiedrowski, 1996).

Table 3-1 Types of nutrition

Type	Chemotrophy	Autotrophy: Photosynthesis	Heterotrophic Absorption	Heterotrophic: Digestion
Characteristics and the source of free energy	**Chemical substances,** carrying free energy: free hydrogen H_2 and its compounds CH_4 H_2S, and Fe^{+2} salts	**Solar light,** carrying free energy, for the photosynthesis of substances carrying free energy (sugar). Materials for photosynthesis: $H_2O + CO_2$	**'Organic' matter:** substances of biotic origin, living or dead, carrying free energy) by slow absorption through the surface of the living being	**'Organic' matter** by means of the very fast swallowing of 'food', followed by rather slow digestion inside the body.
Type of organisms	Unicellular organisms (prokaryotes)	Cyano-bacteria (prokaryotes), plants (eukaryotes)	Uni-and multicellular fungi (prokaryotes and eukaryotes)	Some unicellular organisms and for all animals (eukaryotes)
Speed of nutritional processes	Slow acting nutritional processes: chemotrophy, autotrophy (photosynthesis) and heterotrophy of the absorption type. A living being of this type must have patience, and cannot depend upon a fast reaction rate in the nutrition source, and cannot evade attack of predators.			Very fast swallowing process, followed by a slow digestion process. Reaction in the short time available for catching prey or evading attack of predators.
Energy flux	Relatively slow; < 0.1 watt/kg			Relatively high; 1 watt/kg. Nervous system: 12 watt/kg

The answer to the question of why there is so much interest in these details is of the highest importance. Biotic indirectly acting information processing (Section 1.2.3), in the true sense of the term, is related to the means of utilising free energy. The method of nutrition determines the emergence and evolution of biotic information processing—the nervous systems. The following is a short list of the four different methods of 'nutrition' ('trophic': connected with nutrition) of terrestrial living beings (Table 3-1). It is an open question, how far these principles are valid in extraterrestrial biospheres. Do they have a universal character? Or are they only of local, terrestrial validity?

About 3.7 gigayears ago the first living organisms appeared on the Earth. They were small, single-celled microbes (prokaryotes). not very different from some present-day bacteria, with a diameter of one micrometre, that is with mass of 1 picogram (Ch de Duve, 1996).

The multicellular organisms were based on eukaryotic cells, with a typical range of 10 micrometre, that is with mass of 1 nanogram. Eukaryotic cells include numerous organelles: highly structured chromosomes made of strings of DNA, skeletal elements (microtubules) several thousand specialized structures such as peroxisomes, mitochondria, lysosome, Golgi apparatus and other.

These rather simple principles of the organization of living beings have resulted in at least ten million different species of micro-organisms, fungi, plants, and animals. We have mentioned Nature's strategy, of 'overabundant production'. Evolution is not just 'survival of the fittest' but primarily 'survival of the luckiest' because of the decisive impact of chance (such as a meteoritic impact 65 megayears ago) (Fig. 3-2 and Table 3-3). What could be more specific to this planet than the species of life it contains? If the term 'terrestriality' has any meaning at all, then it does in this area above all others. All species are terrestrial. No species of a universal character exists. Neither does universal genetic information exists, even if the principles could be universal.

3.2 ANIMALS AND INDIRECT INFORMATION PROCESSING

3.2.1 *Material carriers of biotic indirect information*

What is the 'purpose' of indirect biotic somatic information processing? What is the benefit of having it, for a living being, or, more precisely, for an animal? Why is a more highly-developed, more complex, and, thereby, heavier information processing system, using much more energy, better able to adapt to a changing environment? Why is a more complex nervous system more efficient? It must not be forgotten that worms or snails, in spite of having very simple nervous systems, are relatively well adapted to the present terrestrial environment, and have been successful throughout the whole history of the Earth for the past 500 megayears. The relatively highly-developed nervous system of the mastodon was not much help in its survival, which lasted for only some hundreds thousand years. The large brain alone was not enough to ensure success.

There exist many ways of improving a creature's chance of survival. One advantage, among many others, is to save energy required for obtaining food, for faster flight or for running distances. From our point of view, the advantage saving energy as the result of more effective information processing is also attractive. A better brain can save energy, but this simple statement cannot be the full answer; only a partial one. The sexual selection is probably the much stronger motor.

Assuming that biotic information processing can be successful in some cases, from what raw materials can and must a system be constructed? Also, this question will probably have a universal answer, because the materials are on the molecular level. (Table 3-2)

The most appropriate solution is to use the same raw materials for the construction of the living being and for its information processing system—the nervous system. In our opinion, anywhere in the whole Universe and at any time, living beings are made of chemical macromolecules and molecules, molecular ions, metallic and non-metallic ions of the most abundant elements, and electrons, joined by chemical forces, which are a manifestation of the elementary electromagnetic force.

Electrons, ions, molecules and macromolecules, immersed in an aqueous solution, and acting through electromagnetic force, are able to carry very large amounts of information and to process them with a not too high energy requirement. Probably throughout the whole Universe, spontaneously-evolving biotic information processing systems are all built of such bricks and mortar. The products of a technical civilisation, its artifacts, deliberately constructed for information processing, can be built of other raw materials, but must always be based on the action of the electromagnetic force and on appropriate materials, such as non-conductors (insulators) and conductors, semiconductors and super-conductors.

3.2.2 *The emergence of the nervous system*

One of the greatest 'inventions of Nature' seems to be the nervous system—the biotic information processing system. (The other, an abiotic information processing system is the man-made machine system, the computer). Physical signals from the environment, as well as from the body, are transformed into internal representation inside the body, then transformed, stored, related, otherwise processed, and, finally, converted into nervous signals, which put into action mechanical or chemical devices outside or within the body.

Protists (eukaryotic, unicellular organisms, comprising the algae, protozoans, and lower fungi) also have a simple information processing system including some sensors and simple motors, such as flagellar motors. Indeed, protists had some of the basic building blocks of a nervous system well before animals and plants evolved. For example, protists make use of acetylcholine (neurotransmitter: Section 4.5.2) and rhodopsin (part of optical sensors in eyes).

The nervous system being itself a multicellular system, is of course, to be found only in multicellular organisms. The most simple nervous system is the diffuse one, which is found in coelenterates, such as hydroids, jellyfish, sea anemones, and corals. In diffuse systems, nerve cells are distributed throughout the organism, usually beneath the surface. Some local concentrations of nerve cells are called ganglia.

The common hydra has a nervous system that consists entirely of a nerve net—a meshlike system of separate nerve cells dispersed over the organism, but more concentrated near the 'mouth'. Some cells play the role of sensory cells; others that of motor cells. Some of the nerve cells are capable of secretion.

During the 500 megayears of biotic evolution the nervous system was based on the following steps; here some selected examples:

In sea anemones, the neurons are more highly concentrated near the mouth. Although local concentration of neurons reaches its highest degree in the jellyfish, it is not sufficient to be considered a central nervous system.

In echinoderms (starfish, sea cucumber, sea lilies), the nervous system is sufficiently concentrated to be called a central nervous system, but there is no true brain. The major components of the nervous system are a nerve ring around the mouth, radial nerves extending into the arms, and a nerve network beneath the body covering. Many starfish have eyespots and sensory tentacles at the tip of each arm. Neurons are often gathered into larger discrete structures, the so-called 'ganglia' linked by axon tracts.

The worm Caenorhabditis elegans is an animal about a millimetre long, full-grown; a free-living species of soil nematode, found world-wide, that subsists as a filter feeder, mainly on bacteria. It is one of the best investigated animals. Some facts about it are:

–The adult animal has 959 somatic cells, plus those of the germ cells; another 131 cells died during its the growth,

–The entire genome is about 100 million base pairs long (only 5 per cent directly carries the information for synthesis of functional and structural proteins, more then 95 per cent is so called 'introns' which are only indirectly involved in the synthesis of proteins),

Table 3-2 Velocity of the information processing

	Carriers of free energy	
	Chemical molecules *Chemotrophy*	Solar photons *Phototrophy*
Very slow (without nervous system) direct through metabolic system	*Homotrophy,* **Unicellular organisms**	*Homotrophy,* **Unicellular organisms and multicellular plants**
	Heterotrophy by absorption **multicellular fungi**	—
Very fast (with nervous system) indirect, through nervous system	*Heterotrophy* by digestion **multicellular animals**	—

–The nervous system, when fully developed has 407 neurons. During maturation, some of neurons perish and 302 neurons remain in the hermaphrodite, plus 56 of the neural support cells that tend and maintain the wiring, and 381 neurons in the male. Its repertoire of behaviour is rather modest it can recognize chemical compounds, including its own pheromones; the worm can migrate to a familiar temperature; and it prudently avoids bumps on the nose. It is worth remarking, that 80 million base pairs are necessary in Caenorhabditis elegans for steering the evolution and existence of such a small body, with less than one thousand cells.

The only 40 times larger human genome, with about 3 giga of base pairs (from which about 97 per cent is not directly carrying a genetic information) is responsible for the human body, with some 10 thousand giga of cells, including 100 giga of neurons in the central nervous system. The nervous system changes less during evolution than the motor systems (muscles). For example, it has been proposed that the neural circuits underlying locomotion are similar in fish, amphibians, reptiles and mammals, even though the muscles and limbs used are very different. The reason that the nervous system is well conserved during evolution could be because it is complex and highly interconnected; it may be difficult to change one component without causing major disruptions in other components.

3.2.3 *The breakthrough: the emergence of the brain*
The trend towards greater centralization and cephalization (that is, the

Table 3-3 Mass explosions (good genes) and mass extinction (good luck)

Mass explosion; cause	Megayear	Period	Mass extinction; cause
Supernova explosion	4 600		
Origin of **Sun** 30% less luminosity	4 600		Jupiter as scavenger
Origin of planet **Earth**	4 560		
Catastrophic origin of **Moon**	4 100	*Hadean*	
Possible emergence of **life**	4 100-	*Archean eon*	Great Bombardment
Pre RNA and **RNA World**	3 900		
Emergence of **prokaryotes**	3 500	*Proterozoic*	
Emergence of **eukaryotes**	2,000		Huronian ice age
Explosion of **cyanobacteria**	1 700		Oxygen-rich atmosphere,
	800	*Precambrium*	Decline of stromatolites
Emergence of multicellular marine fauna **'Ediacaran'**	700		Ediacaran extinction, no descendants
Cambrian explosion: 40 phyla	545	*Cambrium*	
Emergence of Supercontinent: **Gondwana**, Laurentia	450	*Ordovicium*	1st extinction of 95% marine species, trilobites
Emergence of **first land plants, fishes**		*Silur*	
	438	*Devon*	2nd extinction; fishes, trilobites
Explosion of **Coal** forests	408	*Carbon*	
Emergence of **Pangea supercontinent**, Siberian flood basalt	250	*Permian*	3rd extinction of about 90% of genera. Anoxic ocean, CO_2
Ocean **Tethys** Begin of **Atlantic Ocean**	180	*Trias*	4th extinction: reptiles, ammonites
Explosion of dinosaurs. Emergence of mammals (with **neocortex**)			
Explosion of **mammalian**	65 asteroids	*Cretaceous/ tertiary*	5th extinction: dinosaurs, ammonites:
Emergence of **apes**		*Miocene*	
Explosion of hoofed animals	10		Climate drying savannah
Emergence of **Australopithecus**	3	*Pliocene*	Extinction of Australopithecs
Emergence of **Homo habilis**	2		
Emergence of **Homo erectus**	1.5	*Pleistocene*	Extinction of Homo habilis
Emergence of **Homo sap. neanderthal**	.0.4		
Emergence of **Homo sapiens sapiens**	0.120		Warm Eemian period
	0.045		Extinction of Neanderthals
Explosion of global **human population**	0.010	*Holocene*	6th extinction of large animals; over-killing by human hunters
Explosion of **technical civilisation**	0	*Present*	7th extinction of some species and biotops: tropical forests

emergence of a central nervous system, the primitive brain, where '*cephal*' means: concerning the head) continues within annelid worms, molluscs and arthropods.

Arthropods are crustaceans (shrimps, lobsters, scorpions), spiders, and insects. They have a complex, compartmentalized nervous system consisting of three main regions. Central nervous systems of insects have remained remarkably constant during evolution, while the external body form and the details of behaviour have diversified in profusion. Neural evolution of insects has proceeded primarily by tinkering with the synaptic relationships (changes in connectivity) among existing cells rather than by altering the neuron population.

Neurosecretory cells, specialized types of biotic effectors, which have been identified in all major invertebrate groups, reach their highest degree of development. The nervous system of vertebrates consists of brain, spinal cord and peripheral nervous system. The higher stage of biotic information processing began about 100 megayears ago, when certain gene lines enabled animals to learn from their elders during their lives, rather than inheriting behaviour patterns at conception.

New uses for old anatomy is probably more important in brain evolution than in any other organ, simply because neurons have a common currency (measured in units: millivolts and microamperes) with which to compare very different things. Recent findings suggest that in spite of enormous differences in the evolution path, the processing of pattern vision in insects and higher vertebrates, including humans are based on similar principles (M.V. Srinivasan et al, 1993).

Some megayears ago, our primate ancestors began to rely on tools made of bones, sticks, and stone, and accelerated again, with the harnessing of fire and the development of complex language, some hundreds of thousands of years ago.

3.3   BRAIN, BODY AND ENVIRONMENT

3.3.1   *Inputs of information: internal and external-detectors*

Some remarks concerning the transformation of signals (internal and external) into sensor activity, that is nervous activity, being carrier of information. Figure 3-3 gives appropriate data.

The responsibility for smell sense are carried by around 1000 genes (of a total human genome with 80,000 genes including about 60,000 genes for brain´s structures and functions). Humans are able to differentiate about 10,000 different odorous substances, stinks and scents.

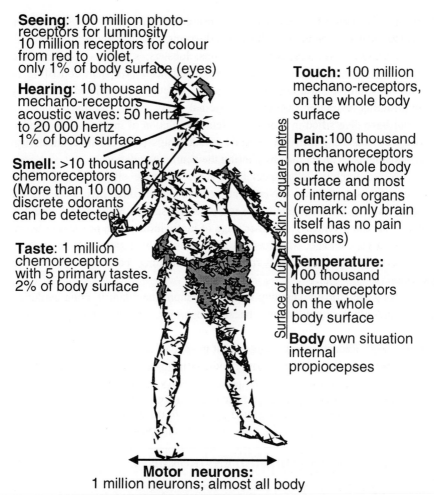

Figure 3-3 Human detectors and effectors

3.3.2 *Brain size and body size*

Our intuition tells us that a larger brain is the carrier of larger, or even higher, intelligence. However, it is a mistake to believe that the largest mammals, such as dolphins or elephants, having a large brain, even significantly larger than that of a human being, have a brain as complex as ours.

The dolphin's brain is significantly less developed than the brain of a primate, and its cell architecture is rather primitive. The size results from the larger body, the greater number of muscles (effectors), and the larger number of some receptors (detectors), but not a larger number of interneuronal connections; and only this is evidence of the complexity of a nervous system.

The other assumption, that the ratio of the masses of brain and body can be a good measure of the complexity and degree of development of an information processing system, is also not correct. The best measure of the 'encephalization quotient' is the relation of the mass of the brain (grams) to body mass (kilograms) to 0.76 power (R.D.Martin, 1981)

A whale has the largest brain at the present time. Its mass is around 6 kg, which corresponds to 1/10000 of its body. An African elephant has a brain of 5.7 kg, or 1/600 of its body mass. An average human brain is 1.35 kg, or 1/40 of his body mass; and small monkeys (macaques) have a brain of 0.08 kg, corresponding to 1/12 of their body. It has been claimed that significant associations exist between diet and encephalization; frugivorous and insectivorous primates are more encephalized than folivores. But it seems that a reverse association is also valid; that with a larger brain it is easier to catch insects than with a smaller one. The domestication of different animals results in a decrease in relative brain mass of around one tenth to even one fifth, in relation to the wild variety. A wild pig has a brain of 0.18 kg, with a ratio of brain mass to body mass of 1/315, while the domestic pig has a brain of 0.11 kg and brain/body ratio of 1/1300 (including fat!). The cerebral cortex is even one-third smaller than that of the wild animal. After thousands of years of domestication, the result is: the 'stupid goose', 'stupid donkey', etc. Domestic cats had only about half as many neurons in the ganglia (nerve clusters) as a wildcats. The way the domestic cat got a smaller brain was by losing more neurons during the first month of life rather than by producing in embryo fewer neurons. Only the racing horse and hunting dog have rather well-developed brains, but a lap dog has deliberately been selected by humans even possessing a small brain and low 'intelligence'.

However, the most significant property of the human brain in relation to the animal brain, even the largest, is its high complexity of the macro- and microarchitecture (cellular, chemical), the complexity of the neuronal interconnections, and the plasticity of function of some brain´s areas.

3.3.3 *The evolution of mammals' brains*

The evolution of the brain of mammals does not, of course, mean a proportional increase of all components. On the contrary, the brains of lower mammals, marsupials, such as the opossum, is deprived of corpus callosum—the connection between the left and right hemisphere (Fig. 3-4).

From the mouse, through the bush baby, and rhesus monkey to the chimpanzee, the relative size of different brain lobes—temporal, frontal and parietal—increases. A significant increase in the size of the cerebral cortex (neocortex) can be seen in Homo erectus. The lateralization of the hemispheres and the increase of the speech areas is probably the characteristic change of the brain of Homo sapiens neanderthalensis (Neanderthal man), though mainly of Homo sapiens sapiens (modern man).

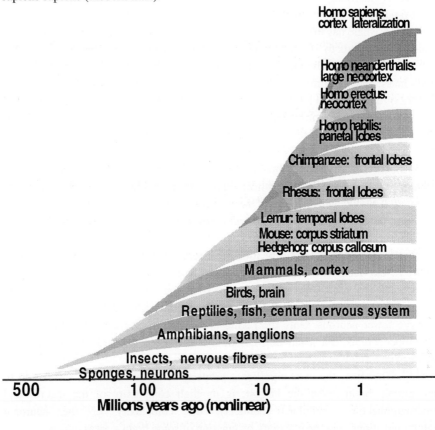

Figure 3-4 Evolution of nervous system

The lower brain stem has a great number of common features in most vertebrate species. In contrast, the forebrain and midbrain show a high degree of diversity. However, the functions of different brain regions change significantly during phylogeny. The brain is the most variable of all complex organ systems and the forebrain is the most variable of the brain regions; one speaks about high plasticity.

3.3.4 Brain and body connections

To illustrate the relation between the nervous system and some features and structures of an animal's body, let us look at the relationship between the means of locomotion of an animal and the level of organization of its nervous system (Fig. 3-5).

J. von Uexküll (1920): "When a dog runs, it moves its legs. When a sea urchin runs, it is moved by its legs."

Worms, such as Caenorhabditis elegans move by contacting the surface with their ventral side. Similar locomotion is also employed by the sea slug 'Aplysia', which has about 100,000, neurons clumped into 100 ganglia. Its stage of information processing corresponds to the lowest—the reflex.

Cephalopods, such as octopus, squid, and cuttlefish, have their hard shells and are the most mobile of the invertebrates. They have the largest nervous system, with a design principally different from that of a vertebrate (Section 4.2.3). Their brain has eight ganglia, arranged in a ring around the oesophagus. Small octopus can learn to solve problems, such as how to open a container of food. This ability corresponds, in our terms to, the stage of instinct.

Decapoda (ten-legged creatures), belong to the class of Crustacea (phylum Arthropoda). Their nervous system is partially centralized in the form of ganglia. However, spiders have eight legs and their nervous system is at the stage of instinct (Fig. 3-5).

Insects have six legs and, some of them two wings. For around 250 megayears, insects conquered water, land and, as the first flying animals, also the air. Their nervous system is rather complex. The nervous system of a bee consists of one million neurons, corresponding to a mass of some milligrams, and a clearly differentiated brain. The corresponding stage of biotic information processing is that of instinct.

Fish are very mobile and have brains larger than most invertebrates, but smaller than other vertebrates; probably because water is a simple environment. Their corresponding stage of information processing is that of instinct.

Reptiles, four-legged with a large and strong tail, emerged 280 megayears ago on land, a rather complex environment, and have larger nervous systems. However, all cold-blooded animals have a slow metabolic rate, and, as the central nervous system requires a relatively high energy inflow per unit of mass, cold-blooded vertebrates cannot have larger brains.

They should be classified as possessing a rather low stage of biotic information processing called instinct. Birds, four-legged (including two wings) and with a tail, although closely related to reptiles, are warm-blooded and can have larger brains; but too large a mass of brain makes flight impossible. The hummingbird has the smallest brain, with a mass of approx. 0.1 grams and 10 mega of neurons. However, some bird species are at a relatively high stage of biotic information processing, at called in this book 'discent'. For example, some parrot species are called 'flying primates'.

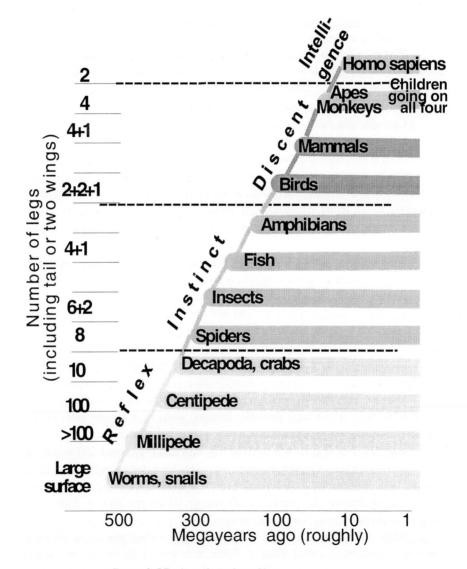

Figure 3-5 Brain and number of legs

Four-legged mammals, often with a tail, are the representatives of the relatively high stage of information processing, we call discent.

Two-legged man, having two hands and no tail, represents the highest stage of information processing, which we call intelligence.

3.3.5 Brain's limits of growth; lateralization

Biotic evolution shows a 'tendency' to improve information processing in a rather monotonic fashion, by increasing the complexity of the central nervous system. Over a very long period, this corresponds to an increase in the mass of the brain, in relation to the mass of the body. The energy requirement of the brain is relatively very high. Specific energy consumption, that is, the amount of energy consumed per unit mass of brain and unit time reaches a value ten times larger than the mean energy consumption of the rest of the body.

For the sake of illustration: the human brain uses about 18 watt per 1.4 kg continuously, day and night, during all life, even during a longer period of hunger. It is easy to see that, when the brain having more than 2 per cent of the weight of the whole body, needs energy inflow, of almost 20 per cent of total body energy consumption. It seems that further evolution in Nature must look for another solution if biotic information processing is to go further. The resources for further evolution can be found in one of the specific properties of the brain. Brains are built in a clear symmetrical way, containing two hemispheres—left and right. In higher animals, the left hemisphere controls the right side of the body and the right hemisphere controls the left side.

This is true for the lower stages of biotic information processing. At the higher stage of animal information processing, called in this book 'discent', the brain areas carrying the highest processes, the so-called tertiary areas (for the human brain: Section 4.1.5 and Fig. 4-3) seem to be involved in all processes of co-ordination of the activity of the body as a whole. It can be said that, at this high level of information processing, there exists a redundancy in some areas of the left and right hemispheres.

The mass of the brain has reached its limit and further growth is prohibited. Here are resources for further improvement in information processing, by removing redundancy and using the left hemisphere for one activity and the right one for another. Thus, the efficiency of the whole brain can be significantly increased. However, redundancy, with all its benefits, is lost. This process is called 'lateralization'. Its significance for our considerations concerning the human intelligence cannot be overestimated.

3.4 BIOTIC INFORMATION PROCESSING

3.4.1 *Reflex, the simplest genetically wired system*

In the literature there is no exact classification of the different levels, or stages, of biotic information processing. Only in very few publications can one find some thoughts in this direction.

In this book we differentiate four stages of biotic information processing: reflex, instinct, discent and intelligence (Table 1-6).

We will begin with the lowest and simplest stage of biotic information processing: reflex. Reflex (Latin *reflexus* reflected), according to Webster's dictionary, is defined as an involuntary action, resulting when a stimulus is carried by an afferent (input) nerve to a nerve centre, and the response is reflected along an efferent (output) nerve to a muscle or gland (effectors). Reflex stage was reached around 700–800 megayears ago, when large multicellular marine organisms, with soft tissue, and without hard exoskeletons, emerged: so-called 'Ediacaran' fauna (Table 3-3).

Typically the stimulus from a receptor cell passes through one or more intermediate neurons, which modify the nervous signal and direct it to the output neuronal cell which results in movement of muscles or activity in internal organs.

The system of co-ordinating neurons is such that several different kinds of inputs (stimuli) may produce the same output (result, action).

Reflex is the simplest wiring in the information processing and of course is included in all multicellular animals, from the 'lowest' as sponge to the 'highest' as Homo sapiens sapiens.

3.4.2 *Instinct: genetically complex wired system*

According to Webster's dictionary, instinct is "a complex and specific response on the part of an organism to environmental stimuli, that is, largely hereditary and unalterable".

An instinct could be generated either when the animal (typically: worms, insects, spiders) becomes so congenitally predisposed to a stimulus that response cannot be improved through learning. This stage of biotic information processing is here taken as being higher than that of reflex. On the other hand, our definition of

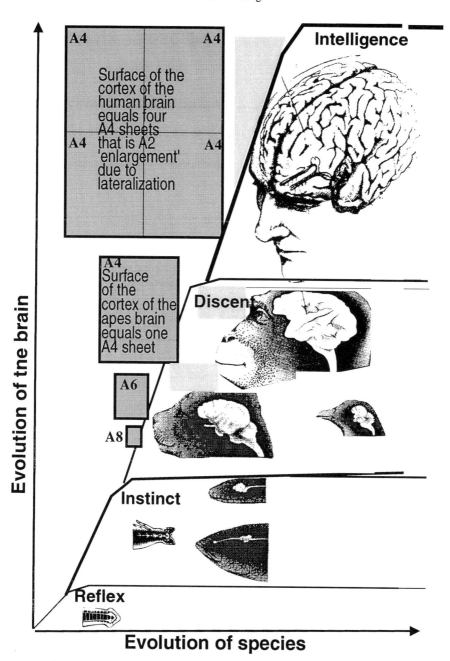

Figure 3-6 Human and animal brains

'instinct' is a complex system, consisting of 'instinct' itself, as well as the lower stage of biotic information processing; that is, of 'reflex'. Five hundred megayears ago, with the emergence of crustaceans and, later, insects, it is probable that the next stage of biotic information processing—instinct—came on the scene, and can be associated particularly with reptiles and fish, that is, with the Permian era, 280 megayears ago. The transition between the lowest stage, the reflex and the higher, that is the instinct, is of course continuous and includes a full spectrum of intermediate states.

Some researchers feel that human behaviours such as aggression and territoriality may have instinctive components. There is some danger of overgeneralizing to human behaviour from animal observations. S.Freud theorized that the sexual drive is essentially instinctive. In our scheme all lower stage of human information processing, including reflex, instinct and discent are significantly influenced and controlled by the highest stage, the intelligence.

Instinct plays a significant role in information processing in all 'higher' animals.

3.4.3   Discent: individual learning of new information

The third stage of biotic information processing is, in this book, called 'discent'. The term 'discent' is not used in English or in any other language. It is proposed here and derived from the Latin *discere* to learn.
This stage of biotic information processing is considered here as being higher than instinct. Discent is, according to our definition, very closely connected to the process of learning. Learning means the memorizing of new information by an individual. The genetic basis for learning is given in the form of larger, more complex, more strongly interconnected 'wiring' in the cortex and in other parts of the brain. However, learning capabilities are strongly limited to some specific kinds of events and activities, characteristic for the given species and the result of the impact of genetic information, of the environment and, last but not least, of the experience of the individual. Discent, as we have called the third stage of biotic information processing, occurred some 150 megayears ago and can be associated with birds and, later, mammals.

On the other hand, in our definition, 'discent' is complex and covers discent itself, as well as the lower stages; that is, instinct and reflex. Even some insects, such as honeybee posses an ability for learning, even if on a rather low level. The transition between the lower stage, the instinct and the discent is of course continuous and includes a full spectrum of intermediate states.

M.Minsky (1986): "You cannot train a beaver to build a termite nest or teach termites to build beaver dams."

It is of interest that the typical bearer of discent, the mammalians, during the period of the recent 65 megayears has been limited to approx. 90 genera only, independent of the fluctuations of climate, migration etc. and a very large diversity of species (Fig. 3-7).

3.4.4 *Intelligence: creator of new information*

According to Webster's dictionary, intelligence derives from the Latin *inter* among and *legere* to gather, or choose, and means the ability to reason, perceive, or understand; the ability to perceive relations, differences, etc., distinguished from 'will', and 'feeling'. As with the previous stages, we use here the term 'intelligence' to describe the whole system, including intelligence, discent, instinct and reflex. This, however, is an arbitrary proposal by the authors. The specific property of intelligence is creation of new, semantically significant, information, concerning the real world and last but not least theory of the intelligence itself.

Our rigid classification system, with the four more-or-less clearly defined stages, from reflex through instinct and discent to intelligence, does not mean that we are neglecting the whole, almost continuous, spectrum from sponge to man.

To some extent, such a classification can be found in the work of other authors, the most important reference being the book of S.E. Luria, S.J. Gould and S. Singer (1981), in which four stages of biotic information processing are considered (Fig. 3-8).

Also, G. Viaud (1960) speaks about four stages of information processing, as follows: 1) reflex, 2) instinct, 3) practical intelligence, without speech ability (typical for higher animals and for human babies, as well as for the practical activity of adults), 4) rational intelligence, including highly developed speech and conceptual thinking.

Intelligence includes a unique ability to produce new information, not only appropriate to the real world, but also to prepare a fantasy world and, what is in everyday life more important, to produce lie, cheat, deception. This gives Homo sapiens the prevalence in the realm of animals and allows to construct generic human social life.

3.4.5 *Hierarchical organisation of the four stages*

One of the most important features of our classification is the principle of the hierarchically organized structure of all four stages, with intelligence at the top and reflex at the bottom. Only such a hierarchical structure ensures that each small step in the development of the nervous system does not climinate the lower stage.

82   Search for terrestrial intelligence

The continuum of evolution is given. This does not mean that, with the increasing complexity of the central nervous system, the lower stages remain in the previous, old state. On the contrary, each higher stage of development influences and even modifies the lower stage, its properties, its functions, and its symbiosis with other stages.

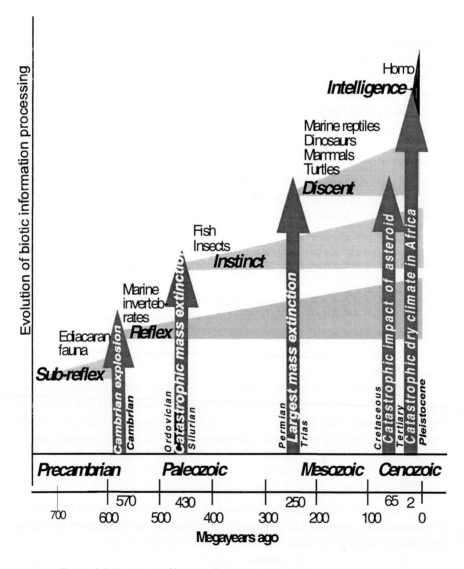

Figure 3-7 Evolution of biotic information processing

3.5 GENUS HOMO EMERGES

3.5.1 *Genus Homo*

In the lowest stage of biotic information processing, reflex, the absolute value of the bonus-malus system is relatively small. Pain is not very great, if such a sentence means anything at all. It is known that lower animals, such as some worms, continue to eat even if the back half of their bodies have been devoured by predators. The bonus-malus system includes entities such as: pain, thirst, hunger and sexual lust. At the stage of instinct, other behaviour, such as defence of territory, social bonds, and the fear of enemies (fear of death?) appear. The corresponding bonus-malus scale increases. Previous labels for pain, thirst, and hunger increase in absolute value, while the features at the instinct stage are given a bonus-malus label with lower absolute values.

Discent is expanded by more complex behaviour, such as the care of offspring, fear of strangers, development of social hierarchy, habitat construction techniques, and finally, sexual activity. All these things are, according to our hypothesis, given appropriate bonus-malus labels, though with absolute values which are not so large. However, the more elementary features of reflex and instinct are now more highly rated, and given higher bonus-malus values.

Consequently, intelligence is associated with the most complex and 'sophisticated' abstract concepts, such as the feeling of freedom, the idea of taboo, hope and moral principles. We will discuss all these problems in Chapter 4. But it cannot be ignored that all these phenomena include not only the ability to produce world-wide culture but also the ability to produce false information: lie, deception, cheat.

3.5.2 *The occurrence of intelligence*

The quality gap between the stages of information processing of the higher primates, apes and modern man is rather large. In this book, we make a clear distinction between the quality of the discent (and not 'animal intelligence') of apes and the intelligence of modern man. If the gap is so large, why not assume that the oldest member of the genus Homo, the species Homo habilis, was at the highest level of discent, which we will arbitrarily call 'super-discent'. Continuing with such non-obligatory terminology, we can assume that the next 'higher' species, Homo erectus, possessed an 'infra-intelligence'. Consequently, a member of the species Homo sapiens, subspecies Homo sapiens neanderthalensis, possessed a 'sub-intelligence'. Only members of the subspecies Homo sapiens sapiens (modern man) achieved the level of full 'intelligence'.

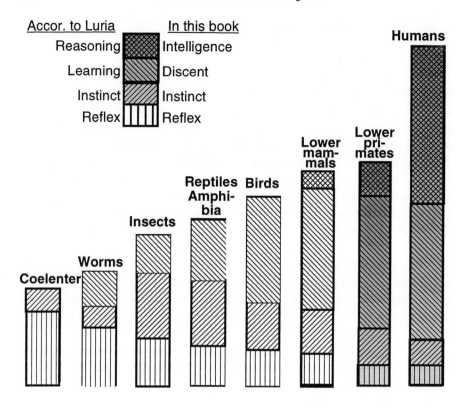

Figure 3-8 Four stages of biotic information processing

Of course, such a classification is very naive and will probably be only of little help in our further discussion. In spite of this, it is worth mentioning that this multi-stage classification gives us an answer to the question why, between present apes and modern man, there exists such a large gap in the system of biotic information processing. The gap results from a missing older subspecies of genus Homo; the lost link in the evolution of humans. Remark: Hominids: family including fossil and recent genus of Homo. Hominoidae: superfamily including family of great apes and family of hominids.

3.5.3 *The occurrence of genus Homo*

Some authors have referred to humans as 'the third chimpanzee' in recognition of our close bio-molecular similarity to the two living species of Pan (chimpanzee and bonobo). Another opinion supports that humans are 'the second gorilla', because of morphological similarity. But, curiously, it is often forgotten that man differs from apes in having only 46 chromosomes, not 48. Recently a hominid called Australopithecus anamensis (probably 4 megayears-old) has been recovered at Aramis, Ethiopia. It is suggested that he represents a long-sought potential root for the Hominidae (Fig. 3-9). Australopithecine brain size increased somewhat during subsequent evolution (that is during 2.5 megayears) but never got above 0.6 kg before this genus became extinct around 1 megayears ago.

Australopithecus—the so-called 'southern ape-man'—evolved in Africa some 3.5 megayears ago. His brain mass was around 0.4 kilograms (within the range of those for living chimpanzees), with a body weight similar to modern man. Australopithecus evidently used his hands habitually to carry loads and make stone tools for cutting meat. By 2 megayears ago, one of the four species of this genus gave rise to the first species of our own genus Homo—Homo habilis. Both genera endured side by side for another megayears, before the last Australopithecus died out, possibly exterminated by its brainier successor.

The hominids, responsible for the manufacture of recently discovered artefacts remain unidentified. The high lithic technology are evidence that the hominids that lived about 2.5 megayears ago were not novices. Some authors ask what were the relative contributions of hand-eye co-ordination, manual dexterity, conceptual facility and opportunity? (B. Wood, 1997).

3.5.4 *Neoteny, the trick?*

There is no doubt that the emergence of genus Homo, from older ancestry is coupled with some biological factors making the tremendous increase in the capacity of the brain possible. One of these factors is probably the impact of the most critical phase during parturition: the passage of the foetal head through the lesser female pelvis. In all nonhuman primates, the diameter of the pelvis substantially exceeds the transverse diameter of the foetal head. In humans, the head of the foetus must be rotated during its passage through the smaller pelvis; and even this is only possible when the foetus is small enough. As a result, a human baby must be born 'premature'. Man is a neotenic ape. It is not clear if the Homo erectus female had the same problem. No female pelvis of this species has been found.

Neoteny is a phenomenon among some salamanders, in which larvae of a large size, while still retaining gills and other larval features, become sexually mature, mate, and produce fertile eggs. In certain lakes of Mexico, only neotenic

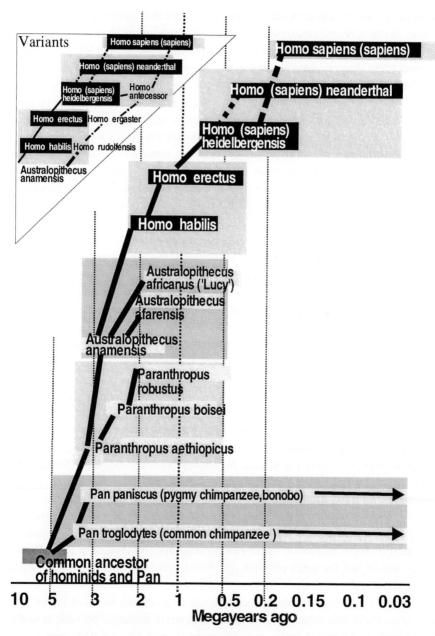

Figure 3-9 Genealogy of the modern man

larvae are present. These are called axolotls.

Adult humans share numerous features found in baby chimpanzees, but which are lost as these grow up. Like humans, baby chimpanzees have a sparse covering of body hair, and a relatively large brain shielded by a bulbous cranium. Like ours, their skull bones are thin.

A juvenile chimpanzee holds its head upright and has almost human features. In the adult, the head juts forward and has ridged brows and a projecting muzzle. As in women, so in young female chimpanzees the vagina faces forwards instead of backwards.

The human brain continues to grow long after birth, longer than in all other mammals, allowing the epigenetic elements to have a significant influence on its evolution quantitatively (the number of neurons and synapses) as well as, and much more importantly, qualitatively (the type of synaptic and neuronal interconnections). The longer human childhood allows a longer period of learning and prolongs our lives beyond that of any other mammal.

The reader must allow an anecdotal remark at this point. Neoteny offers a reasonable explanation for the emergence of a naked ape. A sparse covering of bodily hair has far-reaching consequences. We all know how much time primates need for the care of their bodily hair. The removal of all kinds of insects, worms, ants, and other parasites is a significant activity for efficient survival and, on the other hand, an important stimulus to intensify familial and social bonds. A few hours per day are spent on this activity. Suddenly, thank to neoteny, in the rather short time of maybe ten to twenty thousand generations, there appeared families and whole groups of larger brained, but also almost naked individuals. The time needed for hair care is now free for other activities. But for what activity? Perhaps, some individuals began to use this free time for other social activities, for further tool development, or even for 'thinking'! Who knows what gave the impulse to the development towards intelligence?

3.5.5 *Homo habilis and Homo erectus*

Our own genus, Homo, branched off of australopithecine stock and is associated with a known fossil record that covers approximately the last 2 megayears. Brain size increased much more rapidly in Homo than in Australopithecus, going during 2 megayears from an initial 0.7 kg to its present average of 1.4 kg (Fig. 3-9 and Table 3-4).

Homo habilis ('handy man') stood no more than 1.5 m tall, with a weight of about 50 kg. His brain had a mass of around 0.7 kg, still only half the size of ours and his species lived about 2-1.5 megayears ago.

New discoveries of stone tools are the oldest known artefacts from anywhere in the world. The artefacts show surprisingly sophisticated examples of stone fracture mechanics. This indicates an unexpectedly long period of technological stasis in the Oldovian (Ethiopia) (Semaw S et al, 1997). Artifacts found near its bones suggest that Homo habilis made basic stone tools, built simple shelters, gathered plant foods, scavenged big meaty limbs from the carcasses of animals killed by carnivores, and hunted small game.

The upright posture and obligatory bipedalism of modern humans are unique among living primates. Among the fossil hominids the earliest species to demonstrate the modern human labyrinthine morphology (responsible for unconscious perception of movement) is Homo erectus.

Homo erectus ('upright man') had a bigger brain and body than Homo habilis. He lived about 1.9 megayears–200 thousand years ago. Some adults probably grew to 1.8 m and were at least as heavy as ourselves. Brain mass averaged 1 kg, more than that of Homo habilis, though less than modern man's. Homo in Middle Pleistocene was one-third less encephalized than Recent humans (C.B.Ruff et al, 1997)

Groups evolving in Africa probably spread to Europe, East Asia (including so-called 'Peking man', or 'Sinanthropus') and South-east-Asia (including so-called 'Java man', or 'Pithecanthropus'). Improved technology included standard toolkits, big-game hunting, the use of fire, and improved building methods, enabling this species to invade new habitats and climates. The 'technology' of Homo erectus is called Acheulian—from 300,000-year-old finds at St.Acheul in France. There is evidence that Homo erectus killed and ate large animals, including boar, bison, deer and rhinoceros. He probably used wooden spears with stone projectile points. Recently a 400,000 years old wooden throwing spear, with stone tools etc. has been discovered which strongly suggests that systematic hunting, involving foresight, planning and the use of appropriate technology (in a boreal, cool-temperature climate in central Europe) was part of the repertoire of the pre-modern Homo (H. Thieme, 1997).

Early hominids had a home base—a behavioural innovation—which implied a sexual division of labour—another innovation. Males ranged far in search of scavengeable meat or hunted quarry, females gathered fruits and tubers nearer home and families. Eventually this altruistic behaviour and social co-operation began to select intelligence, language and culture (R.J. Blumenschein, J.A.Cavallo, 1992). In this book, the assumption is also made that Homo erectus used a higher level of acoustic communication including proto-verbs with other individuals, which can be called proto-speech. This assumption is not shared by established science, and is purely a proposal of the author.

3.5.6 *Homo sapiens neanderthalensis*

The first member of the genus Homo and of the species Sapiens was Homo sapiens neanderthalensis, a name based on fossils found in the Neander Valley, near Düsseldorf, Germany. Neanderthal (archaic Homo sapiens) man evolved as much as 200,000 years ago. Typical Neanderthals were extremely muscular and stocky, with large joints and hands. The proportions of the body were compact, which helped to conserve heat in a cold environment. The remains of some hundreds of Neanderthals have been found in more than 70 sites, mostly in caves of Southern, Middle and Eastern Europe. No typically Neanderthal fossils have been found outside Europe and Levant; in Israel and Lebanon (O.Bar-Yosef, B.Vandermeersch, 1993). He had body mass significantly (about 10%) larger than living humans. Relative to body mass, brain mass in Neanderthals was slightly smaller than in early 'anatomically modern' humans (Ch.B.Ruff et al, 1997) However, existence of fully-developed lateralization within his brain, the different features and architecture of his acoustic equipment and the different functions of the left and right hemispheres of his brain are still open questions. (Fig. 3-9 and Table 3-4)

From our point of view the existence of highly-developed lateralization seems to be the prerequisite for the existence of fully-developed speech, with all its cultural consequences. The best assumption seems to be partial lateralization and the associated ability of a low level of speech. Of course, the ability to speak depends upon some important details of anatomical features, muscles, and the appropriate control system. As we will see later, the emergence of fully developed speech results according to one hypothesis from a genetically controlled ability, the so-called 'universal grammar' (N. Chomsky, Section 6.5.4). What properties would a 'missing link' in universal grammar have?

The results of analysis of some locus on chromosome 12 in more than 1600 individuals are consistent with the estimates for the date of the last common mitochondrial DNA ancestor of all modern humans (African and non-African) of 100,000 to 300,000 years before present (out-of-Africa model of human origins) (S.A. Tishkoff, 1996). Scrutiny of the archaeological and palaeontological records pieced together from digs at many sites suggests that hominid creatures migrated out of Africa several times. Africa from the very beginning has been the engine of mainstream innovation in human evolution. Each wave of emigration sent forth a different species onto the world stage—until our own Homo sapiens, eliminated all the others (I. Tattersall, 1997).

Table 3-4 Homo erectus, habilis, sapiens

Genus	Homo			
Species	habilis	erectus	sapiens	
Subspecies			neanderthalensis	sapiens (modern)
Examples, genetic identity (%)		Sinanthropus 99.9%	99.99 %	Cro-Magnon man 100.0 %
Period, years	2–1.6 Myr	1.8-0.9 Myr	0.2 Myr- 35 kyr	150 kyr to present
Geography	Africa	Africa; Caucasus, Asia	Europe, Near East	Europe, Asia, Australia, America:
Body tall	1.4 m	1.65 m	1.7 m, muscular	1.6-1.8 m
Age, average, years			½ reached of 20; 1/20 age of 40	2/3 reached the age of 20
Brain, mass (kg) and structure	0.7 kg	0.8-1.2 kg	1.4 kg without lateralization ?	1.4 kg with lateralization
Encephalization	>3	3.45	4.7	5.288
Male/female dimorphisms			Body 10% larger brain 10% larger	Body 10% larger, brain 10% larger
Vocal apparatus		Undeveloped larynx and pharynx		Full evolved
Walking	Bipedalism	Full bipedalism		
Population	Small	Small	Some 100 kyr	Bottleneck; 150 kyr
Culture	Oldovian	Acheulian	Mousterian, Chatelperron	Aurignacian, Magdalenian,
Material technology	Basic stone tools	Stones from local sides,	Stone from local side, wood tools	Stone (distant) spear thrower, bow,
Shelters	Simple?	Shelters	Fire-heated shelters, clothing	First city: 10,000 years ago
Fire making	No	Conservation of fire	Fire making by iron pyrites	Full-scale fire making
Food	Scavenger,	Scavenger,	Venison, cow, horse	Big game: bow spear-thrower
Labour division	?	Male-hunter / Female-gatherer		
Social behaviour	?	Carry for elderly, sick	Religious intent: burial (doubtful)	Religion (?), social hierarchy
Arts, skin paint, ornaments	No?	Skin painting	Simple ornament, ivory rings	Wall painting, 'cosmetics'
Language Grammar	Rudimentary; no rules	Proto-speech some rules?	Sub-speech, weak grammar	Speech, selfconsciousness, fantasy.
Psychoactive agents use		Only as green plants?		Probably frequently since 40 kyr
Impact on sexual selection	Low	Rather low?	Low impact	Large impact, due to speech, use of psychoactive agents

Neanderthals and Homo sapiens lived in the same area (Mount Carmel) some 40,000 and 75,000 years and used similar tools—yet only one group survived. Some students claim that It must be something else beyond the tools. Maybe they were using their arms differently. In this book the probable 'difference' is the ability to modern speech (cheating) and even use of psychoactive agents (for self-stimulation) (A. Gibbons, 1996).

The Neanderthals were nearing their apogee 100,000 to 80,000 years ago. Their improved technology, especially fire-heated shelters, construction and clothing manufacture enabled them to endure the rigors of winter in the cold climate of the last glaciation period, 70,000–30,000 years ago.

Discoveries suggest that they knew how to start fires by striking sparks from iron pyrites and using dried bracket fungus as tinder. Evidence of clothing is mostly indirect. Yet, about 30,000 years ago, this subspecies apparently died out. Neanderthal tools and implements are termed Mousterian, from finds made at Le Moustier in France, and originated in the cultural period of the Middle Palaeolithic.

A longer scientific dispute concerns the question of whether Neanderthals practised burial rituals, including the use of flowers, etc., although around 70 burial sites, with more that 150 bodies, have been excavated. A clear positive answer about the meaning of their burial practices can help us to understand if Neanderthals had self-awareness. Our definition of intelligence, self-awareness and selfconsciousness are intimately related to human speech. It is, however, still questionable whether the assumption concerning the existence of a lower level of speech is tenable, whether Neanderthals had command of a low level of speech, and, last but not least, if speech and selfconsciousness are really inseparably connected.

Two Spanish caves (in Catalania and Cantabria), one in Israel, and in Arcy-sur-Cure (France) had previously been used by Neanderthals and then by Cro-Magnons—modern Homo sapiens sapiens. There are signs that modern people and Neanderthals about 34,000–40,000 years ago lived side-by-side for a time, probably using intermediate or mixed Mousterian-Aurignacian technology. It is probable that the Neanderthal line of evolution contributed little or nothing to modern European stock, and that all Neanderthals died out between 35,000 and 30,000 years ago.

Neanderthals, despite their large brains, were ineffective hunters living at a low population density. Dangerous prey species, like rhinoceros and pigs and elephants, were not being hunted, because spear thrower and the bow and arrow had not been invented. Also nets and fish-hooks had not been invented.

Neanderthals were not producing the art objects. The lack of inventiveness is the most astonishing feature of the Neanderthals.

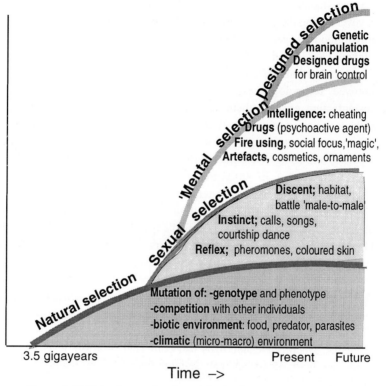

Figure 3-10 Natural, sexual and mental selection

What is the reason for such a fast disappearance of Neanderthals? Genetically Neanderthals have been 99.99% identical to humans today. What ability, what quality of Homo sapiens modern has played the decisive role? The role of Homo sapiens neanderthalensis in the evolution of speech remains an open question (Fig. 6-7 and 6-9). Probably the solution of this question is the emergence of fully developed human speech. Speech, which can easily construct new sentences, including those with an intention to lie and deceive, gives modern Homo sapiens an enormous predominance.

By giving the name Homo sapiens neanderthalensis, modern science has underlined that both sub-species, Homo sapiens sapiens (archaic and modern man) and the Neanderthals, belong to the same species (Homo sapiens) and that sexual reproduction between the two sub-species was possible, in the same way as this takes place among modern human races. At present we have no way of knowing if they were the same species or two different.

3.5.7 Homo sapiens sapiens—modern man

Ch. Darwin (1859): "This leads me to say a few words on what I have called Sexual Selection. This form of selection depends, not on a struggle for existence,.... but on struggle between the individuals of one sex, generally the males, for the possession of the other sex.....Sexual selection is therefore, less rigorous than natural selection. ...But in many cases, victory depends not so much on general vigour, as on having special weapons, confined to the male sex......The war is, perhaps, severest between the males of polygamous animals and these seem oftenest provided with special weapons..... Thus it is, as I believe, that when the males and females of any animal have the same general habits of life, but differ in structure, colour, or ornament, such differences have been mainly caused by sexual selection.....weapons, means of defence or charms."

The similarity between social primates (chimpanzee, bonobo) are numerous. But it should not be ignored that even in domesticated environments including zoo and circus, apes are not able to manipulate or even tolerate fire. All efforts to learn speech, even in rudimentary form, finished without success. There is no evidence that primates in natural environments are using natural psychoactive substances.

Ch. Darwin (1871): "Although a high degree of intelligence is certainly compatible with the existence of complex instincts, as we see in insects just named and in the beaver it is improbable that they may to a certain extent interfere with each other's development. Little is known about the function of the brain, but we can perceive that as the intellectual powers become highly developed, the various parts of the brain must be connected by the most intricate channels of intercommunication. (p.278) Sexual selection acts in less rigorous manner than natural selection. ... We shall further see, and this could never have been anticipated, that the power to charm the female has been in some few instances more important than the power to conquer other males in battle..... (p.384). For my own part I conclude that of all the causes which have led to the differences in external appearance between the races of man, and to a certain extent between man and the lower animals, sexual selection has been by far the most efficient. (p.402) He who admits the principle of sexual selection will be led to the remarkable conclusion that the cerebral system not only regulates most of the existing functions of the body, but has indirectly influenced the progressive development of various bodily structures and of certain mental qualities."

The male and sometimes female efforts to win the approbation of the sexual partner mostly have a rather high energetic cost such as: fight with other partner, long-time display, offering food, preparing a nest, elaborating and especially decorated, the so-called bowers, that are used only for courting and mating. The most effective, that is the most energetically effective methods (lowest energetic input and highest revenue; the mating) is doubtless the art of verbal courting, of curry favour with partner. The lowest price, the shortest way to success, is the verbal communication with mating partner (Fig. 3-10, Table 3-5).

High verbal ability is in some way, but not directly, connected with this property which is known as intelligence. Doubtless humans achieve on this field (verbal speech) the highest level in the whole realm of animals. The effectiveness of intelligence on the field of sexual selection is unique, is truly unique.

An active sexual partner is very generous in his/her promises not only concerning some material goods but also, or even especially often, ideal goods, such as love to the end of life. The astonishing aspect is that the second partner in the depth of her/his soul knows the real value of such promises (mostly being unrealistic), but because of 'inborn' self-deception neglect own doubts.

Table 3-5 Sexual selection activities in all four stages of biotic information processing

	Intelligence	Discent	Instinct	Reflex
Pheromones, attractive	Perfumes	Pheromones		
Coloured skin	Skin painting Cosmetics	Natural coloured skin		
Elaborating habitat	on all three stages			
Bringing food, gifts	on all three stages			
Courtship dance	on all three stages			
Calling	Singing, calling etc.			
Social position	Verbal self-display + non-verbal on all three stages			
Male-to-male battle	Verbal mocking and deception + 'battle'			
'Magic' behaviour	Fire 'using'			
Promises	Verbal			
Psychoactive agents	Deliberated use for self and for partner			

The other extreme is the highest price for sexual selection; the man-to-man battle or even war, paid even with own life. There are some students, claiming that the real selection pressure during the evolution of Homo sapiens comes from the battle from man-to-man having in mind the trophy: the women of his enemy, or in previous two centuries, also 'women' of his duellist-partners.

Intelligence, is on the one hand a product of the natural selection, which as we now know, is determined through contingency, through 'unexpected' catastrophic events and the battle of survival, on the other hand by sexual selection. In the natural selection humans get a unique arm, better as all horns, claws or muscles, namely the highly evolved brain. In the sexual selection humans get instead of colour-full plumes etc. the highly evolved speech, which can convey the most sophisticated lies and deceptions, helping achieving the desired aim: mating.

It is probable that modern humans, Homo sapiens sapiens, a sub-species of Homo sapiens, originated between 200,000 and 120,000 years ago in Africa (A.C.Wilson, R.L.Cann, 1992) and migrated to Europe and Asia by way of the Levant (present-day Israel, Lebanon and Syria). May be it is of significance that 125 to 115 thousand years ago a rather warm Eemian interglacial period were at least as warm as today. If, as many molecular biologists believe, modern Homo sapiens arose as a discrete evolutionary event some 120–150 thousand years ago (the so-called hypothesis of 'mitochondrial Eve', the mother of us all, who lived in Africa), monogenesis of language is likely (F.J.Ayala, 1995).

One hypothesis postulates that for more than 100 thousand years ago the population of Homo sapiens sapiens consisted of only 10,000 individuals ('population bottleneck'). It is also of interest that 73 500 years ago a massive volcanic eruption of Toba, Sumatra (the largest of the recent hundred thousand years, with 2800 cubic kilometres of magma, including some gigatonnes of sulphuric acid) caused a significant cooling of global temperature for 1 or 2 years, of some centigrade, with widespread hard freezes in mid-latitudes. But a major climatic event was this not. Increased snow cover and sea ice could have led to longer term (decadal) cooling, and perhaps decimated early human populations (M.R. Rampino, S. Self, 1993).

New data suggest that anatomically modern people (Homo sapiens sapiens) were present in southern and central Europe at least 43,000 years ago, some recent investigations suggest about 70,000 years. Previously, it was believed that early modern Europeans, or Cro-Magnons (discovered at Cro-Magnon in South France), appeared between 35,000 and 30,000 years ago, using the innovative, more highly-developed 'Aurignacian technology', corresponding to the cultural period known as the Upper Palaeolithic.

His weight was around 70 kg, and his brain around 1.4 kg. Innovations were an upright forehead, a smaller nose, and brow ridges. Cro-Magnon man was probably an early ancestor of today's European. However, Cro-Magnon sites have been found in Borneo with some skeletons also showing similarities to Negroes, Orientals and aboriginal Australians.

The brain of Cro-Magnon man (smaller that Neanderthal's brain) (R.D. Martin, 1995) had all of the properties of a modern brain, without exception. It is very important to mention that his brain had all the features of lateralization— asymmetry of the left and right hemispheres. Among other things, this is an argument for the existence of fully-developed speech and of highly developed mental activity. The existence of genetically carried 'universal grammar' is evident

(Section 7.1). With this, the opportunity for the emergence of selfconsciousness is present. Toolmakers travelled long distances to obtain special materials, such as marine shells and flints.

Homo sapiens sapiens made extensive use of antlers, bone and ivory, not just for tools, but also for ornaments and objects of art. The earliest credible body ornaments are from the Aurignacian period, and date from about 35,000 years ago, or perhaps slightly earlier. Some featured animals and other forms, mostly poorly drawn, on small, portable objects, and are 32,000 years old. Early cave paintings, including hand prints and engraved and painted silhouettes of animals (lions, rhinos, bears), are 25,000 years old. Most paintings lay deep in caves, where artists worked in the light of burning wood. At the end of 1994 in the cave Chauvet at Combe d'Arc, Ardeche, France 300 engravings of highest artistic quality 32,000 years old were discovered. Fertility symbolism could account for human figurines with exaggerated female features. Graves contained refined goods, such as necklaces.

Modern man lived 40,000 years ago in the Near East, and 30,000 years ago in Asia and the Far East. During the coldest period of the last glacial period, probably more than 30,000 years ago, when the ocean surface was at its lowest level (50 metres lower than present), Australia was connected by a land- bridge to Southern Asia, and Siberia was connected to Alaska. The most widely held theory is that the first Americans crossed the Bering Strait from Asia within the last 30,000 years in three distinct waves of migration. First, the Amerind people arrived about 30,000 years ago, followed by the NaDene, 20,000 years later, and finally the Eskimo-Aleuts, who arrived in northern Canada and Alaska within the last 5,000 to 7,000 years. Modern man walked to North America, and then in a relatively short time, crossed Central America and reached South America.

S.J. Gould (1995): "Why is Homo sapiens here?—..We are here because the death roster of anatomical products of the Cambrian explosion did not include a small and 'unpromising' chordate group ..(in this sense, any group alive today owes its existence to contingent fortune.)... No late-Cretaceous bolide and dinosaurs would still be dominating the world of terrestrial vertebrates, with mammals probably still restricted to rat-size creatures in the interstices of their world (dinosaurs had so dominated mammals for more than 100 million preceding years so why not for an additional 65 million?) Step down to a lineage of apes 10 million years ago in African forests. On this replay, climatic drying does not occur, forests do not convert to savannahs and grasslands. The lineage stays in the persistent forest as apes—doing quite well—in an alternate today. .... Nature's laws and history's contingency must work as equal partners in our quest to answer ..(why is Homo sapiens here?)."

Migration towards Beringia (Siberia-Alaska continental bridge), migration towards Polynesia, 'conquering' the Andes by Incas have been 'motivated' not only by 'inborn' curiosity of Homo sapiens, or by search for new habitats and food sources, but probably strength by the 'fantastic' reports of the reconnaissance agents, resulting from human 'inborn' tendency to lie, to rave, to exaggerate, especially when the impact of psychoactive drugs, (hallucinogens cause false perception that take place in the absence of an environmental stimulus; stimulants changes seeing, hearing and thinking giving a feeling of being 'superhuman'; other produce delusions, that is false ideas) deliberate or not, was of significance.

3.6   BIBLIOGRAPHY

Ayala FJ (1995) The myth of Eve: molecular biology and human origins. *Science* **270** 1930
Bar-Yosef O, Vandermeersch B (1993) Modern humans in the Levant. *Sci Am* **4** 64
Calvin WH (1994) The emergence of intelligence. *Sci Am* **271** 4 78
Darwin Ch (1871) *The descent of man, and selection in relation to sex.* J.Murray, London, (photocopy Princeton Univ Press, Princeton, N.J.,1981)
Douglas K RJ, Martin AC (1996) The information superflyway. *Nature* **379** 584
Duve de Ch (1995) *Vital dust. Life as a cosmic imperative.* Basic Books, N.York
Duve de Ch (1996) The birth of complex cells. *Sci Am* **274** 4,38
Erwin DH (1996) The mother of mass extinction. *Sci Am* **275** 56
Ferris JP et al (1996) Synthesis of long prebiotic oligomers on mineral surfaces, *Nature* **381** 59
Gibbons A (1996) Did Neanderthals lose an evolutionary 'arms' race? *Science* **272** 1586
Gould SJ (1994) The evolution of life on the Earth. *Sci Am* **271** 4 62
Gould SJ (1995) What is life? As a problem in history, in MP Murhy, LAJ O'Neill (edit)

What is life?. The next fifty years. Cambridge Univ Press, Cambridge UK
Hublin J-J at al (1996) A late Neanderthal associated with Palaeolithic artifacts. *Nature* **381** 224
Israelshwili J, Wennerstrom H (1996) Role of hydration and water structure *Nature* **379** 219.
Kiedrowski G (1996) Primordial soup or crêpes? *Nature* **381** 20
Knoll AH et al (1996) Comparative Earth history and Permian mass extinction. *Science* **273** 452
Leakey R, Lewin R (1995) *The sixth extinction.* Doubleday, N.York
Lee DH et al (1996) A self-replicating peptide. *Nature* **382** 525
Martin RD (1981) Relative brain size and basal metabolic rate. *Nature* **293** 57
Martin RD (1995) Hirngrösse und menschliche Evolution. *Spekt Wissensch* **9** 48
Orgel LE (1994) The origin of life on the Earth. *Sci Am* **271** 4 52
Pagel MD, Harvey PH (1989) Taxonomic differences in the scaling of brain on body weight among mammals. *Science* **244** 1589
Rakic P (1995) Corticogenesis in human and nonhuman primates, in Gazzaniga MS (edit) *'The cognitive neurosciences'*. MIT Press, Cambridge, Mass.
Raymond J.L. et al (1996) The cerebellum: a neuronal learning machine? *Science* **272** 1126
Ruff CB et al (1997) Body mass and encephalization in Pleistocene Homo. *Nature* **387** 173
Ruyter de R.R. et al (1996) The rate of information transfer at graded-potential synapses. *Nature* **379** 642
Schopf JW (1993) Microfossils of the Early Archean Apex Chert: New evidence of the antiquity of life. *Science* **260** 640
Semaw S et al (1997) 2.5-million-year-old stone tools from Gona, Ethiopia. *Nature* **385** 333
Shepherd GM (1995) Toward a molecular basis for sensory perception, in Gazzaniga MS (edit) *'The cognitive neurosciences'*. MIT Press, Cambridge, Mass.
Skinner BF (1984) Selection by consequences. *Behav Brain Sci* **7** 447
Srinivasan MV et al (1993) Is pattern vision in insects mediated by 'cortical' processing? *Nature* **362** 539
Tattersall I (1997) Out of Africa again and again. *Sci Am* **276** 4 46
Taube M (1962) *Hydrogen as carrier of life*. Atom Ener Publ, Warsaw
Taube M (1985) *Evolution of matter and energy on cosmic and planetary scale.* Springer, N.York
Thieme H (1997) Lower Paleolithic hunting spears from Germany. *Nature* **385** 807.
Tishkoff S.A. et al (1996) Global patterns of linkage disequilibrium at the CD4 locus and modern human origins. *Science* **271** 1380
Wald G (1984) Life and mind in the Universe. *Int J Quant Chem* **11** 1
Wood B (1997) The oldest whodunnit in the world. *Nature* **385** 292.

CHAPTER 4
BRAIN CARRIERS INTELLIGENCE

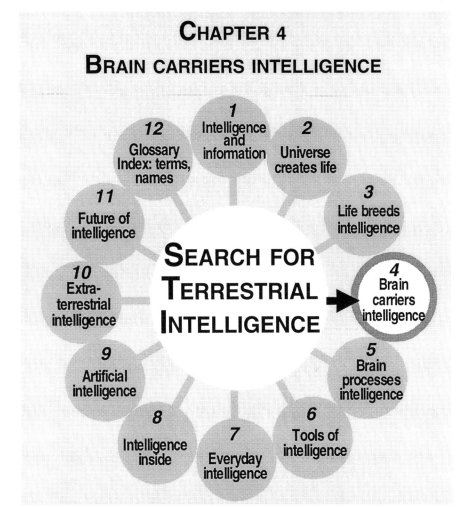

The macro- and micro-structure of human brain is extremely complex. Micro-components: neurons with a complex structure: axons, synapses, microtubules. Development, growth of almost a trillion of neurons. Energy flux through the brain is very high.

Electrical and chemical properties of neurons. Neurons are effective producers of about hundred neurotransmitters, controlling the activity of neuronal networks. Genetics of brain: probably two thirds of all genes of the human genome control the development of the central nervous system.

The asymmetry of the left and right brains is responsible for the increase of effectiveness of the human brain.

4 BRAIN CARRIES INTELLIGENCE 101
4.1 THE HUMAN BRAIN AND ITS MACROSTRUCTURE 101
4.1.1 Intelligence is carried by the human brain *101*
4.1.2 The human brain must have a very complex structure *101*
4.1.3 The brain's architecture *102*
4.1.4 Brain organs: neocortex, thalamus *103*
4.1.5 Other organs: amygdala, hippocampus *105*
4.1.6 Cerebellum, etc. *105*
4.1.7 Brain areas: sensory and motor *109*
4.1.8 Speech processing: Broca's and Wernicke's areas *110*
4.2 THE MICROSTRUCTURE OF THE BRAIN: NEURONS, SYNAPSES 111
4.2.1 Neurons and information processing *111*
4.2.2 Neurons—'brainy' cells *113*
4.2.3 Neuron categories: rather few *114*
4.2.4 The glia cells *116*
4.2.5 The synapse: the neuronal junction *117*
4.2.6 Microtubules play a significant role *119*
4.2.7 Neurons and synapses as electrical and chemical devices *119*
4.2.8 Ion channels *121*
4.3 THE BIRTH AND DEATH OF NEURONS AND SYNAPSES 122
4.3.1 The human brain: from embryo to adult *122*
4.3.2 The birth and death of neurons *123*
4.3.3 The production and elimination of synapses *127*
4.3.4 Modular and laminar organization *128*
4.4 PHYSICAL PHENOMENA IN THE HUMAN BRAIN 128
4.4.1 Do we need a special kind of physics? *128*
4.4.2 Neuroscientists plan an 'atlas' of the brain *129*
4.4.3 The energetics of brain *131*
4.4.4 Electrical activity of the brain *135*
4.5 CHEMICAL PHENOMENA IN THE HUMAN BRAIN 137
4.5.1 Chemical composition seems at first sight simple *137*
4.5.2 Neurotransmitters: 'brainy' molecules *139*
4.5.3 The biochemical basis of the bonus-malus system *141*
4.5.4 The nervous and endocrine systems *142*
4.5.5 Blood-brain barrier *144*
4.6 THE BRAIN AS THE PRODUCT OF GENETIC INFOR-MATION 145
4.6.1 Genetics and the brain *145*
4.6.2 Genetics and epigenetics *147*
4.6.3 Genes and brain normality *149*
4.6.4 Brain plasticity *150*
4.6.5 Genes and brain abnormality *150*
4.7 BIBLIOGRAPHY *151*

4 BRAIN CARRIES INTELLIGENCE

4.1 THE HUMAN BRAIN AND ITS MACROSTRUCTURE

4.1.1 Intelligence is carried by the human brain

Since we consider intelligence to be the highest stage of biotic processing of information, the search for the material basis of this phenomenon is the direct consequence.

In the previous three chapters we have discussed:

–The nature and role of information,

–Material basis of the Universe as carrier of intelligence,

–Life as direct carrier of biotic information processing from lowest stage: reflex, over instinct and discent to the highest stage: intelligence.

The human brain is the seat of the intelligence. Does an intelligence exist without or outside a brain or brain-like structures? Which physical, chemical, and biotic structures and which properties has the intelligence-carrying system?

4.1.2 The human brain must have a very complex structure

The term 'complexity' (Section 1.4.3), opposed to 'simplicity', is in spite of some unsolved problems very useful. Human intelligence is probably the most complex information processing system, and the human brain is probably the most complex structure.

E. Pugh (1965)(not an exact quotation): "If the human brain would be so simple that we could understand it, then it would be so simple that we could not understand it."

J.A. Fodor (1983): "Perhaps solving the riddle of the Universe requires one more neuron than...anyone will ever have. Sad, of course, but surely not out of the question."

Independent of all our hypotheses and opinions about the nature of intelligence, the carrier of intelligence is the human brain. It is also obvious that the only tool available for the investigation of the human brain is human intelligence. This is an old problem illustrated by the German anecdote of Baron Münchhausen, who succeeded in extracting himself from a morass by pulling at his own hair. Is it possible to explain the nature of the intelligence by using the same intelligence? Intelligence, being the highest stage of biotic information processing, must be carried not by a simple, but rather by a very complex, structure, even by the most complex structure which is possible? If the carrier is so complex, the question is then if it is at all comprehensible?

4.1.3 *The brain's architecture*

Our intelligence is probably the most highly developed product of gigayears evolution. It is also generally agreed that the carrier of all or a large part of these phenomena is our brain (Fig. 4-1).

We will begin with a very short description of the most important components of the human brain. The first view under the skull is disappointing for an observer. The grey mass, without clear differentiation, but clearly visible hemispheres, left and right, seems to be everything but the site of our mind.

It is difficult to speak about mental activity, the brain, and biotic information processing, and neglect the new vocabulary which has recently evolved in the realm of computing. Comparisons between the computer and the brain, which is the analogue of computer hardware, are often made. It is not so easy to define which phenomena or components in neurology correspond to computer software. The richness of the phenomena in the human brain responsible for information processing is so large, that the hard/soft distinction is insufficient and inappropriate (Fig 4-2).

L.Boltzmann (1897): "Das Gehirn betrachten wir als den Apparat, das Organ zur Herstellung der Weltbilder, welches sich wegen der grossen Nützlichkeit dieser Weltbilder für die Erhaltung der Art entsprechend der Darwinistischer Theorie bei Menschen geradeso zur besonderen Vollkommenheit herausbildete, wie bei der Giraffe der Hals, beim Storch der Schnabel, zu ungewöhnlicher Länge."

Even if the observer limits himself to investigating the human brain by means of techniques which allow him to use only direct visual observation and discriminate all fine elements of structure, the number of 'organs' in the brain is very large. The phylogeny (the evolutionary history of a group of organisms) of the brain is rather complex. Very roughly, we can assume, that the phylogenetically oldest part of the central nervous system is the spinal cord (as in lower invertebrates).

The old components of the phylogenetically more highly developed brain continue to exist, but may take over new functions, and co-operate with the newly-developed younger 'organs' in the brain. The form, size, and even microstructure of the older 'organs' is also adopted during evolution.

Human brain can be considered from different points of view, discussed in later sections:

–macrocomponents (Fig. 4-3)

–origin (phylogenetic age); young and old

–anterior-posterior differentiation; corresponding to a first approximation to motor and sensory activity (Fig. 4-4)

–hierarchy of activity—primary, secondary, tertiary

–left and right asymmetry
–function in mental activity (for example in speech, vision etc.)
–ontogenic development (from child to elderly age)
–microcomponents.

4.1.4 *Brain organs: neocortex, thalamus*

The cerebral cortex (Latin *cerebrum* brain) is what makes human beings what they are. The cerebral cortex in the human brain, has a surface area of 220,000 square millimetres, a thickness of around 1.5 millimetre, and thus a volume of around 300,000 cubic millimetres; that is, 0.3 litre. The mass of the cortex is 0.3 kg, and it is the largest organ in the human brain. The human cortex is three times thicker than that of the mouse. The area of the cortex of the chimpanzee and gorilla is about 50,000 square millimetres.

The functions of the cerebrum, especially of the cortex, the youngest part of the brain, are very differentiated and all of them are of the greatest importance for our discussion. However, one thing is clear—if the highest level of information processing, intelligence, can be associated with one of the brain's components, then the cortex is this component.

The cortex is macroscopic and microscopic asymmetric. If left-right differences between the brain's hemispheres are measurable, are these differences related to differences in the higher activity of the brain, with different mental activities? In the case of a positive answer, the next question could be: What evolutionary advantages are there in left-right asymmetry? The answer to these questions could e as follows:

Asymmetry of hemispheres, and the probable differentiation of their functions, requires the co-ordination of the distinct halves of the nervous system. There are two ways of realizing such co-ordination—either by preparing a special co-ordination centre, or by continuous, uninterrupted communication between both halves. As we will see, both possibilities are used by the human brain (Section 5.5).

The interconnections between both halves of the brain are called commissures. A commissure (Latin *commissura*, bringing together) is a bundle of nerve fibres passing from one hemisphere, to the other; from left to right and right to left. The largest commissure between the cerebral hemispheres is the corpus callosum (Latin *corpus* body + *callosus* hard).

The significance of the corpus callosum in the emergence and develop-

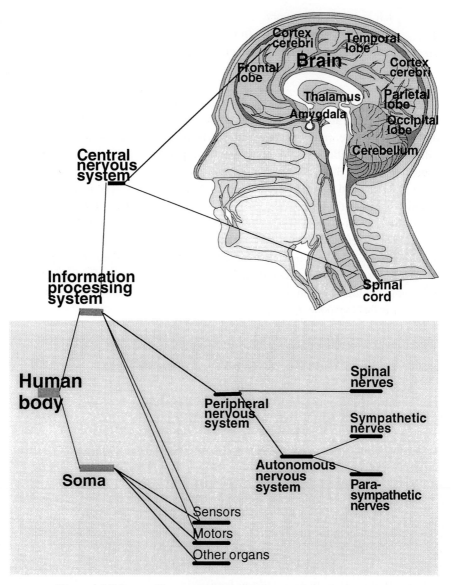

Figure 4-1 Scheme of human nervous system

ment of human intelligence must be stressed. It is a rather young creation of evolution. Marsupials have no corpus callosum (Fig. 3-4), which means that the hemispheres have no, or at least very weak, intercommunication. The higher animals have a corpus callosum proportionally smaller than that in humans. The human hemispheres communicate very intensively, possibly due to significant differences in their functions. The size of corpus callosum may be associated with speed of information processing (J.Lutz, H. Steinmetz, 1994).

The thalamus (Greek *thalamus* inner room) is a large grouping of nuclei situated just anterior to the midbrain, and consists of two ovoids. The thalamus is the final relay station for the major sensory systems that project to the cerebral cortex. Sometimes the thalamus is called the gateway to the cortex, because the main inputs to the cortex have to pass through it. Each thalamic area (about two dozen regions) also receives massive connections from the cortical areas to which it sends information. Probably a specific thalamic nucleus exists for pain.

The hypothalamus is a grouping of small nuclei that lie in the lower portion of the cerebrum, below the thalamus. The hypothalamus interconnects with many regions of the brain. The hypothalamus plays a function which is probably significant in our context—linkage to the bonus-malus system. At least some parts of the hypothalamus seem to play the role of 'pleasure centres'. Of course, pain and pleasure are not purely intellectual phenomena, but they are involved in all stages of biotic information processing.

4.1.5   *Other organs: amygdala, hippocampus*

The amygdala (Latin *amygdala* almond) which receives visual input form other brain organs, contains among others neurons responsible for perception of facial expression of other individuals and therefore carry the ability for appropriate social communication (R. Adolphs et al, 1994).

The hippocampus is a major structure, belonging to both loops in the central nervous systems—the 'own-body system' and the 'world-outside system'. It plays a significant role in the processes of memorizing and is an essential component in the initial registration of episodic and declarative memories (Section 5.2.1).

The group of large nuclei lying in the central regions of the cerebral hemispheres are called the basal ganglia. These nuclei (the main one is called striatum) play among other functions a role in the control of the motor system. The nucleus accumbens is a part of the basal ganglia and plays a significant role in the bonus-malus system: primary drug-reward circuitry (e.g. endogenous opioid peptide) (G.Schulteis, G.Koob, 1994).

4.1.6   *Cerebellum, etc.*

The cerebellum (Latin, diminutive form of *cerebrum*) belongs to the oldest part of the brain and is larger in man than in other animals. However, in lower vertebrates the relative weight of the cerebellum is larger. The cerebellum lies at the bottom back of the cerebrum (Fig. 4.3). From outside, it resembles two balls of wool a few centimetres in diameter. The cerebellum has a volume of 1/10 of the brain, however includes almost half of all brain's neurons, that is approx. 30 giga (30 thousand million) of neurons.

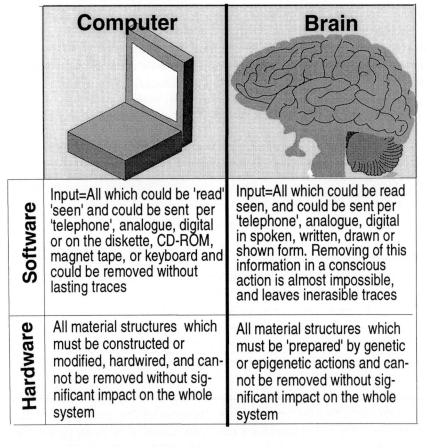

Figure 4-2 Brain and computer

The posterior lobe is the most recent development—the neocerebellum—and is most prominent in primates and man. This part seems to be responsible for precise co-ordination and control of the movement of the body and its various parts. Without the cerebellum, all movement would become clumsy. From our point of view, it must be noted that the cerebellum is not 'crossed-over'. This means that the right half of the cerebellum largely controls the right side of the body and the left half the left side. A bridge of fibres, called the pons, permits co-ordination between the two sides of the body.

The cerebellum performs the same general computation for many motor (and perhaps nonmotor) tasks. The anatomy and physiology of the cerebellum are central to thought about the sites of plasticity and mechanisms of motor learning. Experiments suggest that; 1) plasticity is distributed between the cerebellar cortex and the deep cerebellar nuclei; 2) the cerebellar cortex plays a special role in learning the timing of movement; 3) the cerebellar cortex guides learning in the deep nuclei (J.L. Raymond, 1996). Both, long-term depression and long-term potentiation may be some of many cellular mechanisms of plasticity (Section 4.6.4) in cerebellar cortex, as it is elsewhere in the brain. Also, with this mechanism, some sites of learning and short-term memory may not be sites of long-term memory. Cerebellum participates in functions ranging from analysis of sensory information, to telling time, to solving puzzles.

The enlarged cerebellum is related to the acquisition and discrimination of sensory information and to the adaptive co-ordination of movement through a complex three-dimensional environment (T. Rowe, 1996).

The cerebellum seems to be responsible for 'automatic', actions, especially for co-ordination of the subtle movements of hands, legs, etc. When one is learning a new skill, such as cycling, or the sophisticated movements of the trained typist or musician, one must initially think through each action in detail, and control is performed by the cerebral cortex. When the skill has been mastered, it is the cerebellum that takes over control, and the skill becomes 'second nature'. This relieves the tertiary areas of the cortex cerebri from automatic functions, making available more resources for higher mental activity.

Recent work suggests that a wide variety of classical, conditioned-learned responses may be stored in the cerebellum. Does this mean that the cerebellum is not involved in the highest level of information processing, in intelligence, but is mostly concerned with the lower stages of information processing, such as instinct or discent? It has been proposed that cerebellum is connected also with spatial working memory, that is cognitive functions. However, this is controversial. Learning of new movement trajectories involves the cerebellum, while overlearned trajectorial movements engage the premotor cortex (R. Seitz et al, 1994).

In the pons-stem brain, a small nucleus, (small pigmented region) the locus coeruleus plays the role of a funnelling mechanism that integrates all sensory input. Some of its axons can have an enormous number of synapses, often extending over

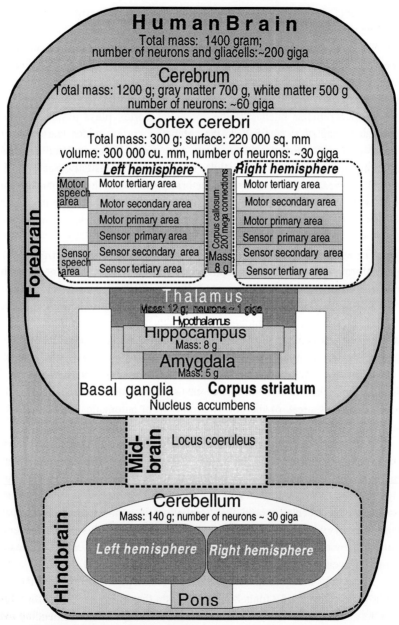

Figure 4-3 Scheme of human brain

a large region of the cerebral cortex, and making millions of connections to the other nerve cells on the way. It lumps all types of sensory messages, from sight, hearing, tactile pressures, smell and taste, into a generalized excitation response within the brain, providing emotional colouring, which plays a significant role in the bonus-malus system (Section 5.3). It must be of interest that in this nucleus the neurons tend to vanish with brain ageing (D.J. Selkoe, 1992).

4.1.7 *Brain areas: sensory and motor*

The processing of incoming information takes place in 'sensory areas'. Nerve signals from detectors, after processing in the sensory apparatus itself, proceed to the 'primary sensory areas'. In the brain the information is further processed before passing to the 'secondary sensory areas'. Some information reaches the tertiary sensory area.

The output information from a brain is mostly, but not only, in the form of signals going to muscles: that is, to effectors. The cortex areas which directly send the executive signals are called 'primary motor areas'. More complex signals have been prepared at a 'higher level', in the secondary motor area. The 'highest level' of processing is the tertiary motor area. The motor areas are in the anterior cortex, that is, in the frontal lobe. There exists asymmetry between the sensory and motor activities. The sensory inputs are very heterogeneous and not always orchestrated. Different detectors (sensory input channels), such as vision (colour, form, distance, etc.), hearing (frequency, intensity, speech, etc.), taste (salty, acid, bitter, sweet), smell (with its spectrum), and touch (temperature, pain, etc.), are on the one hand, rather heterogeneous but, on the other hand, are very exactly co-ordinated and directed towards different effectors (muscles or glands). Of course, the differentiation of effector outputs results from the enormous spectrum of human motor activities. These range from subtle movements of hands and fingers, extremely rich vocal apparatus (human speech), the differentiated expression of emotions by facial muscles and by 'body language', to the enormous spectrum of artifacts and technical devices which are activated by the movements of human fingers, feet, and vocal signals (e.g. voice-operated computers).

However, the territories are not fully differentiated; electrical stimulation of the brain of rhesus monkey, suggests, that territories controlling different fingers overlap. Despite the adult brain´s fixed anatomical wiring, recent results show that the topographic maps in sensory and motor cortices show remarkable plasticity. Such flexibility could provide neurobiological explanations for well-known observations such as learning. The topographic reorganization within the cortex was largely due to synaptic changes intrinsic to the cortex (C.D. Gilbert, T.N.

Wiesel, 1992).

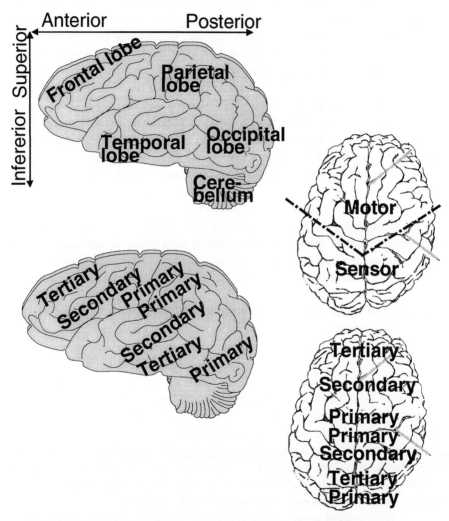

Figure 4-4 Human brain: motor and sensor, primary, secondary, tertiary

4.1.8 *Speech processing: Broca's and Wernicke's areas*

The clearest evidence of both asymmetry and direct coupling between macrocomponents of the human brain and higher mental activities, which we call 'intellect', are the two areas called Broca's and Wernicke's areas. Both are only in the left hemisphere especially in right handed subjects (at least, for the majority of people).

Wernicke's area is at the posterior (back), that is, the sensory half of the human brain, on the temporal lobe, in the region which is classified as the tertiary part. Wernicke's area is directly associated with the perception of human speech, and will be discussed in more detail later.

Broca's area is in the anterior (front); that is, in the motor half of the human brain, on the frontal lobe, in a region which is also classified as a tertiary or associative part. Broca´s area is directly associated with the active production of human speech, and will be described in further detail later on. Broca´s area clearly is not the human language organ. It probably should be regarded as multipurpose higher-level association area that is specialized to access subroutines for certain sequential operations. However, a fluent repetition of known words does not activate the two 'speech' areas, neither Broca´s nor Wernicke´s.

Our linguistic capacity is distributed rather more widely across the brain. For grammatical comprehension, as opposed to production, Wernicke´s area is at least as important as the Broca´s area, and the two areas are separated by some centimetres and a major lobal boundary.

Human neocortex is regarded as a six-layered structure, although this is a convenient fiction and greater or lesser number of layers are found in some areas. The reptile forebrain is in general only bilaminate.

4.2 THE MICROSTRUCTURE OF THE BRAIN: NEURONS, SYNAPSES

4.2.1 *Neurons and information processing*

The fundamental cellular units of the nervous system are neuronal cells—the neurons (Greek *neuron;* Latin *nervus* string). Neurons are the fundamental units for transmission and processing of information. Neurons and their networks form the basis of our perceptions, actions and memories. The complexity of information processing at the level of a single neuron reveals a previously unimagined complexity and dynamism (Ch.Koch, 1997). Which mechanisms are used for processing of information and what is the maximum amount of information that can be transmitted?

Spiking neurons communicate with other neurons by electrical impulses, in 'digital' form (interval between successive impulses being typically tens of milliseconds) and they could transmit information at a rate of about 300 bits per second (R.J.Douglas, K.A.C. Martin, 1996). The graded responses of the non-spi-

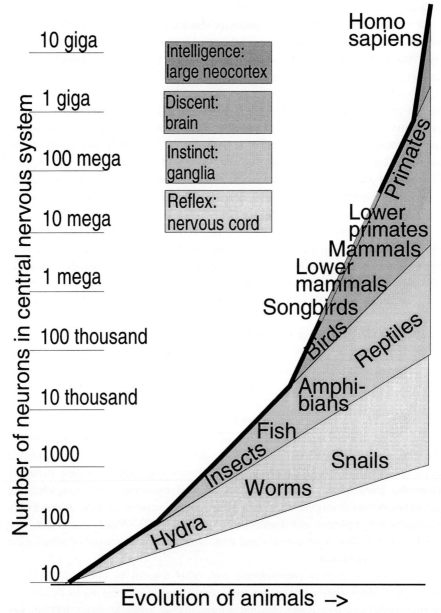

Figure 4-5 Number of neurons in the central nervous system

king neurons transmit as much as 1500 bits per second, five times the highest rates measured in spiking neurons. It was demonstrated that sensory neurons and chemical synapses are remarkably effective in this respect. Thus measurement of information capacity provides a benchmark for understanding the performance and design of neural systems at cellular and molecular levels (R.R.Ruyter de et al, 1996). Non-spiking neurons, do not communicate with other neurons by nerve impulses, but instead transmit their messages across synapses as analogue signals coded in form of graded voltages. Assuming that a primate's retina includes 10 mega of photoreceptors, and assuming a high rate of information transportation, the total information flux would be a gigabyte per second. The importance of active mechanisms for the analogue processing of information cannot be overestimated.

The optic nerve, which contains about one million fibres, may convey between one and hundred megabits per second—compare this with a quad-speed CD-ROM drive, which transfers information at 4.8 megabits per second (Ch.Koch, 1997).

4.2.2 Neurons—'brainy' cells

The human body consists of approximately 1 tera (10^{12}) of cells. Each day about 10 giga (10^{10}) of cells die; that equals 1 per cent of all cells. On average a cell in the human body exists 3 months. The dead cells must in general be exchanged by newly produced ones. However, the neuronal cells, 10 per cent of all cells, 100 giga (10^{11}), are 'immortal'.

Once they are produced in the first months of the existence of the human being (neurogenesis) they may stay alive for over almost 100 years. Part of the neurons die in the early period of development of the human child and some of them die in older humans (Section 4.3.1). Only a direct impact may damage the neurons. Neurons share some basic features with virtually every other cell in a living being.

However, in other mammals, for example in rodents neurogenesis occurs in the hippocampus throughout life, but the function of these neurons and the mechanisms that regulate their birth are unknown. Significantly more new neurons exist in the hippocampus of mice exposed to an enriched environment compared with littermates housed in standard cages (G. Kempermann et al, 1997).

Neurons make up only 15 per cent of brain cells; the other 85 per cent are glial cells (Section 4.2.4). Common components are: a plasma membrane (the boundary between its interior and the external environment) and the cytoplasm containing organelles (mitochondria, microtubules) and the cell nucleus. The membrane contains numerous pumps, channels, and receptors. An average neuron has a diameter of some micrometres, and a volume of one hundred micrometre.

Into one cubic millimetre (around 1 milligram) some 100,000 neurons are packed, which includes about 4 km of axonal wiring, 0.5 km of dendrites and close to one giga (10^9) synapses. The number of synapses in the cerebral cortex is around one thousand million million (1 peta) in the human cortex. Even counting a thousand per second, one would need about 10 thousand years to count them.

The most characteristic and unique component of neuronal cells is the axon (Greek *axon* axis), by means of which they send nerve signals to other partner neuronal cells.

The other protruding elements of the neuron are called dendrites (Greek *dendron* tree), which, because of their shape, resemble the branching of a leafless tree. They are the receiving zones for information (nerve signals) from other neurons. A large number of appendages, about 0.1 micrometre long, about 1 micrometre in diameter, cover the dendrites. These are called spines. In humans, one can count an average of at least 20,000 spines per neuron of the pyramidal cell type. Spines are discrete sites where ions can enter and activate cellular mechanisms modulating neuronal responses and thus information processing. Spines are the major postsynaptic target of excitatory synaptic input. Spines are a possible anatomical substrate for long-term memory.

Dendrites have additional levels of complexity that we are just beginning to understand. Dendrites can transmit information about spike timing. Differences in spike timing of a few milliseconds are crucial for synaptic plasticity (T.J. Sejnowski, 1997).

Recently an intriguing finding has been made: people who had pursued intellectually vigorous life-styles—going to college, holding intellectually demanding jobs—had dendrites with a length larger by as much as 40 per cent than those who had not. Especially in the Wernicke's area, which plays an important role in understanding language, this was found.

4.2.3   Neuron categories: rather few

The emergence and evolution of neuronal cells probably began 700 megayears ago and, as we can see, has been very successful. Two different strategies in the evolution of the nervous system can be discerned:

–Differentiation: by increasing the number of neuron types, rather than the number of similar neurons. This requires a rather large number of genes, at least one gene for each neuron type.

–Redundancy: by increasing the number of similar neurons belonging to the same type, and limiting the number of types. This requires a rather limited number of genes, but, consequently, strong influence of epigenetic mechanisms.

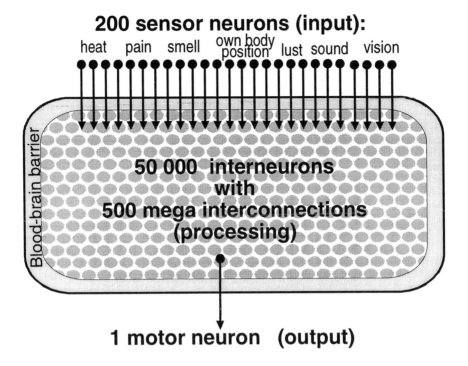

Figure 4-6 Interneurons are very numerous

A good example of the first strategy is the sea slug, Aplysia, in whose nervous system practically every neuron differs from its neighbours in its state of differentiation. Interesting exceptions are octopoda, such as the octopus, which has a nervous system built according to the 'redundancy strategy'.

The second strategy reaches its culmination in man. The number of neuronal cell types is relatively small (perhaps some tens or up to a hundred), but the number of neurons of the same type is very high (order of magnitude: hundred million). It must be pointed out that the human brain does not contain a special, exclusively, 'human' type of neuron. All neuron types existing in the human brain have been found in apes, monkeys, and rats. The neurons are the unique immortal cells in living beings. The mass of the cerebral cortex in man is 300 grams, contained in a volume of 300,000 cubic millimetres. With a density of 100,000 neurons per cubic millimetre, the whole cerebral cortex thus contains around 30 giga of neurons. Cerebellum also consists of 30 giga of granule cells, 30 mega of Purkinje cells, and many others. For comparison, the Milky Way consists of 100 giga of stars, and the present number of human beings on this planet is around 6 giga. A rat, with 400 to 500 square millimetre of cortex surface, would have only 65 million cortical neurons. Almost all neurons in the cerebral cortex are so-called 'interneurons'; that is, they connect one part of the cortex to another. Only 500 mega neurons are 'sensory neurons'—coming from sensors—and only 2-3 mega are 'motor neurons'—sending from the spinal cord nerve signals to the muscles (Fig. 4-6). The signals to or from the cortex are relayed at the level of the brain stem or spinal cord.

The classification of all neuron categories is rather extensive and, for our purpose, irrelevant. It seems that the list will eventually extend to tens, or even hundreds, of neuronal categories. However, compared with the number of neurons in the human brain, the number of categories is astonishingly small. We will speak mostly about only two categories: pyramidal neurons and stellate neurons.

A neuron, on average, has more dendrites but only one axon (Fig. 4-7). The axon may make contact, by means of 10,000 synapses, with the same number of other neurons. The dendrites of one neuron may link with up to 10,000 synapses from 10,000 other neurons. The average number of synapses in the human cortex is around 6,000 per neuron, rising in some cases probably even to 20,000.

4.2.4   *The glia cells*

Eighty five per cent of brain cells consists of glia cells or neuroglia ('nerve glue'), but constitutes only 50% of cerebral mass. Glial cells were once thought to be little more than passive support elements. They now appear to nourish (by transporting glucose from the blood and preparing it for the neurons), protect, and 'listen' to neurons. They may even 'talk back'. They do not send out the long projections called axons that terminate in synapses on neurons or other cells. They send out shorter, branchlike projections, some of which envelop the synapses while other contact other glia cells. They clean up the neuronal environment and recycle some neuronal products. Some glia cells (the oligodendrocytes) form the myelin that wraps the axons of neurons in the white matter of the brain, giving the white matter its characteristic colouring and distinguishing it from the grey matter. Some other glia cells, so-called astrocytes (named for their star-like shape) wrap the cerebral capillaries and form the structural link between the blood supply to the brain and the neurons. Many structural features of the mammalian and human central nervous system can be explained by a morphogenetic mechanism that involves mechanical tension along axons, dendrites and glials. In the human cerebral cortex tension along axons in the white matter can explain how and why the cortex folds. This results in a cortical surface of about 1,600 cm^2, nearby three

4.2.5 *The synapse: the neuronal junction*

All neurons are mutually connected to an enormous number of other neurons (Fig. 4-7). The synapse is the junction between two neurons. On one side, on the axon, is the synaptic terminal. A small cleft separates the terminal from the other side, the postsynaptic membrane, belonging to another neuron and lying on a dendrite or neuronal cell body and occasionally on the axon of another neuron. The synaptic cleft is only a few millionths of a millimetre wide, and the whole synapse is just a few thousandths of a millimetre in size (Fig. 4-7). Synapses look even more uniform in structure than neurons. They play an important role in the growth and death of neurons. When neurons communicate, the information is carried not only by electrical pulse but also by means of specific chemical messengers—the synaptic transmitters. A given neuron uses the same neurotransmitter for all of its numerous synapses.

Information on synapses is transmitted in one direction only—from the axon terminals of the sending neuron (the presynaptic neuron), across the synapse cleft, to the receptive surface (postsynaptic membrane) of the receiving, or postsynaptic neuron.

Neurons respond to stimuli by producing action potentials, the rapidly travelling electrical impulses which trigger neurotransmitter release.

In general terms, there are only two functional kinds of synaptic mess a neuron is analogue (continuous values between 0 and 1), and a neuron's output is discrete either it spikes or it does not), though some neurons may have analogue outputs. An action potential (spike) lasts about 1 millisecond ages: excitation, in which one neuron (presynaptic) commands another (postsynaptic) to electrical activity (to start firing), and inhibition, in which the recipient neuron (postsynaptic) is prevented from firing. A postsynaptic neuron, receiving an excitation signal in the form of a chemical message through the synaptic cleft, may or may not fire in response. One factor in determining the threshold for firing is the amount of neurotransmitter released onto the receiving neuron.

A single release from a presynaptic neuron may not be enough to change the receiving neuron's resting potential to an action potential. Often, a summation of many impulses is required to produce the potential change necessary. The summation can come either in firing more times in rapid succession by one neuron or through the firing of more neurons simultaneously.

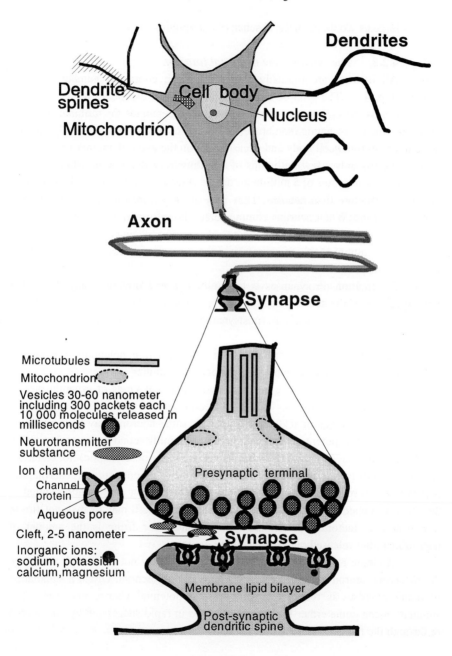

Figure 4-7 Scheme of neuron and synapse

4.2.6 *Microtubules play a significant role*

Neurons, like other cells, must strike a balance between structural stability and plasticity. One agent of this compromise is the intracellular system of polymeric protein filaments, called the cytoskeleton. In the neuron, these filaments contain microtubules.

These microtubules are very small tubes, with a diameter of 20 millionths of a millimetre and a length of 20 thousandths of a millimetre, consisting of the protein tubuline (Fig. 4-8).

Neurons use the cytoskeleton to maintain their shape and position, as well as to change them—to shuttle their internal contents about, or to exclude these from certain regions. Microtubules and microfilaments do not translocate in the axon, but undergo assembly and disassembly there (N.Hirokawa, 1992).

Microtubule bundles are observed to elongate at rates of 0.2 micrometres per second. They play the role of rails, for the transportation of neurotransmitters. Some students (R.Penrose, 1994) claim that microtubules are the vary carrier of the quantum mechanical effects, which are responsible for the highest intellectual activity, the consciousness.

Recently the role of the neurofibrillary tangles in the contribution to the death of neurons has been explained referring to the microtubules.

4.2.7 *Neurons and synapses as electrical and chemical devices*

Neurons are primarily electro-chemical devices. They perform their work of receiving and transmitting nerve signals by means of electro-chemical energy. The energy is generated by the flow of charged chemical entities—ions—through the neuron's membrane. The fluid that surrounds all the cells of the body, neurons included, is an aqueous solution in which positively and negatively charged ions are distributed freely in equal numbers, neutralizing each other's charge.

The positive charged ions are atoms deprived of one electron—the carrier of negative charge. In neuronal processes, ions of the following metallic elements are present: sodium, potassium, calcium, and magnesium. The ions of non-metallic elements such as chlorine, or compounds of such elements as phosphorus bound with oxygen carry an additional electron (therefore having a negative charge), in the form of chloride, phosphate, carbonate. The inside of a neuron is negatively charged, or 'negatively polarized', with respect to the outside.

When the neuron is at rest, the negative polarity is maintained. The plasma membrane's selectivity is possible because it has, along its length, openings, or channels, that are specialized to allow the passage of specific substances. There are channels for potassium, for sodium, and so on. Parallel to the channels there are ion pumps, powered by energy generated metabolically. These pumps help to remove

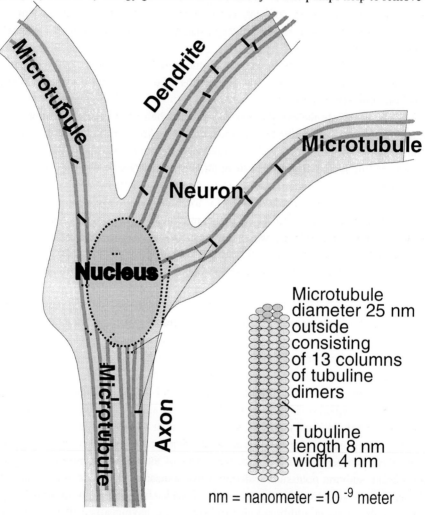

Figure 4-8 Microtubules: significant component of neurons

excess sodium ions that enter during periods of excitation. This electrical state of an unstimulated neuron is called its resting potential. In resting neurons, the transmembrane potential is 0.07 volts, with the inside of the cell negative relative to the outside. Nerve cell culture experiments including neurons and glial cells (astrocytes) of rats show that the electrical signals are transmitted not only to neighbouring neurons but also to other astrocytes (M. Nedergaard, 1994).

A neuron is stimulated to fire, that is, to conduct an electrical impulse, by receiving certain chemical signals at excitatory synapses. A rather complex mechanism of transportation of sodium and potassium ions, inside and outside the neuronal membranes, spreads along the axon. Each such reversal of the interior membrane's electrical state, from negative to positive, is called an action potential, or depolarization. Such electrical impulses, which last less than a thousandth of a second, are essentially momentary, self-limiting reversals of the transmembrane potential.

Long-term potentiation (LTP) is a chemical mechanism of highest importance for some extremely significant mechanism in brain. Long-term potentiation is a stable and long-lasting potentiation of synaptic activity. Changes in synaptic strength may be critical for learning either as a mechanism for the direct storage of memories, or as a process that transforms information making it suitable for long-term memory. R. Malinov (1994) writes that over the past 20 years many properties of LTP have been identified: the title of his paper 'LTP: desperately seeking resolution'.

It is of highest importance to see the synapses as a scene for very complex chemical events including non-electrical synaptic integration.

4.2.8 *Ion channels*

Cells (also neurons) respond to external signals by transducing the information into internal messengers such as calcium Ca^{2+}. The information-processing capability of the ionic signalling system is enhanced by the well-known engineering principles of amplitude (low-high concentration) and frequency (temporal) modulation (M.J.Berridge, 1997). The membrane of nerve cells are made of lipid bilayers (approx. 0.004 micrometres thick) and are not permeable for ions (Fig. 4-9). As special organs, the ion channels take care of ion transport.

The transport of simple positively charged ions, such as sodium Na^+, calcium Ca^{2+}, potassium K^+, magnesium Mg^{2+} is the decisive mechanism of electrical properties of neuronal activity. Negatively charged ions are chloride Cl^-, and sulphate SO_4^{2-}. All ion channels must perform a delicate energetic balancing act to allow ion permeation. The flux of ions through the channels achieves one thousand to a few thousand ions per millisecond. The ion channels are built of one protein molecule folded into four distinct parts, or of five separate protein subunits.The existence of two different types of ion channels: chemically-gated (that is activated by neurotransmitters) and electrically gated (that is activated by an electrical potential) illustrates well how neurons are electro-chemical devices.

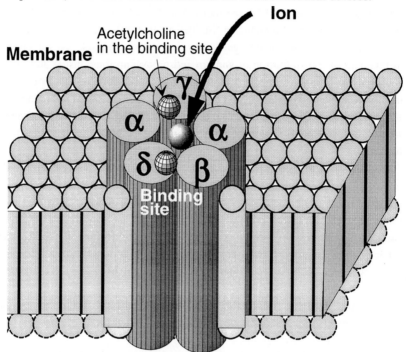

Figure 4-9 Ion channels

4.3 THE BIRTH AND DEATH OF NEURONS AND SYNAPSES

4.3.1 *The human brain: from embryo to adult*

The formation of the nervous system, in humans, begins about two weeks after fertilization. During the second month, the three major divisions of the brain—hindbrain, midbrain and forebrain—are formed. The cerebral cortex begins to form before the brain differentiates sexually. Already by the sixth week of human embryonic life, the most forward vesicle of the neural tube divides into two compartments, each of which will form a cerebral hemisphere. At times, the neuron production rate reaches thousand neurons per second. At around 150 days after fertilization, nerve cell division probably stops. The male hormone testosterone, secreted by male embryos very early on, has an organizational effect on brain development. The prenatal brain's development is probably influenced by the body of the mother and by environmental events (Fig. 4-10).

The maximum number of cortical neurons is attained well before birth. Although humans are born with almost all the neurons they will ever have, the mass of the brain at birth is only about one fourth that of the adult brain. The brain becomes bigger because neurons grow in size, and the number of axons and dendrites increases (C.J.Shatz, 1992). We are born with a brain in which the number of neurons can only diminish. Massive numbers of neurons die in utero, and the dying continues during the first two years before levelling off at age seven. But head size, brain mass, and thickness of the cerebral cortex continue to increase rapidly in the year after birth. Long distance connections (white matter) are not complete until nine months, and they continue to grow their speed-inducing myelin insulation throughout childhood. Synapses continue to develop, peaking in number between nine month and two years, depending on the brain region. At that time the child has fifty per cent more synapses than the adult (probably 100 tera, that is 10^{14} synapses). Metabolic activity in the brain reaches adult levels by nine to ten months, and soon exceeds it, peaking around the age of four. At birth, after 270 days of embryonic development, the brain has achieved 25 per cent of its adult weight; that is about 350 grams. The next 350-gram increase occurs during the following 6 months. A further 350 grams, are added in the next 18 months; that is, by the age of two years. The remainder is added during the following 12 to 14 years (Fig. 4-10).

4.3.2 *The birth and death of neurons*

To produce 30 billion neurons in an adult human cortex and taking into account the overabundant production in early period of life, the production rate must be about 700 neurons per second, with a maximum rate of 4,000 per second.

Neurons have very complex and diverse lives. It can be said that each neuron has its own individual life—at least, in some microscopic details. Almost all neurons in the human brain are 'born' during the foetal period. A large number of them die in the first year of infancy (Fig. 4-10). The overproduction of neurons and glial cells is enormous and achieve a factor two. Only a small number of neurons live some tens of years, and a minority die at the moment of the natural death of the body. The following phases of a neuron's life are of importance: overabundant proliferation (overproduction), migration, aggregation, differentiation, selective elimination, stabilized activity, apoptosis (programmed death of cells), and then death of whole organism.

All neurons in a human brain are born in the first phases of the development of the human being, in the prenatal period. The inability of neurons to divide makes the nervous system the most susceptible of any organ to damage of all kinds. Neurons cannot be replaced, in contrast to many other types of human cells. This is the high price to be paid for being intelligent.

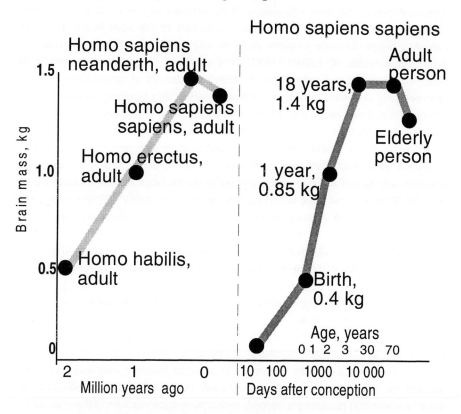

Figure 4-10 From embryo to adult-brain's mass

For example, in the brain of the new-born monkey the measured rate of elimination of axons equals 8 million per day, that is 100 per second (P. Rakic, 1995). New-born neurons must migrate from the region in which they proliferated to their final destination in the brain, to its different organs, tissues, layers, fibres, etc. The neurons must be adapted to particular modes of transmitting messages (excitatory or inhibitory), as well as specific size, shape, location, and connections.

The connectivity of the nervous system becomes gradually established during embryonic development when each neuron as it differentiates, extends a thin axon to appropriate target cells to form synaptic connections. Axonal trajectories are highly reproducible from embryo to embryo, as growth cones rarely make errors of projection despite the large distances—as much as several centimetres—some of them travel. How do growth cones achieve these navigational feats? Current evidence suggests that they find their way in the embryo by detecting and interpreting directional information encoded in specific molecules (Nerve Growth Factor, NGF) in their environment.

The next step is the establishment of synaptic connections with other neurons, through the growth of axons and dendrites. The axons have specialized tips, called growth cones, that can recognize the proper pathways. Once axons have arrived at their targets, they still need to select the correct address. Unlike pathway and target selection, address selection is not direct, in fact, it involves the correction of many initial errors. The necessity for neural activity to complete the development of the brain has distinct advantages. The first is that, within the limits, the maturing nervous system can be modified and fine-tuned by own experience, thereby providing a certain degree of adaptability. It can begin in utero and, as in the primate visual system, continue well in neonatal life (C.J. Shatz, 1992).

The number of newly generated neurons and newly established synapses in a foetus, and then in a child, is much larger than that in an adult's brain. The superfluous and excessive neurons must be eliminated—they must die. In some small neuron circuits, the ratio of neurons which die is between 15 and 80 per cent of the initial number of created neurons. Also superfluous and excessive synapses must be eliminated—they too must die. The process of the elimination of neurons and synapses is one of the most important ways of learning about the real world and establishing a good internal representation of it—a good semantic correspondence.

The number of neurons, and synapses, in these brain nuclei does influence the skill with which a particular canary sings. Although neurogenesis has been reported in some adult mammals, its occurrence there seems to be much more limited and more controversial. Why is it so prominent in song birds? It may have something to do with their relatively long life-span.

There is no doubt that the processes of overabundant production and the death of neurons (selective elimination) are of the highest significance for our discussion about human intelligence. However, it must be stressed that, as yet, there is no clear evidence that neurogenesis occurs during the rest of life, in man or in other primates.

There is strong evidence that at least some categories of 'immortal' neurons in the human brain are lost during life. During the development of the nervous system, up to 50 per cent or more of many types of neurons normally die soon after they form synaptic connections with their target cells. The massive neuronal death is thought to reflect the failure of these neurons to obtain an adequate amount of specific neurotrophic factors that are produced by the target neurons and that are required for the neurons to survive. These survival signals seem to act through an intrinsic cell suicide program, the protein components of which apparently are expressed in most cell types (M.C. Raff, 1993). The simplest rate of the continuous loss (that is, 'death') of neurons in the human brain is probably one half of one per cent per year. This means that, after seventy years of life, the brain has only a half or a quarter of its initial number of neurons. This value is probably on the high side. The brain decreases somewhat in size with age, and the ventricles enlarge. Naturally occurring cell death ('apoptosis') reduces many neural populations by between 25% and 75%. The weight of the brain declines with age: from 1,5 kg for age of 20 years to around 1,1 kg for age over 80 years (D.J.Selkoe, 1992).

From our point of view, it must be said that this phenomenon suggests that, if an individual lives long enough, all reserve capacity will eventually disappear. The minimum number of cells required to maintain normal function will no longer remain, and decompensation will occur. Disease produced by 'normal' ageing will appear.

In this or another way, neurons are the real carriers of biotic information processing, including the highest stage—intelligence. The question may really be: Why do neurons live so much longer than other types of cells? Perhaps the answer to this question is the reason for the 'longevity' of individuals who maintain their full mental capability.

A simple hypothesis concerning the birth and death of neurons in humans is as follows: The rate of formation of neurons in a developing embryo increases very rapidly, and probably reaches a value of more than four thousand neurons per second some weeks after fertilization. After birth, the rate decreases and falls to zero at the age of a few years. Recent data seem to suggest that the number of neurons in the human brain may not decrease with time. The volume of the brain decreases by round 10 per cent, but the reason for this would not be a decrease in the number of neurons, instead only a real volumetric decrease of the liquids. In our hypothesis, we postulate that the number of neurons slowly, but clearly, decreases (Fig. 4-11).

Because of the similarity to the strategy observed in the evolution of species (Section 3.1.6), G.M.Edelman (1987) called the phenomenon of overabundant production and selective elimination, 'neuronal Darwinism'. No other method can exist for the generation of around a hundred billion neurons, using the genetic information from some tens of thousands of genes, devoted to the generation of the human central nervous system. Only overabundant, more-or-less 'chaotic', generation, ruled by genetic factors and followed by epigenetic selective elimination, can be responsible for this extremely complex phenomenon. At present it is a routine to grow neurons, taken from the brain, in the laboratory, outside of living organisms.

4.3.3 The production and elimination of synapses

The excellent strategy of overabundant production and selective elimination is used with enormous effectiveness in the development of the human brain.

All the tactics, and 'tricks', of this strategy are used for the birth and death of around a million billions synapses in the human central nervous system. We must not forget that the number of genes controlling this phenomenon equals some tens of thousands, from the total number of approx. 80,000 genes in the human genome. That means that the human genome provides the blueprint for about 80,000 different proteins. The process seems to be very wasteful and 'chaotic', but appears to be the only feasible strategy available. A well functioning repair system contracts the spontaneously emerging error rate of DNA replication. But the neurons are 'immortal cells' and do not replicate itself.

In our hypothesis, which is weak and naive but gives some quantitative suggestions, the formation and death of synapses, called synaptogenesis, continues throughout the entire life of a human being. In the embryonic period, synaptogenesis is probably zero, increasing very rapidly towards the moment of birth. Synaptogenesis is active during the first few years and reaches a growth rate of more than half a million synapses per second.

This is the basis for the very intensive learning of what we called 'common sense' (Chapter 7). During this period, the rate of death of synapses must also be enormous, because the excess of synapses—that is, the number of synapses being generated—is five to ten times larger than the number which remain after childhood. This must mean that, during the first five to seven years, the death rate of synapses must reach the same order of magnitude as the synaptogenesis—that is, around half a million per second. Later, both processes come to equilibrium. In old age, the death of synapses exceeds the formation rate. However, synaptogenesis is, in spite of this, a significant process up to the end of life. This is probably the basis for a learning ability which, in some cases, continues into old age. While growing up until around 80 years of age, the number of synapses probably decreases, particularly in the sensoric part of the cortex, and especially when the neuronal network is less used.

Recently, the mechanism of the formation of synapses under the influence of a peptide called synapsin has been further unravelled. Synapsin may contribute to both short-term and long-term synaptic plasticity.

4.3.4 *Modular and laminar organization*

However, the evolution of the brain did not proceed by the simple addition of parts to the brain of pre-existing species. The human brain is not that of an reptilian brain or ancestral ape with addition of a bit of frontal lobe and some language areas.

While this is easy to state, it is much more difficult to articulate the subtle organization changes that must have taken place during the evolution of the human brain. It now seems fairly clear that neither new neuronal types, nor columns, modules or even laminae are to be found in the human brain.

M. Kinsbourne (1995): "The traditional, serial hierarchical centred brain model conflicts with findings in brain structure, physiology and focal lesion effects on behaviour. An alternative model is proposed, in which multiple representations are active in parallel....The 'penetrability' of modules to each other's influences is consistent with a parallel neuronal architecture."

4.4 PHYSICAL PHENOMENA IN THE HUMAN BRAIN

4.4.1 *Do we need a special kind of physics?*

The investigation of the carrier of intelligence by means of the scientific method requires from time to time, answers to the following questions: Are our scientific methods or more exactly, our physical methods, sufficient to investigate the human brain? Do we need a new, as yet unknown, kind of physics, new methods, new hypotheses or theories, new concepts of elementary particles and elementary forces, or even new laws of nature? Do we expect, or sense, the existence of an insuperable barrier, which will prevent further progress in the investigation of the human brain?

Of course, these questions provoke various answers, which, in turn, generate further questions. Some prominent scientists have clear answers. For example:

R.Penrose (1994): "They require that the biological systems that are our brains have somehow contrived to harness the details of a physics that yet are unknown to human physicists! This physics is the missing 'objective reduction' theory that straddles the quantum and classical level and, as I am arguing, replaces the stop-gap 'State vector reduction'-procedure by a highly subtle non-computational (but undoubtedly still mathematical) physical scheme. That human physicists are, as yet, largely ignorant of this missing theory is, of course, no argument against Nature having made use of it in biology."

F. Wilczek (1994): "Thus he (Penrose) concludes that mental feats of humans cannot be explained within conventional laws of physics. Finally, he argues that the required new law of physics will explain....The quantum theory of gravitation (M.T. postulated by Penrose) is fraught with difficulties ... that may or may not be resolved by superstring theory. My own tentative conclusion is that the predicted effects are exceedingly small and are likely to be overwhelmed. So (Penrose's) speculations about a spectacular computational ability of microtubules based on quantum coherence and central to human consciousness appear quite bold at this time.... It appears to me that Penrose's argument....that quantum theory is incomplete is unconvincing and his proposed remedy implausible."

F. Crick (1994): "I feel that no-one has shown, properly, how quantum mechanics would solve the binding problem; what quantum coherence could effect and how that is going to do the binding."

Some authors have made the extreme suggestion that the human brain is a quantum computer. It is naive to hope that the existence of quantum mechanics by itself is sufficient to explain the phenomenon of human selfconsciousness.

4.4.2 Neuroscientists plan an 'atlas' of the brain

Parallel to the development of the Theory of Everything (the recent variant is called 'Duality Theory') (Section 2.1.2) which is a theory of the most elementary phenomena, must be considered the development of the theory of the most complex issue—the physics of the brain (and mind?) of physicists themselves. The application of physical methods in the field of brain research brought very positive and encouraging results.

Scientists plan to prepare a series of computerised maps of the brain's anatomy, function, physiology, biochemistry, and molecular biology (Fig. 4-12). The macro-architecture of the brain contains 120 different structures, such as the thalamus and amygdala. Neuroscientists say that computerised catalogues would provide a 'common language' for new specialities emerging from new imaging techniques such as positron emission tomography (PET), magnetic resonance imaging (MRI), computerised X-ray tomography (CT) and the recently developed magnetoencephalography (MEG), measuring the magnetic currents that are generated by electrical impulses in the brain. MEG-EEG machines will offer a millisecond by millisecond picture of brain activity (J. McCrone, 1995).

Some of these maps will ascribe 'geographic addresses' to such traits as fear, sexual desire, anxiety, depression, vigilance, language processing. A larger objective of the planned atlas is to understand how the brain works as a whole. Organizing a neural database will be much more complicated than setting up the human genome database. The human genome is a finite project, it involves

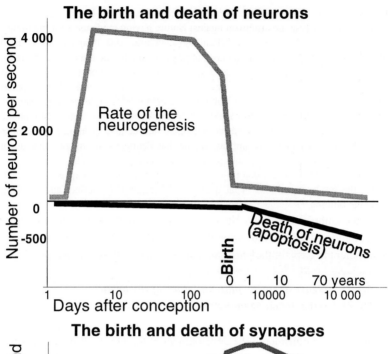
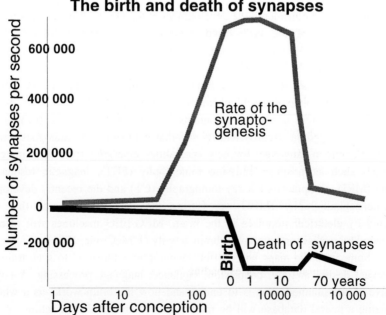

Figure 4-11 The birth and death of neurons and synapses

sequences and mapping data which are relatively straightforward.

The power of neuro-imaging is also permitting in vivo measures of circuits and mechanisms of the human brain. These advances have created an era in which a scientific psychopathology that links mind and brain has become a reality. Brain-based cognitive models of illnesses, such as schizophrenia and depression have been tested with a variety of techniques and a relatively sophisticated picture is emerging that conceptualized mental illnesses as disorders of mind arising in the brain. These can be understood as dysfunctions in specific neural circuits and their functions and dysfunctions can be influenced or altered by a variety of cognitive and pharmacological factors (N.C. Andreasen, 1997).

Using PET and MRI an area of abnormally decreased activity in the prefrontal cortex (reduction in mean grey matter) and parts of the corpus callosum in depressives has been localized. These regions have previously been implicated in the mediation of emotional and autonomic responses to socially significant or provocative stimuli, and in the modulation of the neurotransmitter system targeted by antidepressant drugs (W.C.Drevets et al, 1997).

4.4.3   *The energetics of brain*

Information processing requires the consumption of a rather large, more or less continuous, flow of free energy. It is a process producing order, in the form of correct and good information, and competing against disorder— the destruction of information. High-level order can be maintained only by the use of well-ordered energy—free energy. Non-free, that is bound energy represents disorder.

The energy consumption of the central nervous system in relation to the 'basal metabolism' (that is mean total energy flux) of most vertebrate species equals between 2% and 8%. The comparable figures for the rhesus monkey and the humans are 13% and 20% respectively. And the human children's brain uses about 60% of the total energy flow. The effect of age is rather significant showing continuous decrease of energy flow (K.L.Leenders et al, 1986, K.L. Leenders et al, 1990) (Fig 4-13).

The human brain is the largest consumer of fresh blood of all human organs. The human heart has a pulse of something less than 1 second pumping 60 millilitres blood per pulse. About 1/5 of this amount goes into the adult's brain. Each millilitre of blood includes 1.2 milligrams of glucose, the primary carrier of free energy for the brain. The corresponding flux of oxygen equals 0.8 millilitre per second. The energy flux is around 20 watts; that is, some 20 per cent of the consumption of the whole body (100 watts corresponds to a daily consumption of

8.64 megajoules, or, in other units, around 2 100 kilocalories).

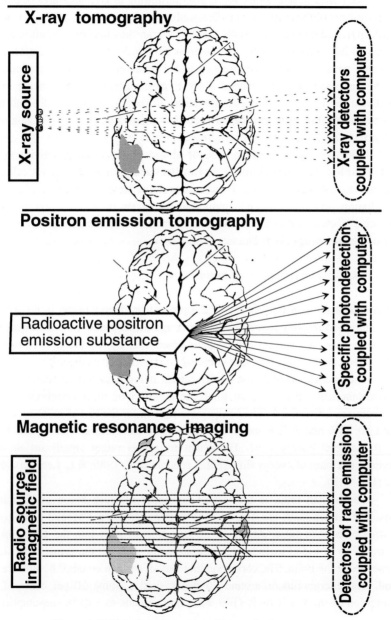

Figure 4-12 How to make a map of human brain

The specific energy flow of 20 watt per brain mass of 1.4 kg corresponds to a consumption of around 14 watt per kilogram (Fig. 4-13 and 4-14). If we do not take the whole mass of the human brain, but only the 'dry' mass (that is, without water), the specific energy flow is around 65 watts per kilogram of dry brain mass. This value must be compared with values of other specific flows in nature, in order to understand the uniqueness of the human brain, from this point of view. For example, let us compare it with some typical rates of energy transformation.

The Sun produces free energy with an average specific energy flow of only 0.0002 watt per kilogram of solar matter. The terrestrial biosphere, taken as a whole, with all micro-organisms, green plants and animals, has an average energy flow of 0.075 watt per kilogram of dry matter. Bacteria's specific energy flow achieves some kilowatts per kg. The horse's muscle achieve a specific power of 100 watt/kg in maximum and less than 10 watt/kg during working time, decreasing to some watts/kg during rest. The human body, as a whole, requires a specific energy flow of 1.5 watt/kg (100 watts, and 65 kg body weight). Now, the energy flux of a 6-year-old child's brain is 21 watt/kg. The brain oxygen consumption decreases with age, approximately 0.5% per year between 20 and 80 years of age, This indicates constant loss of dendritic synaptic density or reduced neuronal firing. Probably the first option is true (K.L.Leenders et al, 1990). Adults of 50 years of age require, for their brain, only 2/3 of the above-mentioned flux of 20 watts, and, at 70 years of age, only 1/2. An older man, with sclerosis, uses only 10 watt/kg of his brain (Fig. 4-13). Effect of age (0.6% per year loss of dopamine D2 receptors) in normal health (without obvious loss of function) is demonstrable also for specific neurotransmitter functions. (A. Antonini, K.L.Leenders et al, 1993). But it must be stressed that the energy transformed in the brain is always coupled with consumption of glucose but not of oxygen, the so-called 'anaerobic metabolism'. This means that the brain relies on tactics similar to those used by sprinter's muscles. And this occurs despite the presence of abundant oxygen in the normal brain. It remains a mystery.

As we can see, the human brain is an enormously intensive energy-consuming 'machine', also during sleep. Information processing is not a negligible energy consumer, as we also know from the cooling of a personal computer. The free energy is carried by the energy-rich molecules of adenosine triphosphate ATP. One mol that is $6 \cdot 10^{23}$ of ATP-molecules contains 33.5 kilojoules of free energy. Human brain needs around 20 joule per second, which corresponds to 0.6 millimol of ATP. Assuming that each of around 50 giga of neurons continuously use equal amount of free energy the number of ATP molecules used during one second by one neuron equals around 10 giga. On average each synapse gets around one million ATP molecules per second. The human brain has a blood circulation rate of 0.013 litre per second, carrying 0.0013 grams of glucose. The burning of this amount of glucose uses 0.0008 litres of oxygen and produces around 20 joules per second; that is, 20 watt of free energy. A neuron uses roughly one femtojoule (one millionth billionth) of energy per operation (e.g. one neuron activating another at a synapse). One femtojoule corresponds to 10,000 electron volts; the energy exchange between one molecule reacting with another requires around 1 electron volt). By contrast, the most efficient silicon technology currently requires about 0.1 microjoule per operation (multiply, add, etc.), that is 10 to 100 million times more than a neuron.

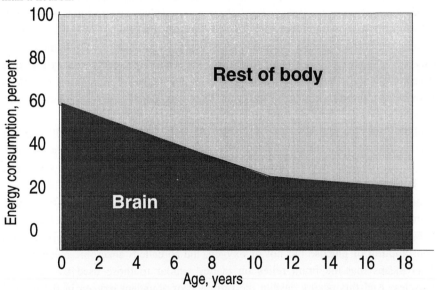

Figure 4-13 Energy flow in a child's brain

An interruption in the inflow of fresh blood into the human brain of only six seconds is sufficient to cause an oxygen deficiency which results in loss of consciousness and even irreversible brain injuries, ending in death.

A direct consequence of the high energy efficiency of neurons is that brain can perform many more operations per second that even the largest supercomputers. The present fastest digital computers are capable of around teraflops (10^{12} = tera floating point operations per second); (Section 9.1.3) the brain of the common housefly, for example, performs about 0.1 tera operations per second when merely resting.

And it must be remembered, that events in an electronic computer happen in the nanosecond (one billionth) range, whereas events in neurons happen in the millisecond (one thousandth) range.

4.4.4 *Electrical activity of the brain*

It is useful to measure the integrated electrical activity of human brain during appropriate intellectual tasks. This technique is know as electroencephalography, in short EEG.

E. Callaway (1978): "Berger, the father of electroencephalography, was a psychiatrist with some interest in extrasensory perception. One can speculate that Berger, like some of the rest of us, must have felt frustrated that minds are locked each into its own bony prison. We are limited in our communications with fellow prisoners to a prison telegraph developed out of the sensory and motor systems. For mystics, the appeal of extrasensory perception lies in its promise of direct mind-to-mind communication. For one with scientific training and sceptical nature, brain (electrical) potentials hold a similar fascination in their elusive promise of some additional window on the mind."

Some characteristic peaks on the curve of the electroencephalogram are observed, among others the so-called 'P300' peak of measured current, which is a positive peak emerging as a response around 300 milliseconds after task-related stimuli.

The most important fact from our point of view, is that some intellectual tasks, such as the selection of a sequence of three even digits or three odd digits in a pseudo-random sequence of single digits (Ch. Michel, 1988), has been correlated with measurement of the changes in the P300 component of EEG. The influence of different exogenous factors, such as smoking, or the use of drugs, has been observed and interpreted. It is widely assumed that the hippocampus and cortex are responsible for the P300 peaks, but chemical processes in the synapses have also been invoked to explain some features of the changes seen in EEG waves.

A recent advance in this field is the high-resolution electroencephalogram system device which records, from 124 sensors on the subject's scalp, the rapid pace of brain activity in real-time. In the split second of anticipation before the subject receives signal to which they will have to respond, activity occurs in brain areas that will be activated a moment later during the execution of the task. This can be called 'an act of mental preparation'.

Event-related potentials, called ERP are small fluctuations in the spontaneous electrical activity measured by means of electroencephalogram, or EEG.

136 Search for terrestrial intelligence

A dramatic and significant change in electric activity of brain in pathological cases is seen in patients suffering from epilepsy (S.J.Schiff 1994).

Also by means of measurement of magnetic activity, connected with the electrical currents so called 'magnetoencephalography' (MEG) significant knowledge about human intellectual activity could be achieved.

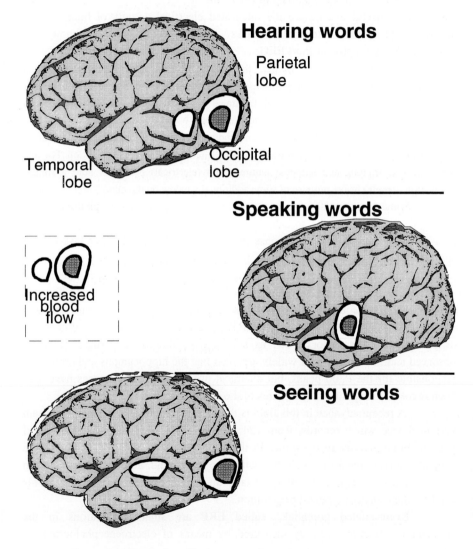

Figure 4-14 Energetics of brain and intellectual activity

4.5 CHEMICAL PHENOMENA IN THE HUMAN BRAIN

4.5.1 *Chemical composition seems at first sight simple*

The human brain has been declared to be the spatially most complex structure in our region of the Galaxy. However, from the point of view of elementary chemistry the same object seems to be very simple, even extremely simple. If we distinguish only two classes of chemical compounds, water and the rest, then we will probably be astonished to learn that the human brain contains 77 per cent by weight of water. The 'dry' rest, consisting of proteins, lipids, etc. comprises only 23 per cent by weight. It is reasonable to compare this figure with the average content of water in the whole terrestrial biosphere, which is 56 per cent by weight.

Cerebral ventricles, five in numbers, are cavities in the brain filled with cerebrospinal fluid, with a total volume of approximately 0.05 litre in 30-year-old individuals to 0.120 litre in 80-year-old males. The proud carrier of human intelligence is thus three quarters water, and its other chemical constituents are not very exotic. The chemical composition of the human brain does not differ from that of the brains of other higher animals. There is nothing mysterious or unique in the brain's chemistry.

If we consider not the proportion by weight, but the number of different molecules, then we see that around 90 per cent of all molecules are water. We are made not from ashes but from water. 'You are made of water, and to water you shall return'.

It is true that water molecules are the most abundant molecules in our Galaxy (and probably in the entire Universe) and hydrogen and oxygen are the most abundant elements (excluding helium, which is unable to react chemically). This enables us to say that the human brain is a good sample of cosmic matter, which may better satisfy our vanity! On the other hand, we cannot neglect the fact that atoms and molecules are only 10 per cent of the entire cosmic matter—the remaining 90 per cent are neutrinos or other more exotic elementary particles (Fig. 4-15).

The other components of the human brain, apart from water, are the following:

-Lipids, which are fat-like substances, and are about 12 per cent by weight. Included here is cholesterol,

-Proteins, macromolecules built of 20 different, simpler molecules of amino acids, which together make up around 7 per cent by weight,

-Other organic substances, totalling 2 per cent by weight,

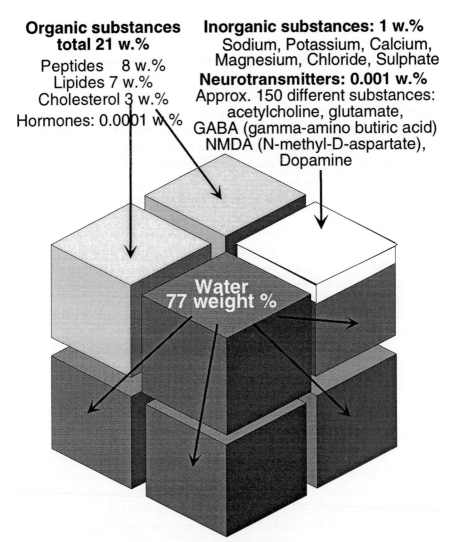

Figure 4-15 Chemical composition of human brain

–Inorganic salts, mostly of sodium, potassium, magnesium, calcium, plus chlorides, phosphates and some organic acids; The human body consists of 26 elements (including hydrogen, oxygen, carbon) and the bacterium Escherichia coli of 17 elements,

-Specific substances which play an extremely important role in the activity of the brain, and which form the so-called neurotransmitters. Neurotransmitters in the human brain are probably 0.001 per cent of brain weight. The number of different neurotransmitters which have so far been discovered is around 60-70, and some scientists believe that the number will increase to around 100-150. The brain's chemical processes were recognized by prominent scientists at the end of the nineteenth century:

Ch. Darwin (1858): "Why is thought, being a secretion of the brain, more wonderful than Gravity, a property of matter?"

S. Freud (1930?): "We must remember that all our provisional ideas in psychology will one day be explained on the basis of organic substrates. It seems then probable that there are particular chemical substances and processes that produce the effects of sexuality and permit the perpetuation of individual life."

4.5.2 *Neurotransmitters: 'brainy' molecules*

There was a period when everyone thought that synaptic transmission was electrical, whereas now we know that it is mainly chemical. Neurotransmitters are substances which are produced directly in neurons—more precisely— in the pre-synaptic nerve terminal. There are other substances which have a direct effect on the activity of synapses, but are produced outside neurons, for example, some hormones such as angiotensin (see below). The neurotransmitters are packed in so-called synaptic vesicles (Table 4-1).

It is no exaggeration to say that all mental activity, every act of information processing, each expression of our mind and of our free will (if such a phenomenon really exists) is directly associated with the production, transportation, chemical activity, and metabolic destruction of numerous neurotransmitters. At the level of synapses they are the decisive actors.

All known neurotransmitters in the human nervous system produce one or other of the following actions: they either excite the post-synaptic cell membranes, causing them to fire, that is, to emit nervous signals, or inhibit them from firing, or modifying their excitability. The general characteristics of neurotransmitters are important to know, not only because they perform such a significant role in the normal working brain, but also because their absence or excess production can play a major role in brain disease and behavioural disorders. The brain´s reaction to drugs also depends on the transportation of neurotransmitters through the synaptic cleft.

In order for the neural network to function smoothly, these transmitters, after being released into the synaptic cleft (the gap between neurons), must be removed. Some of the transmitter molecules are carried into astrocytes (see glia cells, Section 4.2.4). In the astrocytes the transmitter can be metabolized to form amino acids and other small molecules. Subsequently these amino acids can serve as raw material for making more neurotransmitters.

To summarise, we can say, that higher mental activity is directly associated with the extremely small number of around 100 small and rather simple molecules, built of hydrogen, oxygen, carbon, nitrogen, and, in some cases, sulphur, called neurotransmitter, but also including such simple molecules as nitric oxide (NO), carbon monoxide (CO) called secondary inter-neuronal messenger and also such omnipresent as the energy-carrier molecule: adenosine triphosphate (ATP) (F.A.Edwards at al, 1992).

At the simplest chemical level are the amino acids that act as transmitters. The main amino acids that have been established as neurotransmitters are glutamate and aspartate, which act as excitatory agents, and gamma-aminobutyric acid (commonly called GABA), which is inhibitory. The monoamine neurotransmitters are slightly more complicated chemically than amino acids. Neurons themselves synthesize these monoamines from amino acids, using enzymes present only in these neurons to make the modifications. Among the most important monoamines are: acetylcholine, dopamine, epinephrine, and norepinephrine. Each of these has a special function; for example dopamine plays a role in the control of complex movements.

Much more complex are other neurotransmitters, such as endorphin (a shortened version of their original name 'endogenous morphines') or enkephalin. They can relieve pain and slow the release of other neurotransmitters. Their activity corresponds to the activity of well-known drugs, such as morphine and heroin. Some scientists assume that the brain's own opiates have a role in organising social attachment. It must be clear that the role of neurotransmitters has been much better investigated at the lower stages of biotic information processing, such as discent, instinct and reflex, because of the limitation of performing experiments with humans.

Cocaine's reinforcing properties result from its ability to prolong the synaptic action of the neurotransmitter dopamine by blocking its re-uptake into dopamine neurons in brain reward regions such as the nucleus accumbens and prefrontal cortex.

Around one third of all synapses in the brain, especially in the cerebral cortex, are influenced by this neurotransmitter. The modern techniques of measuring living human brain, such as positron emission tomography (PET) allow to demonstrate details of pre-synaptic and post-synaptic mechanisms of neurotransmitter interactions (for example of the dopaminergic system, K.L. Leenders et al, 1984)

J-P. Changeux (1986): "The temptation is even stronger when one considers the great variety of neurotransmitters found in the brain ... of man. .. So why not a thirst substance and another for pain or for pleasure? In general terms, is there a chemical coding of behaviour? The case of thirst can serve as an example. We drink when we have lost water. This water loss causes a reduction in the blood volume and a change in its salt concentration. This variation in physico-chemical properties provokes a desire to drink, through the intermediary of the nervous system. Only a few neurons are involved. Recording their activity has helped to identify the substances that stimulate them a hypothetical 'thirst transmitter'. It is one of the numerous peptides that serve as hormones in some circumstances and as neurotransmitters in others. Called 'angiotensin II', when it is injected into the blood .. the animal will soon begin to drink. Strictly speaking, angiotensin is not a neurotransmitter because it is not released by nerve terminals."

Some types of neurotransmitters, but not all, and their related enzymes show a marked decline with age. Losses of neurotransmitter indices are frequently greater than can be accounted for by neuron death alone.

So often in the history of mankind, at least up to the present time, has knowledge and science been used for military purposes. Chemical weapons, especially the so-called nerve gases, are substances which, in one way or another, influence the activity of neurotransmitters and therefore result in very severe injuries to the nervous system, generally having irreversible or even fatal consequences.

Neurotransmitters, being rather small chemical molecules, have an appropriate counterpart: the transmitter receptors. These receptors can be grouped into two large and growing families based on their amino acid sequences and on presumptions about the shape that the molecules assume as part of the cell membranes in which they are embedded. One receptor family consists of ion channels, proteins that can form aqueous pores through which ions cross the membrane. They act rather rapid. As an example: dopamine (the first dopamine receptor gene was isolated four years ago). The other family of receptors does not form channels. Instead its members interact with membrane proteins, and initiate a cascade of biochemical reactions. Therefore they mediate slowly.

The components of the whole transmitter cycle are influenced significantly by the age of individuals, what may be directly measurable by PET (A. Antonini, K.L.Leenders et al, 1993).

4.5.3 *The biochemical basis of the bonus-malus system*

The efficacy of biotic information processes, in all four stages of reflex, instinct, discent and intelligence, depends directly on the mechanism of labelling all information in terms of quality. The mechanism of quality labelling : the bonus-malus system (Section 1.3.2).

A. Bain (1855, in his book 'The senses and the intellect'): "Every state of pleasure replies to an increase, every state of pain to a decrease of part of or all the vital functions." (According to JP. Changeux, 1985)

R.A. Wise (1980): "The cold information regarding the physical dimensions of a stimulus is translated into the warm experience of pleasure."

The involvement of the frontal cortex in brain reward has been demonstrated in monkeys as well as in rats. The great differences among species in the size of the frontal cortex creates great variation in the significance to a given species of the reward system. Brain reward can be found not only in the frontal cortex and the hypothalamus (that is, in the forebrain) but also in the midbrain and the hindbrain. It is present at all levels of the brain (A. Routtenberg, 1978).

Some neurotransmitters, such as norepinephrine and dopamine, have been identified as active agents in the reward system. Dopamine transmission is enhanced by drugs that increase its release (amphetamine, cocaine), facilitate receptor activation (apomorphine) or inhibit reintake of dopamine into the post-synaptic knob (amphetamine).

This raises the possibility that the highest stage of information processing, intelligence, is influenced by evolutionary primitive norepinephrine and dopamine reward systems. A connection between brain reward and learning, that is memory consolidation, has been recognized for about 30 years.

The positive part of the bonus-malus system, the 'bonus', is probably labelled by other substances and other networks than the negative part, the 'malus' (Section 5.3.2).

Dopamine plays a key role in reward signalling and addiction (A.H. Glassman, G.F. Koob, 1996). Dopamine neurons are rather activated by appetitive than aversive stimuli (J. Mirenowicz, W. Schultz, 1996).

4.5.4 *The nervous and endocrine systems*

Neurons secrete their chemical messengers, neurotransmitters, into the synaptic gap to regulate the activity of their synaptic target neurons. This secretory activity influences only the very near neighbourhood, and this only momentarily. Endocrine organs, called glands, secrete their chemical messengers, hormones, into the bloodstream.

This carries them to all the cells that have receptors for the particular hormone. The target cells are often very distant and the whole action extends over a long time period. Some identical substances work in both systems, and are hormones secreted by certain endocrine cells and transmitters secreted by certain

Table 4-1 Neurotransmitters, brainy molecules

Chemical properties	Specific activity	Synthesis etc.
Amino acid transmitters		
Glycine The simplest amino acid	Inhibitory in brain stem in small neurons in the spinal cord and	The smallest of all 'natural' 20 amino acids
GABA Gamma amino-butyric acid	Inhibitor for 1/3 of all synapses, acts as an inhibitor; most abundant transmitter	Influences memory
Aminergic transmitters		
ACh: Acetylcholine Monoamine	Identified as active agents in the reward system. Excited neurons secrete large concentrations of acetylcholine in discrete bursts at synapses	
Dopamine Monoamine		Is concentrated in the basal ganglia
Epinephrine Monoamine		
Norepinephrine Monoamine	Identified as active agent in the reward system.	Concentration in the hypothalamus
Serotonin	Decrease the propensity for aggression and violence	Drugs boost serotonin in synaptic cleft
NMDA N-methyl-D-aspartate	Closing and opening calcium ions	The importance for the emergence of memory
PEA Phenyl-ethyl-amine	Endogenous amphetamine	Acts on the bonus-malus system
Endorphin (Enkephalin)	Endogenous morphine Relieve the pain	
Small molecules		
NO Nitric oxide, short-lived radical	In low concentrations acts as a interneuronal second-messenger.	Toxic
CO Carbon oxide	Retrograde secondary messenger	Noxious gas, synthesized on demand
ATP Adenosine-triphosphate	Not merely a neurotransmitter but is one of the second-messengers	Omnipresent as energy currency
AA Arachidonic acid Poly-carboxylic acid	Retrograde secondary messenger	Basic mechanisms in long-term potentiation
Other unknown neurotransmitters		
Other 40 neuro-transmitters	With well-known properties	
Probably tens of 'new' neurotransmitters	With unknown properties	

neuronal cells. Examples are norepinephrine and vasopressin. The importance of the endocrine system, from our point of view, is not in its regulating role over the brain but in the feedback observed in the influence of the hormones on the brain's activity, not only by sometimes acting as neurotransmitters but also directly. Hormones can affect the nervous system during its growth, in the following ways:

–Altering the direction of growth of hormone-sensitive neurons,

–Affecting the growth rate of axons and dendrites. The hormone-sensitive neurons would end up making different connections in male and female brains,

–Triggering some neurons either to form synaptic connections or not,

–Preventing cell death of sensitive neurons in competition with others.

All these possible mechanisms can be responsible for the differentiation between male and female brains. Steroid sex hormones, androgens, oestrogens and progesterone, can all influence behaviour by direct action on the brain (J.B. Hutchison, 1991).

Stress makes the body more vulnerable to some physical illness; immune responses can contribute to depression and fatigue. Even though the brain and the immune system differ in their function and organization the brain is the centralized command centre. The immune system is decentralized and its organs are located throughout the body and linked at a biochemical level. This fact suggests that drugs traditionally used to treat neurological problems might help against inflammatory maladies and vice versa. A corollary of the communication between brain and immune system is that psychoactive drugs may in some cases be used to treat inflammatory diseases, and drugs that affect the immune system may be used in treating some psychiatric disorders (E.M.Sternberg, P.W.Gold, 1997).

4.5.5 *Blood-brain barrier*

One essential activity of the central nervous system is based on two contrary requirements:

–Separation from the rest of the body by means of a special barrier, to prevent the ingress of undesirable materials which could disturb the regular function of information processing,

–Importing necessary components from other organs, in particular biotic fuel glucose and the appropriate amount of oxygen, and then removing the combustion products, primarily carbon dioxide.

Both functions, (the separation and the exchange of matter) are realized by means of a very sophisticated blood-brain barrier. This is a complex of membranes and blood vessel walls that prevent blood from coming into contact with the tissues of the central nervous system and the cerebrospinal fluid.

The transport of glucose from blood towards neurons is rather complex.

Partially the anaerobic glycolysis is taking place in the astrocytes, the cells surrounding the neurons (P.J. Magistretti, L. Pellerin, 1996).

It is very important to see that a number of synthetic, man-made or natural, but man-prepared, substances have the ability to cross the brain-blood barrier and to influence numerous chemical and physical processes going on in the neurons and synapses. The results manifest themselves in the form of clearly observable changes in the functions of the central nervous system. Some of these substances are alcohol, and some drugs.

A specific receptor in the granular cells of cerebellum has been discovered, which is the target for alcohol. Alcohol induces lack of motor co-ordination and the inability to co-ordinate voluntary muscular movements.

The problem of how the blood-brain barrier is crossed by the organism's own substances, such as hormones, can only be mentioned here. The barrier protects the brain from noxious chemicals that circulate in the blood, and keeps electrically charged molecules from passing from the blood into the brain (Section 5.3.3). Most of the 70,000 chemicals in commercial use have not yet been tested for neurological effects.

4.6 THE BRAIN AS THE PRODUCT OF GENETIC INFORMATION

4.6.1 *Genetics and the brain*

The human genome consists of 3 giga of base pairs (do not forget the maternal messages stored in the cytoplasm of the egg). Only about 10 per cent of genome is active for coding protein synthesis. Humans differ by 1.6 per cent of the active genome from the chimpanzees. But our brain is about four times larger than the chimpanzee brain and with more expressed laterality. Clearly, the information needed to provide a unique specification of the 200 tera ($2 \cdot 10^{14}$) of synapses in human cerebral cortex cannot be encoded in the human gnome, including at maximum 60,000 genes responsible for central nervous system organization.

Even the most complex and most highly developed human brain is only the product of, at most, two components: the genetic and the epigenetic (Greek *epi* besides + *genesis* birth). The height of an individual is both genetically determined and dependent on nutrition (epigenetic). The total DNA content per cell nucleus has evolved from bacteria to man. The amount of DNA for one cell is in picograms (picogram: one thousand-billionth of a gram) (Table 4-2). The next relationship is illustrated in Table 4-3.

A map of the human genome should become available within the next few years. The task is to determine the complete 3 giga nucleotides of the human DNA sequence. At the end of 1996 less than 20,000 genes were mapped, which may be represent 1/4 to 1/5 of all protein-coding genes in our genome (G.D. Schuler, 1996). Only about 3% of the 3 giga individual units known as 'bases' that make up DNA actually code for proteins, which is the simplest definition of a gene. Until the international Human Genome Project is completed, scientists will not know for sure the number of genes in human chromosomes. Some are sure that we know almost exactly: 60,000 genes, that there is not much room for much more. In fact, it may be less than that. According to another guess 80,000 to 100,000 genes, and even 120,000 to 150,000 are possible. 'People like to have a lot of genes' (J. Cohen, 1997).

Table 4-2 Genetics and brain

Organism	Caenorhabditis elegans (worm)	Homo sapiens sapiens (man)
Body dimension	1 millimetre	1.7 metre
Body weight	< 1 milligram	70 kilogram
Number of somatic cells	959	1000 giga
Number of neurons	302	30 giga
Number of genes for whole individual (genome)	15,000	80,000
Number of genes for nervous system	2,000	60,000
Genome, number of base pairs	0.1 giga	3 giga

Probably 2/3 of all human genes (that is approx. 60,000 from the total of 80,000 genes) are involved in the control of human brain structure and activity, that is round 2 giga (thousand million) of base pairs. For the rest of the body remain 1 giga (thousand million) of base pairs. About 97 per cent of human genome does not carry the genetic information for protein synthesis, but probably contain another kind of organized information.

Some of our genes come in multiple versions, or alleles, so that an offspring may have two versions to choose between; furthermore, the unexpressed 'junk DNA' is full of DNA sequences that look like fragments of known alleles, so that there is the possibility of a new allele being expressed on the rare occasions when the two usual copies are inactivated. Although DNA is the carrier of genetic information, it has limited chemical stability. Hydrolysis (action of water), oxidation and methylation of DNA occur at a significant rate, and are counteracted by specific DNA repair processes.

The self-repair-mechanism of DNA is enormously effective. The spontaneous decay of DNA is likely to be a major factor in mutagenesis, carcinogenesis and ageing (T. Lindahl, 1993).

No simple relationship exists between the complexity of organization of the genome and that of the central nervous system. The aphorism 'one gene—one protein' can in no way become 'one gene—one neuron', and certainly not 'one gene—one synapse'. More than one-half of human structural genes code for proteins in the nervous system. Many of these are related to the ion channels, neurotransmitters, and receptors that determine synaptic function.

4.6.2 *Genetics and epigenetics*

The features of the anatomical and functional organization of the nervous system depends, generally speaking, upon two different mechanisms:

–Genetically controlled mechanisms (preserved from one generation to another) which govern the proliferation, categorization, migration to the ultimate location, and formation of widespread and excessive connections (axons, dendrites and synapses). It must not be underestimated that genetic mechanism are responsible only for a limited number of instructions. Even in genetics, chance plays an important role. Genetics gives the general rules, chance results in a spectrum of possibilities

Table 4-3 DNA size and number of neurons

Organism	Mass of DNA in one cell (picogram)	Genome size, mega of base pairs	Number of neurons in organism
Bacteria, prokaryote Escherichia coli,	0.01	4	0
Yeast, eukaryote, unicellular		12	0
Worm, Caenorhabditis elegans	0.01	90	302
Fruit-fly, Drosophila	0.24	180	0.1 mega
Chicken	2.5		2-3 mega
Mouse	few		5 mega
Man	6	3,000	100 giga

–Epigenetic mechanisms, acting individually, which depend upon experience within the environment, the body itself, and the current state. Epigenetic mechanisms cannot produce new neurons, or even new synapses, but they can very effectively kill (that is, cause the death of) excessive neurons and synapses, and therefore influence the ultimate state of neuronal networks, modules, etc.

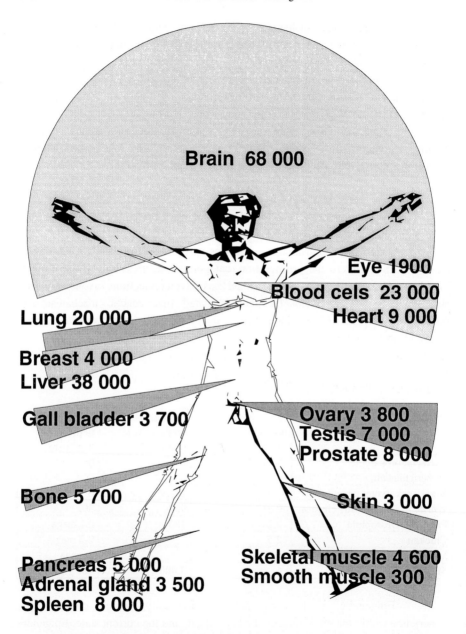

Figure 4-16 Number of genes for human brain

The chance of survival of a synapse depends upon the more-or-less effective 'use' of the neuronal circuit and the quality of its internal representation—the bonus-malus system. Certain pre-existing connections are selected by activity or 'experience', without inducing the formation of new structures or the synthesis of new molecular species. Epigenesis is the controlling mechanism of self-learning neuronal circuits.

It would be far too naive to assume that genetic information results in one unique brain structure, and to neglect the parameters of freedom contained within the genetic codes. Mankind seems to consist of an enormous number of different individuals, without any being repeated twice. The unlimited spectrum of human individuals reflects directly or indirectly the unlimited spectrum of the human brain.

4.6.3 *Genes and brain normality*

The genetic system is not as rigid as a first glance might suggest. There are many mechanisms which could result in significant changes in the products. It has long been considered possible that the central nervous system, which superficially shares much in terms of complexity and cell-cell interactions with the immune system, may use somatic recombination to increase its ability to process information, and an increasing number of shared molecules or molecular motifs are known to be shared by the two systems. Some of these mechanisms are listed below, even if it is not certain that they are directly linked with the development and modification of the brain's components:

– Principal mutability of genomes,

– So-called 'jumping genes' (transposon),

– Methylation of some components of deoxyribonucleic acid, in which small molecular groups are chemically linked to the large DNA. Methylation is one of the possible mechanisms of the 'parental genome imprinting' of genes. Imprints change as genes pass from one sex to the other. During ageing subtle elimination occurs of some methyl groups from certain parts of DNA molecules,

– Plasticity of genes, similar to that playing such an important role in the immunological system, among other things resulting in the role of the recombinational activating genes,

– Formation and death of neurons,

– Plasticity of neuronal micro-components, especially of microtubules,

– Formation and death of synapses,

The epigenetic influence corresponds to cultural influence, the most characteristic feature of the development of human intelligence.

W. Singer (1990): "During early phases of brain development gene expression and post-translational modifications of gene products are controlled in the cellular microenvironment. Late, however, brain development starts to differ radically from the development of other organs because electrical activity is added to the biochemical messengers as a further signalling system in the self-organizing dialogue between the genes and their respective environments. It has to be assumed that these electrical signals are capable of influencing gene expression and post-translational modifications."

4.6.4 Brain plasticity

There is a lot of evidence in support of the existence of brain plasticity, suggesting that in many instances, the human brain is able to compensate for early damage by shifting function from brain areas traditionally believed to subserve certain cognitive skills, to other regions.

Blind subjects showed activation of primary and secondary visual cortical areas during tactile tasks, whereas normal controls showed deactivation. Thus in blind subjects, cortical areas normally reserved for vision may be activated by other sensory modalities. There is a 'critical period', or brief window of opportunity, early in childhood, during which major alterations in primary visual cortex can take place after a given eye manipulation (N. Sadato et al, 1996).

4.6.5 Genes and brain abnormality

Is the brain of a genius a product of genetic mutation? The answer is doubtless negative. There is no gene for intelligence, any more than there is for genius. Nobody expects that any single gene will have a major effect on intelligence.

But there are some hundreds genetic mutations known at present time which cause pathological manifestations in the central nervous system (Section 7.7.7).

Recently (1996) some DNA sequences on chromosome six are under suspicion as responsible for schizophrenia, which may well be the most mysterious of all mental illnesses. Schizophrenic symptoms include: paranoia, delusions, social withdrawal, auditory and visual hallucinations, and disorganized thoughts—often do not surface until adulthood. About 1 per cent of the population world-wide is affected by this mental illness. If there are problems with which future mankind will have difficulties, then among them will doubtless be the question of whether Homo sapiens sapiens should manipulate its genome, especially that part which contains the genes regulating the formation of the brain.

4.7 BIBLIOGRAPHY

Adolphs R, et al (1994) Impaired recognition of emotion in facial expressions following bilateral damage to the human amygdala. *Nature* **372** 669

Alttwel D et al (1993) Arachidonic acid as a messenger in the central nervous system. *Semin Neurosci* **5** 3 159

Andreasen NC (1997) Linking mind and brain in the study of mental illness: a project for a scientific psycho pathology. *Science* **275** 1586

Antonini A, Leenders KL et al (1993) Effect of age on D(2)-dopamine receptors in normal human brain measured by PET and C-11-raclopride. *Archiv of Neurology* **50** 474

Barinaga M (1991) Is nitric oxide the 'retrograde messenger'? *Science* **254** 1296

Belliveau JW et al (1991) Functional mapping of the human visual cortex MRI. *Science* **254** 716

Bliss TVP, Collingridge GL (1993) A synaptic model of memory: long-term potentiation in hippocampus. *Nature* **361** 31

Braitenberg V, Schütz A (1989) Cortex: hohe Ordnung oder grösst-möglicher Durcheinander. *Spekt Wiss* **5** 74

Changeux J-P (1993) Chemical signalling in the brain. *Sci Am* **269** 5 30

Cohen J (1997) The genomics gamble. *Science* **275** 767.

Crease RP (1993) Biomedicine in the age of imaging. *Science* **261** 554

Drevets WC et al (1997) Subgenual prefrontal cortex abnormalities in mood disorders. *Nature* **386** 824

Edelman GM (1987) *Neural Darwinism. The theory of neuronal group selection*. Basic Books, N.York

Edwards FA et al (1992) ATP receptor-mediated synaptic currents in the central nervous system. *Nature* **359** 144

Eigen M (1987) *Stufen zum Leben*. Piper, München

Engel AK, et al (1993) Why does the cortex oscillate? *Curr Biology* **2** 332

Gilbert CD, Wiesel TN (1992) Receptive field dynamics in adult primary visual cortex. *Nature* **356** 150

Glassman AH, Koob GF (1996) Psychoactive smoke. *Nature* **379** 677

Garthwaite J (1993) Nitric oxide signalling in the nervous system. *Semin Neurosci* **5** 171

Hirokawa N, Okabe S (1992) Microtubules on the move? *Curr Biology* **2** 4 193

Hubel DH (1988) *Eye, brain, and vision*. Sci Am Libr, N.York

Ingvar DH (1990) On ideation and 'ideography'. in Eccles JC, Creutzfeldt O (eds) '*The principles of design and operation of the brain*'. Springer, Berlin

Kalil RE (1989) Synapse formation in the developing brain. *Sci Am* **261** 6 3

Kempermann G, et al (1997) More hippocampal neurons in adult mice living in an en riched environment. *Nature* **386** 493

Kinsbourne M (1995) Models of consciousness: serial or parallel in the brain? in Gazza niga MS (edit) '*The cognitive neurosciences*'. MIT Press, Cambridge, Mass

Koch Ch (1997) Computation and the single neuron. *Nature* **385** 207.

Leenders KL et al (1984) Pre-synaptic and post-synaptic dopaminergic system in human brain. *Lancet* **II**, 110

Leenders KL et al (1986) Inhibition of L-(18)flurodopa uptake into human brain by amino

acids demonstrated by positron emission tomography. *Ann Neurol* **20** 258
Leenders KL et al (1990) Cerebral blood flow, blood volume and oxygen utilisation: normal values and effect of age. *Brain* **113** 24
Lutz J, Steinmetz H (1994) Interhemispheric transfer time and corpus callosum size. *Neuro Report* **5** 2385
Magistretti PJ, Pellerin L (1996) Cellular mechanisms of brain energy metabolism. Relevance to functional imaging and to neurodegenerative disorders. *Ann N.Y.Acad Sci* **777** 380
McCrone J (1995) Maps of the mind. *New Sci* 7.01 30
Michel Ch (1988) Electroencephalographic correlates of human information processing and psycho-pharmacological influences, *Diss.ETH,* No **8502**, Zürich
Mirenowicz J, Schultz W (1996) Preferential activation of midbrain dopamine neurons by appetitive rather than aversive stimuli. *Nature* **379** 449
Nottebohm F (1989) Vom Vogelsang zur Bildung neuer Nervenzellen. *Spek Wiss* **4** 112
Petersen SE et al (1988) Positron emission tomographic studies of the cortical anatomy of single-word processing. *Nature* **331** 585
Posner MI, Raichle ME (1994) *Images of mind.* Sci Am Library, N.York
Raff MC et al (1993) Programmed cell death and the control of cell survival. *Science* **262** 695
Raichle ME (1994) Visualizing the mind. Sci *Am* **270** 4 36
Routtenberg A (1978) The reward system of the brain. *Sci Am* **239** 5 122
Sadato N et al (1996) Activation of the primary visual cortex by Braille reading in blind subjects. *Nature* **380** 526
Schiff SJ (1994) Controlling chaos in the brain. *Nature* **370** 615
Schuler GD et al (1996) A gene map of the human gnome. *Science* **274** 540
Schulteis G, Koob G (1994) Dark side of drug dependence. *Nature* **371** 108
Sejnowski TJ (1997) The year of the dendrite. *Science* **275** 178
Seitz RJ et al (1994) Successive roles of the cerebellum and premotor cortices in trajectional learning. *Neuro Report* **5** 2541
Selkoe DJ (1992) Aging brain, aging mind. *Sci Am* **267** 3 97
Shatz CJ (1992) The developing brain. *Sci Am* **267** 3 35
Singer W (1990) Self-organization of cognitive structures. in J C Eccles, O Creutzfeldt, (eds) *'The principles of design and operation of the brain'.* Springer, Berlin.
Snyder SH (1986) *Drugs and the brain.* Sci Am Books, N.York
Snyder SH, Bredt DS (1992) Biological roles of nitric oxide. *Sci Am* **267** 5 28
Sperry RW (1976) A unifying approach to mind and brain: ten years perspective, in Corner MA, Swaab DF, (eds) *'Perspectives in brain research'.* Elsevier, Amsterdam
Sternberg EM, Gold PW (1997) The mind-body interaction in disease. *Sci Am Spec.Iss. Mysteries of the mind* Vol **7** 8
Travis J (1994) Glia: the brains other cells. *Science* **266** 970
Van Essen D.C. (1997) A tension-based theory of morhogenesis and compact wiring in the central nervous system. *Nature* **385** 313
Wilczek F (1994) A call for a new physics. *Science* **226** 1737

CHAPTER 5
BRAIN PROCESSES INTELLIGENCE

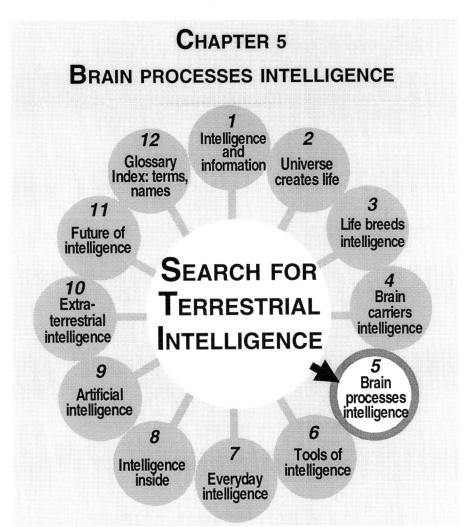

Human brain as carrier of intelligence. Time and space perception. Brain as a permanently changing complex system. Different types of memory. Chunks, the units of memory. Where is the memory stored? The bonus-malus system: emotions, pain, psychoactive agents. Sleep, one third of life. Cognitive operations: visual information processing as an example. Decomposition, synchronization of processed information. Importance of attention. Asymmetry of human brain; lateralization. Female and male peculiarities of the brain. Handedness and brain asymmetry.

5 BRAIN PROCESSES INTELLIGENCE 154
5.1 THE NETWORK OF NEURONS AND SYNAPSES 154
5.1.1 Brain with a complex structure and function: carrier of intelligence 154
5.1.2 Dimensions of the information processes 155
5.1.3 Time perception 155
5.1.4 Spatial pictures or linear structures 156
5.1.5 The brain: a permanently changing system 159
5.2 MEMORY; INFORMATION STORAGE 161
5.2.1 Memory, different types, etc. 161
5.2.2 Memory overabundance and selective forgetfulness 164
5.2.3 Chunks, the units of memory 165
5.2.4 Where is memory stored? 167
5.3 EMOTIONS, DRUGS AND THE BONUS-MALUS SYSTEM 169
5.3.1 Neural functions; own body and the world outside 169
5.3.2 The bonus-malus system and emotions 170
5.3.3 Psychoactive agents, pain and the bonus-malus system 173
5.3.4 Sleep, one third of life 177
5.4 COGNITIVE OPERATIONS AND THEIR PHYSICAL BASIS 178
5.4.1 Cognitive operations could be physically measured 178
5.4.2 Visual information processing; the best known system 178
5.4.3 Visual illusions as gate to cognitive activity 179
5.4.4 Complex information is initially decomposed 180
5.4.5 Synchronization of neural networks 182
5.4.6 The processing of just one word is measurable 185
5.4.7 Attention! Attention is very important 186
5.4.8 Planning and execution 188
5.5 THE HUMAN BRAIN AND ITS FUNCTIONS ARE ASYMMETRIC 189
5.5.1 Macro- and micro-asymmetry 189
5.5.2 Emergence of the asymmetric human brain 189
5.5.3 Lateralization—significant or not? 191
5.5.4 Female and male peculiarities of the brain 194
5.5.5 Handedness 198
5.6 BIBLIOGRAPHY 199

5 BRAIN PROCESSES INTELLIGENCE

5.1 THE NETWORK OF NEURONS AND SYNAPSES

5.1.1 *Brain with a complex structure and function: carrier of intelligence*

The complex macro and microstructure of the brain has been discussed in the previous chapter. The unique properties of the human information processing result from the multiconnected functional interaction of all these components. This is the content of this chapter.

5.1.2 Dimensions of the information processes

A primary cerebral map may be considered to be the translation that makes it possible to sample and preserve selected coherent portions of the overall topographical order of the external world in time and space, in the form of an inner representation.

The internal representation of space, which is three-dimensional, seems to be simple. We should not forget that brain organs can act as three-dimensional devices, such as the thalamus, or as a two-dimensional projection on its surface, as in the case of the primary sensory and motor cortex—the so-called 'homunculus'. In this statement is hidden the assumption that the dimensionality of both space and the brain's organs are the same, or at least are connected via projection from three to two dimensions (Fig. 5-1).

5.1.3 Time perception

At all stages of biotic information processing, the perception of time and space is the most important element. Consequently, the internal representation of space and time must be adequate. Survival of the internal representation during the next minute must be much more complex. This is because time is a one-dimensional phenomenon, which has only one direction, instead of the two directions which are possible in a one-dimensional physical structure. As the brain's organs are two- or three-dimensional, and they have to represent one-dimensional and uni-directional time, it seems that a valid scientific explanation of how this is achieved will take some time to obtain.

R.Penrose (1994): "One of the most striking and immediate features of conscious perception is the passage of time. It is something so familiar to us that is comes as a shock to learn that our wonderfully precise theories of the behaviour of the physical world have had, up to this point, virtually nothing to say about it. Worse than this, what our best physical theories do say is almost in flat contradiction with what our perception seems to tell us about time."

E. Pöppel (1976): "There is no such thing as time perception. If one deals with problems of subjective time, it is suggested that one begins with event perception."

G.M. Edelman wrote a book in 1989 entitled "The remembered present. A biological theory of consciousness", which clearly demonstrates the significance of the perception of time in the origin of consciousness, or more precisely of selfconsciousness. Note that, in that book, Edelman does not use the term 'intelligence'.

Edelman is of the opinion that almost all animals, with the possible exception of chimpanzees, have not achieved the level of primary consciousness (in this book called proto-consciousness), which is able to free animals from the tyranny of ongoing events. In higher-order consciousness, the freedom is greater. The emergence of concepts allows the use of memory to develop a coherent picture or an internal model of the present, past, and future.

G.M.Edelman (1989): "Primary consciousness (present in all humans and perhaps in some animals) may be considered to be composed of certain phenomenal experiences such as mental images, but in contrast to higher-order consciousness (in humans having language and reportable subjective life), it is supposed to be bound to a time around the measurable present, to lack a concept of self and a concept of past and future, and to be beyond direct individual report." (Remark: This quotation is a 'mosaic' of two sentences.).

When does the present end and the future begin? What does it mean that two events are simultaneous? And everyone knows how time can sometimes appear to pass more quickly in some circumstances than in others.

It is known that two sounds must be minimally 0.03 seconds (30 milliseconds) apart before their sequence can be perceived, that syllables are typically 0.3 seconds in duration, and that 3 seconds is 'the length of the human present moment'. Our eyes usually move three or four times a second (round 300 milliseconds). Clearly, it takes a certain amount of time to experience a conscious percept. Some experiments led to the conclusion that the processing period for perception is about 60 to 70 milliseconds (F.Crick, Ch.Koch, 1992).

It can be assumed that biotic time-perception is quantified into 'time-chunks'. It seems that we cannot differentiate between events occurring within less than 40 milliseconds of each other. However, a pianist must make around 30 decisions per second. It seems that his finger movements are controlled not by conscious decisions but by means of unconscious, automatic signals emanating from the cerebellum. We can say that such rapid decisions are made below the level of intellect, at the level of well-learned movements; that is, at the level of discent.

5.1.4 *Spatial pictures or linear structures*

The correspondence of external objects and processes within the real world, and the internal representations of these objects and processes in the brain must be sufficiently efficacious. The internal representations are performed by means of physical, chemical, and cellular mechanisms, and processed by different brain organs.

There are numerous hypotheses as to how the internal representations are realized, but we will try to simplify the situation as far as possible. A very old, but always current, question is: Do we have the possibility of investigating the highest mental activities, such as consciousness and free will, by studying what occurs in synapses, neurons and modules (or 'columns'), as the reductionists claim, or must we consider consciousness as an entity without looking into its components, details and material carriers, as the holists claim?

How do we recognize the face of a friend? Do we analyze, more or less consciously, each detail and compare with stored information about the face, or do we recognize and distinguish the face by means of one, complete, indivisible glance, without consciously knowing how the process has been performed (Section 5.4.2)? Both our individual experience and objective scientific experiments confirm that the latter, the holistic, strategy is used.

To look at or reach for what we see, spatial information from the visual system must be transformed into a motor plan. The posterior parietal cortex is well placed to perform this function, because it lies between visual areas, which encode spatial information, and motor cortical areas. The observation of L.H. Snyder et al (1997) indicates that the brain organizes its information in a framework that is defined by its repertoire of motor functions. The idea is that the brain naturally queues up several motor plans, only a subset of which will be executed. So, for the brain, spatial location is not a mathematical abstraction or a property of a map, but involves the issue of how the body navigates its hand or gaze.

Global mapping includes both sensory input and the results of motor activities. The dynamic structure of global mapping is maintained, refreshed, and altered by continual motor activity and rehearsal. Each region of the cortex is organized in such a way that it could support different maps.

The maps are two-dimensional cortical sheets, representing three-dimensional real-world objects. The motor generator sends signals to the muscles, which do their job in a three-dimensional real world.

One-dimensional speech (linear structure of information) must be processed in two-dimensional cortex sheets, and retransformed into three-dimensional muscle activity.

There are some doubts about the interpretation of the 'maps' prepared by means of any technique. V. Mountcastle is concerned that blood flow (measured by positron emission tomography, PET) does not always indicate an increase in neural activity—that it is too crude a measure for the subtle processes neurologists are trying to trace. This technique is just not fast enough, and changes in neural activity happen long before there is any change in blood flow. Mapping tells us where the parts are but not how they work.

At this moment we are interested in the other sense of the term 'maps'. Maps are a structural level in the partially hierarchically organized structure of the brain, particularly in the human brain. In Fig. 5-1 and Fig. 5-2 some principles of this structure are given in a very crude and schematic form. This classification and the hierarchical relationship are far from being accepted, even far from being well described, but these terms are generally used in many current theories.

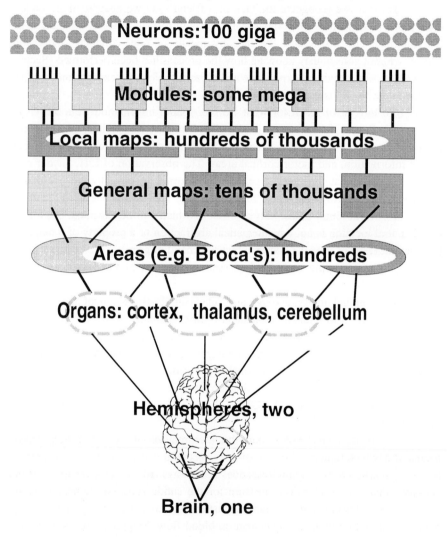

Figure 5-1 Maps: local and general

All these structures are directly controlled by the genetic system. However, the lowest levels, especially the neurons and modules (or 'columns'), are mutually connected in a way which depends on epigenetic factors, on the influence of the experiences of the individual after birth.

Cortical maps are subject to constant modification based on the use of sensory pathways; the architecture of each of our brains is modified in slightly different ways, which contributes to the biological basis for the expression of individuality.

Why do we pay so much attention to the details of brain structures? Does a link exist between the extremely complex structure of the brain and intelligence? The brain is not only complex, but is predominantly changeable, at the microscopic level (modules), at the cellular level (birth and death of neurons), at a sub-cellular level (synaptogenesis), on a macromolecular level (microtubules), on a molecular level (neurotransmitters) and, last but not least, on the atomic level (metallic ion concentration).

Neuronal variability is the basis for plasticity (according to G.M.Edelman, 1987):

–Variation in genetic traits and developmental primary processes; neuron division, neuron migration, mutual adhesion, differentiation of functions, death at different times. The rate of neuron death in an adult can be assumed to be some hundred per second. These latter processes are probably strongly influenced by epigenetic mechanisms,

–Variation in neuron structure: neuron shape and size, axonal and dendritic arborizations (both processes resulting in different spatial distribution), branching order, length and number of spines,

–Variation in connection pattern: number of inputs and outputs, connection order with other neurons, local versus long- range connections,

–Variation of micro-architectonics; number and density of neurons, position of somata.

–Variation in chemical activity: type and number of neurotransmitters, spatial and time differentiation,

–Variation in dynamic response: synaptic chemistry, electrical properties, excitatory and/or inhibitory synapses, short-term and long-term synaptic alteration,

–Variation in neuronal transport,

–Variation in interactions with glia.

5.1.5 *The brain: a permanently changing system*

Some mechanisms which are of significance:

–Neurons, belonging to one or a couple of modules, could change from belonging to one module to moving to another,

–Changes in the positions of some neurons during their migration to their final positions in the brain. Of course, this type of plasticity is active mainly in babies,

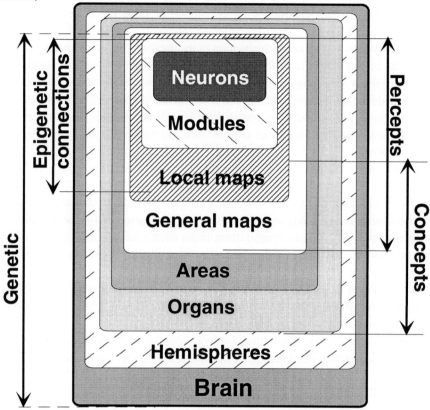

Figure 5-2 Hierarchy and functions of maps

–Variability of axon growth, resulting in the emergence of different neuronal networks, modules, and fibres,
–Differentiation of growing dendrites,
–Birth and death (or even short distance migration) of synapses, throughout the whole life of an individual. The rate of synapse death in an adult is about a hundred thousand per second,
–Formation and decay of microtubules inside neurons, with a characteristic time period of the order of minutes or hours (Section 4.2.6),
–Production and decay of neurotransmitters inside and outside of neurons,

with a characteristic time period of seconds or minutes.

All these changes probably have a not-insignificant impact upon the so-called 'spontaneous activity' of the brain. Observation indicates that neuronal networks are able to exhibit spontaneous activity in the form of the unexpected and untriggered emergence of nervous signals. A question arises, of whether these spontaneous firings could be a material basis for the emergence of so-called 'pre-concepts' (Section 6.6.1), of the phenomenon called intuition (Section 8.1.4) and of free will (Section 8.3.2). The reason for the spontaneous firing of neurons is probably, on the one hand, so-called 'chaotic phenomena', and, on the other hand, possibly due to spontaneous radioactive decay of radioisotopes in the human body or external ionizing radiation.

This is also a good moment to repeat an old question: Where are the components, systems, cells, and chemicals hidden which are uniquely present in the human brain? Has the human brain such components, at a macroscopic or microscopic level, which exist only there? And if not (and the answer is categorically not), where is the mystery of the human brain hidden? The most probable answer is that formulated by M.Minsky (1986): "What is the trick of intelligence? The trick is, that there is no trick." The answer seems to be hidden in the complexity of the human brain, in a unique co-operation and mutual feedback of all its parts and components, on the macro-scale and the micro-scale. Unlike the molecular building blocks and the functions of individual neurons, it cannot be assumed that the intricacies of cortical connectivity will conserved in different species. The intricacy of this network, after all, is what distinguishes Homo sapiens from all other forms of life (G.D. Fischbach, 1992).

In a previous chapter we discussed the macroscopic and microscopic structures of the brain. To the former belong the large brain organs, such as the cortical lobes, thalamus, hypothalamus, and hippocampus. As microscopic structures we mean the cellular components, such as neurons and their synapses. The question arises, of whether an intermediate structural level exists between both these levels. It is possible that such structures, so-called 'modules', could exist. Plasticity is the tendency of synapses and neural circuits to change as a result of own activity.

G. Fischbach (1992): "Plasticity weaves the tapestry on which the continuity of mental life depends."

5.2 MEMORY; INFORMATION STORAGE

5.2.1 *Memory, different types, etc.*

One of the most curious phenomena of the higher activity of the nervous system is that of memory—the storage of new information and the system of the retrieval of the memorized (acquired) information. Memory is a gift of nature, the ability of living organisms to retain and utilize acquired information or knowledge.

Memory is a trick that evolution has invented to allow its creatures to compress physical time. Owners of biological memory systems are capable of behaving more appropriately at a later time because of their experiences at an earlier time (E.Tulving, 1995).

The curiosity of the phenomenon of memory results from the following statements, which are almost trivial:

–Processes in living cells are of a chemical nature. All chemical processes involve an enormous number of molecules transforming in two contrary directions. Chemical equilibrium is a steady-state of two contrary flows of chemical entities. The mean life-time of so-called stable molecules in the brain is in the order of minutes. Even large macromolecules, such as proteins, achieve a longevity of only 10 days,

–Memory has a longevity from seconds up to a hundred years, at least in some extreme cases, when an individual lives this long,

–Memory, at least the long-term type, because it occurs in living cells, cannot be carried only by simple chemical means. Another mechanism must be used,

–Memory is a phenomenon whose significance increases from the lower stages of information processing up to the highest stage. Reflex does not require a memory; instinct also not. Discent is the next higher stage of information processes, and gets its name from the Latin word, *discere*, meaning learning, and learning is memorizing. The highest stage, intelligence, consists of three different types of memory, one of which—long-term memory—exists successfully throughout the whole life of the individual. Memorizing and learning are different aspects of the same process (Section 6-4). It is a well established belief that memory can be classified according to the time period over which it is active in these three following categories (Table 5-1, 5-2 and Fig. 5-3):

Working memory is an erasable mental black-board that allows to hold briefly in your mind, and manipulate that information—whether it be words, menu prices, or a map of the surrounding—essential for comprehension, reasoning, and planning. Working memory is causally related to success in higher thinking, in verbal and non-verbal problem-solving (I. Wickelgreen, 1997). Working memory operates over mere seconds, but is a central element in the organization of behaviour, language, and thinking. Deficits in this function is common in normal ageing and brain disorders such as Parkinson's disease and schizophrenia. The prefrontal cortex appears to assume the lion's share of the working-memory duties, both holding relevant information on-line and performing complex processing functions for brief periods of time.

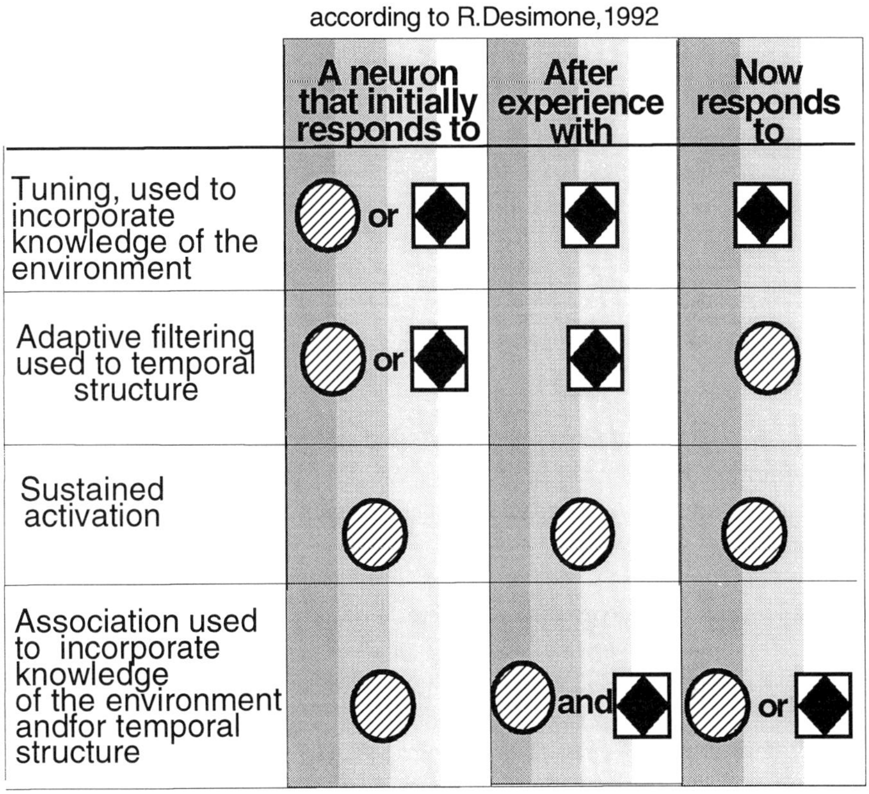

according to R.Desimone, 1992

	A neuron that initially responds to	After experience with	Now responds to
Tuning, used to incorporate knowledge of the environment	◎ or ◆	◆	◆
Adaptive filtering used to temporal structure	◎ or ◆	◆	◎
Sustained activation	◎	◎	◎
Association used to incorporate knowledge of the environment and/or temporal structure	◎	◎ and ◆	◎ or ◆

Figure 5-3 Different levels of memory

The working memory has also been studied by positron emission tomography (PET). These results begin to uncover the circuitry of working memory systems in humans (J. Jonides et al, 1993).

The long-term changes are stored at the same site as the short-term memory, but they require something entirely new: the activation of genes, the expression of new proteins and the growth of new connections (E.R.Kandel, R.D.Hawkins, 1992). How these categorizations of knowledge are realized in neuronal substrates is still a mystery (Y.Dudai, 1989). It is plausible to assume that the first 'how' systems preceded the first 'that' systems in evolution, because knowing 'how' is more basic for survival than knowing 'that'.

Reflexive systems, such as modifiable defensive reflexes in invertebrates and vertebrates, are primitive 'learning how' systems. It must once more be repeated that in this book, genetically stored information is not classified as memory. It is not clear at what phylogenetic stage the so-called declarative memory (Table 5-2) emerged. What is clear, though, is that knowing 'that' is a key to the mammalian level of information processing, to discent and especially to human intellect. 'Learning that' and 'learning how' systems probably share molecular and cellular mechanisms and operational rules. 'That' systems, with their extensive use of working memory, may have developed novel neuronal mechanisms to deal with strings of temporarily retrieved representations.

Interference in retrieval of information from long-term memory can produce 'memory blocking' effects, which can be commonly observed when information is on the 'tip-of-the-tongue'.

5.2.2 *Memory overabundance and selective forgetfulness*

How much information is really stored in our memory? Do we store only that information which is of significance for later use? How large is the redundancy of memorized information compared with that which is really used, at least once in a lifetime?

Forgetfulness is a special kind of process of elimination of memorized information. But it is very uncertain if forgetfulness is a kind of selective elimination. Even if some memorized information is not used for tens of years, it can be recalled when needed; it has not been eliminated, even by lack of use over such a long time.

It is often forgotten that this other side of the memorizing process—forgetfulness—is a very important component of effective information processing. Short-term forgetfulness is very similar to the process of selective attention. What is, for us, insignificant must be forgotten in the next few seconds. This is a well-known process.

Also of the highest importance is the removal of memorized information from our long-term memory. There are different causes which influence this ability to forget. Some of them are of a rather trivial nature, such as the biotic disintegration of the memory carrier. Another mechanism must be called a 'psychological' one; we tend to forget all that information which has an unpleasant, painful, and fearful, subjective character. All this information is probably labelled with a clear malus-value on our bonus-malus scale. Of highest importance is the property of human memory to change along the time step by step in such a way that loss of the 'appropriate' content is consciously undetectable. We use the old and partially wrongly memorized information being convinced that this is the true information.

Table 5-1 Short, working and long memory

Short-term memory acting over tens of second. Common for all animals of the stages of instinct and higher. Involved in the processing of recently acquired information	Working memory operating over minutes and longer, which is the basis for higher mental activities, such as calculation, logical thought, etc. It requires the simultaneous storage and processing of information. Following three subcomponents: 1) the central executive, which is assumed to be an attentional-controlling system 2) the viso-spatial sketch pad and 3) the phonological loop, which stores and retrieves speech-based information and is necessary for acquisition Working memory in humans fundamental to language comprehension, to learning to reason and problem solving *)	Long-term memory which acts over years, even throughout the whole lifetime. It is principally associated with the information processing stage called 'discent'. Some hypotheses claim that dendritic spines are a possible anatomical substrate for long-term memory

*)A. Baddeley, 1995; E.E. Smith, J. Jonides, 1995)

The objective uncertainty and the subjective certainty of our memory is the most principal property of the human memory system.

It is important here to note the well-known property of a computer—the impossibility and inability of forgetting stored information. Here is ignored the physically limited life of all electronic information carriers: tape, disk, CD-ROM, etc. The human characteristic of forgetfulness (full and/or partial) and 'adaptation' to the own bonus-malus system is doubtless a very important aspect of intelligence. To be intelligent is to be able to forget. The ability to make errors is another human characteristic which is not shared by the computer.

5.2.3 Chunks, the units of memory

How is the process of 'writing' in memory storage performed and how does the retrieval of information from memory take place? In what units does memory work? In some cases it is possible to explain memory mechanisms if a term called a 'chunk' is introduced. What is a 'chunk'?

If we want to remember a number, say 89561918, for most readers this task will be rather hard. For one of us (M.T.), however, the task is very easy. Why? Because M.T. recognizes 8956 as the postal code of his private address and 1918 as the year of his birth. For him the number 89561918 is one chunk, or, in the worst case, two chunks. M.T. can remember this number not only for the short time of minutes but, with a little effort, for years. For the reader this number is not a chunk and he must pay a high price for remembering it and then using it. Another chunk can be, for many readers, the word 'Washington George', but not 'Leenders Klaus', the name of one of the authors of this book; but both are perfect chunks for K.L..

Table 5-2 Procedural and declarative memory

Memory of 'that', 'declarative' (explicit) memory, e.g. dogs name, which is conscious and consists of two types: episodic memory, which is short-term remembrance of the recent past and of events in an individual's life: and semantic, also called reference memory, which is knowledge that is independent of events in the individual's life (e.g. long-term references to knowledge). Can be called up within a variety of contexts. Connected with hippocampus. Explicit memory is the deliberate recollection of an experience, whereas implicit memory is the influencing of a response by a previous experience without the person knowing that he or she is being influenced.				Memory of 'how', 'procedural' (implicit) memory, which is unconscious and concerns skills. There are many different types of learning which are currently dubbed as 'procedural', such as classical conditioning, and motor and perceptual skill acquisition by operand and incidental learning. Can be evolved only in highly restricted range of circumstances. Is connected with some structures of right cerebral hemisphere, and for 'automatic' action in cerebellum .			
Facts		Events		Skills, habits	Priming	Classic conditioning	Non associative learning

A good example of the movement of some chunks from working memory to long-term memory is the technique of unspoken 'repeating to myself' of the chunk. Using the notion of chunks it is easy to demonstrate that our short-term memory has a capacity limited to five, or at most seven, chunks. The contents of each chunk depends on individual abilities and experience with this chunk. The capacity of short-time memory is evidently very limited and is often the bottle-neck of our ability to learn (Section 6-4-1).

Sometimes memory capacity or memory rating is measured in bytes, the units of information theory. It seems that this is not an exact description of human memory mechanisms, for the reason that information theory has to do exclusively with linear, sequential information, contrary to human memory, which also employs picture-like and map-like notations. During the normal daily life, we must use, or partially learn anew, some tens of thousands of chunks per day. A student of civil or criminal law, at the end of his study must know 'only' 200,000 statements, mostly transformed in his memory into chunks.

5.2.4 Where is memory stored?

'Engram' is a term used for the long-lived material carrier of the memory. However, the definition of engram leaves a large question mark concerning the real nature of this phenomenon. It can be a molecule, a macromolecule, a circulating electro-chemical current, or a local concentration of ions. It can be situated on synapses (synaptic efficacy might be altered; transmitter release can be enhanced by a small increase in the amount of calcium that enters a nerve terminal with each action potential), short-term synaptic changes associated with molecular modification of proteins (but proteins are degraded on a time scale that ranges from minutes to days), in the neuronal soma, a point-like event (maintenance of memories that may last a lifetime requires more stable alterations, such as those associated with persistent changes in gene expressions, (so called 'immediate early genes', which encode proteins that regulate the expression of other genes) or changes in a whole neuronal network.

The problem begins with the question: where is the memory located in the human brain? Is there a special memory store, such as in a present-day computer? Or is the memory dispersed throughout those parts of the brain which are involved in the acquisition and processing of the memorized information?

The important point is that memory and learning are not dispersed throughout the whole brain, but depend much more on a specific 'organ' of the brain. The search for the real 'organology' of memory, on the basis of the experimental facts, is now in full swing. Memory appears to be at least partly encoded as modifications in the efficacy of synaptic connections. It is clear that a number of different molecular and cellular mechanisms can be expected to function in the acquisition of short-term, long-term and working memory. In simple systems, the locations that change during learning are also those that retain the memory. Long-term memory involves additional locations which are not modified during the learning process.

In male canary brains, especially in the higher vocal centre, the synapses of existing neurons could well supply all the neuronal circuit flexibility needed for the acquisition of new memory.

The hippocampus is probably responsible for memorizing spatial information, especially short-term. Damage to both left and right halves of the hippocampus results in serious learning problems. Such an injured person is unable to store memories of anything, and cannot even remember the name or face of someone encountered only minutes before. The hippocampus operates in some way as a 'working memory'. Injury to the hippocampus prevents the transfer of information from short-term memory to long-term memory. The hippocampus performs a computation needed for storage in memory in a manner that will allow conscious retrieval of an item once it has left current attention. The hippocampus appears to act in the formation and processing of memorized information rather than being the site of permanent storage.

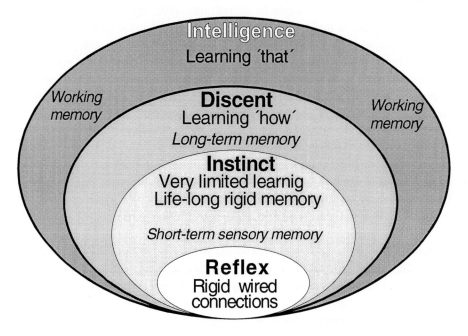

Figure 5-4 Mechanism of memory

The amygdala plays also a significant, role in making sense of experience, something that is very significant in the process of memorizing. The amygdala is the 'heart and soul' of the brain's emotional network. This is a clear example of the bonus-malus system, which is, in this book the principal means of identification giving each piece of information a quality label. The amygdala is not a site of memory, but one element in the process of memorizing.

The cerebellum is responsible for the unconscious control and memorizing of 'automatic' subtle movements; that is, of movements which have been learned, more or less consciously, in the past. Patients who have sustained injuries to the cerebellum report that they must consciously perform each step of a complex movement that they had performed 'automatically' before their injury.

There is no doubt that the cerebral cotex in the human brain is vital to learning and memory, but its complexity makes it difficult to study. Because human thinking and problem-solving usually employ language, animal experiments can offer only the roughest analogy.

H. Teracce (1987);"Though our knowledge of animal representations is embarrassingly meagre, we can be fairly confident that animal representations differ from those generated by human beings in two important respects. Most studies of human memory use verbal stimuli. Even when non-verbal stimuli are used, memory may be facilitated by verbal mnemonics and control processes. It also seems clear that cognitive processes in animals may be more limited by biological constraints than those of their human counterparts."

Memory is a process involving facilitated pathways, not a fixed replica or code. Memory is always altered or enriched by ongoing information processing. It is influenced during this processing by the state of the bonus-malus system.

As an illustration of the complexity of human memory, let us consider prosopagnosia. Prosopagnosia—which literally means 'not knowing faces'—often appears as the aftermath of a stroke or head injury which has caused localised brain damage. Sufferers, who may not have any other problems with their memory, find it hard, or even impossible, to identify a person's face, even after seeing it many times. More men than women seem to suffer from prosopagnosia. Also, sufferers are mostly left-handed and artistic. Some of them have difficulties in naming objects or colours, and in carrying out non-verbal spatial tasks. However, experiments have shown that people with prosopagnosia exhibit a change in the electrical resistance of the skin when presented with a familiar face, even though they are not consciously able to identify the face.

By using magnetic resonance imaging (MRI) data and positron emission tomography (PET) findings indicate that prosopagnosia results only in damage of the right cerebral hemisphere (for right handed prosopagnosics) and no involvement of the left hemisphere. Experiments with macaque monkeys show that surprisingly only a few tens of neurons are sufficient to recognize a face (M.P. Young, S. Yamane, 1992).

5.3 EMOTIONS, DRUGS AND THE BONUS-MALUS SYSTEM

5.3.1 *Neural functions; own body and the world outside*

The central nervous system has two different functions. One is dedicated mainly to the adaptive, homeostatic, and endocrine functions of the individual animal. These functions of the individual relate to its immediate needs for survival. These functions, called by G.M.Edelman 'self', we will call 'own body' (at the stage of intelligence, the concept of 'self', or better 'myself', is also included). The brain components involved are: the hypothalamus, various portions of the brain stem, the amygdala, the hippocampus, and the rest of the limbic system. Of course, the 'own body' system is very strongly influenced by the bonus-malus system. The other function of the central nervous system is dedicated mainly (but not exclusively) to processing signals arising from the outside world which we will call the 'world outside' (G.M. Edelman calls it the 'nonself'). This function is carried out by the cerebral cortex, the thalamus, and the cerebellum.

5.3.2 *The bonus-malus system and emotions*

In our model of intelligence, information processing is intimately and indivisibly associated with the bonus-malus system; that is, with the phenomena of emotions, with reward and punishment, and with pain and pleasure. The most important driving force of behaviour—survival—depends, of course, on doing things that are rewarding or do not lead to punishment. An animal must learn (i.e. memorize) to avoid behaviour which results in its being hurt or frightened. We learn not to repeat behaviour that is accompanied by fear and discomfort; in our terms, having a malus label.

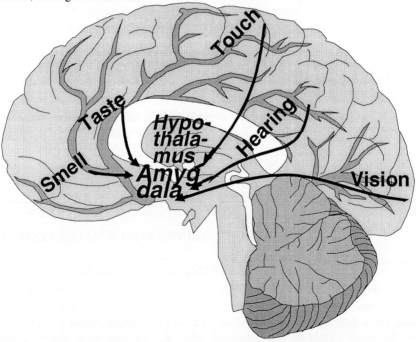

Figure 5-5 Complex information processing

J. Olds and P. Milner (1954) investigated the so-called 'reward centres' and 'punishment centres' in rats. They concluded that about 1/3 of the cells of the rat's brain lie in reward centres, about 1/20 in punishment centres and the rest in neutral areas. Since these historic experiments, numerous other investigations have been undertaken, with more highly developed animals. Opportunities to make observations in conscious humans, during subcortical electrical stimulation have also been presented by hospital patients. Reward centres have been found, but in the rather non-specific form of feelings of well-being, with pleasant sensations ascribed to distinct parts of the body, and with sexual arousal. Punishment areas have also been found, the stimulation of which can elicit terror, anger or pain.

The implication of all these studies is that much learning and decision-making, which presumably takes place chiefly in the cortex, is directed towards stimulating pleasure centres and away from stimulating punishment centres. That many of the functions associated with pleasure and punishment should have survival value for self and the species is, no doubt, of more than trivial importance for our own existence.

Some experiments provide evidence that nicotine can induce sufficient calcium influx into hippocampal neurons to trigger neurotransmitter release, and increases the strength of synaptic communication between neurons in the hippocampus—a centre for learning and memory (D.S. McGehee, 1996).

Enkephalins are endogenous opioid peptides which are thought to be vital in responses to stress, aggression and dominance and modulate responses to painful stimuli (M. König et al, 1996).

There are reports of monkeys (in South America) eating hallucinogenic plants (W.W.Gibbs , 1996).

It is well known, from our general experience, that the conscious perception derived from some sensory input is greatly modified by emotions, feelings, appetite, or sexual drive. Note that, in this book, drives are classified as components of the second stage of biotic information processing—instinct.

The here discussed bonus-malus system corresponds more or less with the 'emotional memory' postulated by J.E. LeDoux (1994).

J.E. LeDoux (1994): "Emotional information may be stored within declarative memory, but it is kept there as a cold declarative fact... These are declarative memories that are dependent on the hippocampus. The individual may also become tense, anxious and depressed, as the emotional memory is reactivated through the amygdalic system. Emotional and declarative memories are stored and retrieved in parallel, and their activities are joined seamlessly in our conscious experience. That does not mean that we have direct conscious access to emotional memory; it means instead that we have access to the consequences— such as the way we behave... Amygdala plays an essential part in modulating the storage and strength of memories.... Thus, emotion or feelings are conscious products of unconscious processes."

A.R. Damasio (1994) is of the opinion that having feelings and emotions is one of the main jobs the brain does. The decision-making process is directly connected with emotions. The amygdala appears necessary both to recognize the basic emotion of fear in facial expressions, and to recognize many of the blends of multiple emotions that the human face can signal. The amygdala may be an important component of the neural systems subserving social cognition in part because fine-grained recognition of the emotions signalled by faces is essential for successful behaviour in a complex social environment (R.Adolphs, 1994).

The bonus-malus system, called also 'reward-punishment' system is recently an object of experiments. When monkeys are trained to work for a particular goal or reward, 'reward expectancy' is processed by prefrontal neurons (M.Watanabe, 1996). Primates process reward information in specific regions of the brain, including amygdala ventral striatum, midbrain dopaminergic neurons and frontal cortex.

In humans measurements of regional cerebral blood flow, an index of neural activity, in healthy volunteers performing a delayed go-no-go task for money examine human reward mechanisms. Behaviour rewarded by money, as compared to a simple 'ok' reinforcement, was most significantly associated with activation of dorso-lateral and orbital frontal cortex and also involved the midbrain and thalamus. These results are consistent with animal research and suggest that assessment of consequences in goal-directed behaviour is mediated by the prefrontal cortex, probably in association with the basal-ganglia-thalamo-cortical system (W.Thut, W.Schultz, K.L.Leenders et al, 1996).

E.T. Rolls (1995): "Some of the factors that enable a very wide range of human emotions to be analyzed include the following: 1) the reinforcement contingency 2) the intensity of the reinforcer 3) any environmental stimulus might have a number of different reinforcement associations. For example, a stimulus might be associated with both the presentation of a reward and the presentation of a punishment, allowing states such as conflict and guilt to arise."

A.Damasio (see A. Bechara, 1997) believes that in humans the prefrontal cortex is part of a system that stores information about past rewards and punishments (in this book: bonus-malus) and triggers the nonconscious emotional responses that normal people may register as intuition or a 'hunch'. Damasio stresses that humans, after all, are set apart from animals by their ability to reason. Human beings are also the sum of all their previous emotional experiences of rewards and punishments—experiences from which we learn, it seems, whether we know it or not (G. Vogel, 1997). The bonus-malus system is partially carried by some transmitters. It is probable that beta-endorphin, the 'endogenic morphine' (enkephalin) (Section 4.5.2), is responsible for the feeling of well-being. A pleasant feeling gives more pleasure, and an unpleasant one is felt to be less bad, when the level of beta-endorphin is high enough.

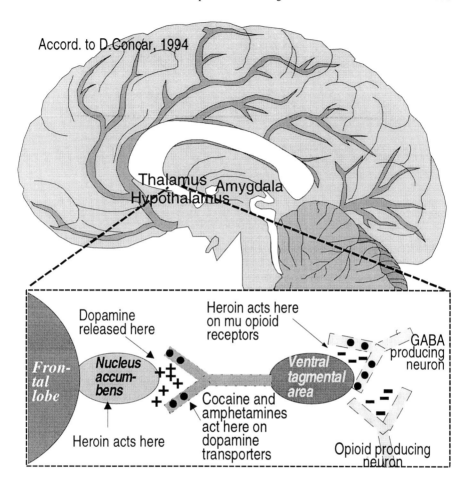

Figure 5-6 Drugs ways in human brain

The brain is able to produce more, or less of these neurotransmitters, depending on its own needs. A functional endorphin-deficit results in the emergence of dysphoria (the opposite of euphoria). One way to acquire the feeling of dysphoria is by the taking drugs.

5.3.3 *Psychoactive agents, pain and the bonus-malus system*

The term 'psychoactive agent' in this book includes following substances: drugs (synthetic, semi-synthesic, and vegetable natural substances) that affect the mind, and is used instead of: 'psychedelics', 'narcotics', 'drugs'. Psychoactive agents include tobacco, coffee, (mental endurance), alcohol, and also myrrh, incense (in most religious ceremonies). Neurotransmitters and other body own substances do not belong to psychoactive agents (endorphin = enkephalin) (Table 5-3, Table 7-7).

Encyclopaedia Britannica: "Not only can nearly anything be called a 'drug', but things so called turn out to have enormous variety of psychological and social functions—not only religious and therapeutic and 'addictive', but political and aesthetic and ideological and aphrodisiac and so on. Indeed, this has been the case since the beginning of human society."

Table 5-3 Drugs, psychoactive agents
(see also Table 7-7)

Sedative	Barbiturates and diazepam (e.g. Valium) are used medically to help people sleep and anti-anxiety, to calm without sleep.
Analgesics Opiates	Reduction of pain, as well as in the generation of euphoria (but of a passive, dreamy type) and other emotional responses. The mass of opiate receptors should constitute no more than one millionth of the brain mass, that is, around one milligram
Antischizophrenics	Used by some mental abnormalities: schizophrenia
Antidepressants	Prozac, enhance the action of serotonin, an amine with a wide variety of functions, by blocking the re-absorption by cells. Lithium (in form of salts) eliminate or diminish the episodes of mania and depression experienced by manic-depressive patients.
Stimulants-Euphoriants	Amphetamine, cocaine, decreasing fatigue. Cocaine facilitates the action of dopamine. Nicotine activates acetylcholine receptors, which are distributed in the cerebral cortex.
Tranquilizers	Having a calming effect.
Alcohol	Ethanol, passing metabolic pathways in humans leads to condensation of two molecules of the neurotransmitter dopamine, resulting in the synthesis of one molecule of tetra-hydro-papaverolin, the first step in the synthesis of opioids. Amount of 1 pro thousand in the blood, corresponds to 100 grams of pure alcohol.
Psychedelics Hallucinogens	LSD, mescaline, psilocybin and 'Ecstasy'. For LSD, the 'dose' is less than a tenth of a milligram, while for mescaline it is around a hundred milligrams.

If we are looking for something of basic importance in the processing of information in the human brain, then we must consider the impact of chemical substances belonging to the category of psychoactive agents. Why are human organisms sensitive to such substances? Why do some of these substances result in addiction? And why do human organisms have drug receptors (opiate receptors) at all? The answer is that nature has designed receptors for some normally occurring transmitters and, presumably, opiate drugs just happened to fit them nicely.

S.G Korenman et al (1993) claim that substance abuse has a neural basis in an endogenous reward centre in the brain that mediates all types of reinforcement, including natural reinforces such as food as well as artificial reinforcement, such as drugs. Some controversy exists as to the precise neuro-anatomic site of this reward centre, called in this book 'bonus-malus system'. Korenman et all give some evidence that this centre is in the ventral tegmental area or in the nucleus accumbens.

Psychedelic drugs act on the locus coeruleus (Section 4.1.6) in humans. One can readily appreciate that stimulation of the locus coeruleus will cause the drug user to feel that sensations are crossing the boundaries between different modes, the so-called 'synesthesia'. The locus coeruleus directs all types of sensory messages—from sight, hearing, touch, smell, and taste—into a generalized excitation system within the brain, and provides the emotional colouring, the feeling response, to sensory inputs. There are probably persons who, for genetic reasons have a lower than normal ability to produce or release these 'bonus-carrying' substances. These persons, deprived of the natural opiates, such as endorphin, may become highly sensitive to pain and lapse into depression, and even are more predisposed to suicide. Coming into contact with drugs they exhibit sensation-seeking behaviour. It is then relatively easy to reach a state of addiction, because there exists a phenomenon which could be called 'addiction memory'.

Addiction results not only from genetic, but also from epigenetic, causes. Some facts concerning the behaviour of narcotic-taking mothers during pregnancy leads to the following conclusions. The intake of some drugs, such as tranquililzers and pain-killers, during pregnancy results in the transport of some of these chemicals to the foetus, whose brain memorizes their 'bonus-making' effect. In adulthood the addiction memory drives the person to take one 'bonus-making' drug or another. If this hypothesis is confirmed, the epigenetic causes of drugs addiction will allow us to derive new contra-measures to tackle the problem.

The quantity of a drug which results in observable effects is of the order of a few milligrams. The mass of opiate receptors should constitute no more than one millionth of the brain mass, that is, around one milligram.

Psychoactive agents can influence the process of synaptic transmission by:

–Causing neurotransmitter molecules to leak out of synaptic vesicles (the small packages full of transmitters). Once out of the vesicles, the transmitters are degraded by enzymes,

–Blocking either the release of neurotransmitters into the synapses or their subsequent reception.

–Inhibiting enzymes that synthesize or degrade neurotransmitters,
–Binding to the receptor of neurotransmitters, because they have very similar molecular architecture,
–Interfering with, or facilitating, second-messenger activity.

Do drugs act at all four stages of biotic information processing? The answer is that they do, because they influence the action of neurotransmitters, which are present at all four stages. Of course, the results of narcotic impact are very different, depending on the stage of the processing (intelligence, discent or instinct).

Some psychoactive agents, such as psychedelic drugs are, from the point of view of pharmacology, of negligible value; but from our point of view they are highly significant, because they affect the very core of the user's consciousness, the highest stage of the biotic information processes.

The first of these effects is a change in sensory perception, especially the visual. According to the discoverer of LSD, A.Hofmann, this drug helps 'make conscious the unconscious'. Chronic users of LSD become irreversibly impaired in their thinking processes.

Many proselytizers of psychedelic drugs have argued that these agents produce a heightened level of awareness, and reveal a world of the mind that is 'more real' than the one we normally experience. Now that pharmacologists are learning how these agents act biochemically, however, they are developing techniques that will almost literally sculpt new drugs, psychoactive agents, to fit their intended sites of action with far greater strength and selectivity than the parent drugs, whose discoveries were accidental. A more potent drug is, by definition, more selective; that is, it binds so specifically and preferentially to its target receptor that it produces its specific effects even when administered in very small doses. It can be assumed that, in the future, perhaps within the next few decades, for each kind of neurotransmitter receptor a corresponding synthetic drug will be synthesized. Particularly tantalizing is the possibility that drugs acting discretely upon some of the newly discovered transmitters will exert effects unlike any produced by the drugs we know today, modulating mental functions in a far more subtle and precise manner than is now conceivable, and thus altering individual psychological symptoms and higher mental activities in a highly selective manner.

In the 'own body and myself' system, the highest positive value—the maximum bonus—is connected with satisfying the simplest needs: thirst, hunger, sexual drive. Only in pathological cases do other emotions acquire higher priority. At the stage of intelligence, in the 'world-outside' system, the striving after success, in social, artistic, or even scientific fields can have a very high value—the maximum bonus in some individual cases.

The representation of pain in human cerebral cortex is less well understood than that of any other sensory system. However, with the use of positron emission tomography (PET) it has been demonstrated that painful heat causes significant activation of secondary somatosensory, and primary somatosensory areas. This could be considered as a first step to objectively measurable pain, which is considered to be an example of subjective perception. (Section 8.4.3: Alcohol and drugs-impact on the bonus-malus system, and Section 8.9.2: Psychoactive agents and culture).

S.H.Barondes (1994) discussed the problem of Prozac (descendant of imipramine) which is now widely prescribed not only for depression but also to help people cope with a range of less serious but highly prevalent behavioural symptoms. This drug mitigates depression, alleviates anxiety and alters temperament. But are these different psychological phenomena due to the drug or is the drug just an expensive placebo?

The power of the molecule-to-mind approach can be illustrated by recent advances in the pharmacological treatment of schizophrenia, the most common and the most devastating of all thought disorders (Section 8.4.4).

5.3.4 Sleep, one third of life

One third of the total living period of human brain is spent asleep. The global energy consumption of the brain does not differ much compared to being awake. Each theory of intelligence must include a discussion of the role of sleep in the activity of brain. The brain spontaneously generates complex pattern of neural activity. This dance of perceptual and motor activity within populations of neurons changes promptly when the brain falls asleep. The role of sleep is therefore of highest significance.

Sleep is characterized by synchronized events of synaptically coupled neurons (giga numbers) in thalalomo-cortical systems. The rapid patterns characteristic of the aroused state are replaced by low-frequency, synchronized rhythms of neuronal activity. The activation of a series of neurotransmitters during awakening blocks low-frequency oscillations, induces fast rhythms, and allows the brain to recover full responsiveness (M.Steriade, 1993). This suggests that sleep oscillations are highly orchestrated and highly regulated. Not only does the brain during sleep exhibit coherent activity at a variety of frequencies, ranging from 1 hertz to over 40 hertz, but the extent of its spatial coherence is also quite variable.

Experiments have shown that a process of human memory consolidation, active during sleep, is strongly dependent on REM (Rapid Eye Movement) sleep

(J.Karni, 1994). An overnight improvement of perceptual skill (memory 'how to do' for example: play the piano, ride a bicycle) has been measured. The findings provide evidence that something important (for example: strength of neural connections) is happening during sleep with regard to learning and memory.

5.4 COGNITIVE OPERATIONS AND THEIR PHYSICAL BASIS

5.4.1 *Cognitive operations could be physically measured*

The problem of material carriers and the localization of cognition in the human brain is an old and difficult one. However, current analyses of the operations involved in cognition, and new techniques for the imaging of physical and chemical processes in the brain during execution of cognitive tasks, are opening up new possibilities for research in this field.

Using the PET-technique it is possible to measure that the brain of a novice computer-game player is much more active, that means needs more energy, compared to someone who has experience. The energy consumption of a mentally retarded patient is significantly larger than that of a normal volunteer. The brain of a clinically depressed person shows less activity (less energy consumption) than that of a healthy person.

5.4.2 *Visual information processing; the best known system*

Honeybees remember the shapes of flowers and are guided by visual landmarks on their foraging trips. How insects recognize visual patterns is poorly understood. Flies store visual images at, or together with, fixed retinal positions and retrieve them from there only. Position invariance, an acknowledged property of human pattern recognition, may not exist as a primary mechanism in insects (M.Dill, 1993). The huge size difference between the insect and primate visual systems suggests that insects may lack some of the features that make the human system so flexible and powerful.

It is a large step from understanding elementary perception processes, such as vision, to understanding information processes terminating in the emergence of internal representation, in concepts, ultimately to memory mechanisms. We are far from understanding the higher mental activities—the cognitive processes.

F.Crick (1994): "I suggest that the best way to approach the problem of consciousness is the study visual awareness, both in man and his near relations."

The picture of an object in the human eye is projected onto its 126 million receptor cells. The receptors, composed of rhodopsin molecules which are sensitive to light, and the next two plate-like layers of neurons, are housed in the retina. The third layer, consisting of so-called ganglion cells, integrates the extensive information received and decreases the number of signals to about a million. Electrical nerve signals travel along the optic nerve, with its 1 million axons, to arrive at 'lateral geniculate bodies'. It must be clear that the number of axons in the optic nerve is more than a hundred times smaller than the number of receptors in the retina. The lateral geniculate bodies contain two types of cells, small cells (parvo-cells) and large cells (magno-cells), each with different properties. The parvo system carries information about colour, and the magno system about luminance. Signals are also analyzed to give information about motion and depth. The magno system combines the visual characteristic of an object in a way that enables it to perceive the image as a whole.

The signal from the eye is transported to the primary visual area of the cortex, which contains two different primary visual areas, numbered 1 and 2. The first area is a folded sheet of neurons, the size of a credit card but three times as thick, situated at the very back of the brain. The astonishing fact is that the signals are analysed by around 1 giga of neurons in the primary visual area; that is, a thousand times more than the number in the optic nerves. The processing of optical information at different stages of analysis requires a different number of cells, differing by a factor of a thousand.

The visual processing needed to perform some highly demanding task can be achieved in under 150 milliseconds (S.Thorpe et al, 1996).

The receptive field is not uniform, but consists of a small circular region surrounded by an annulus, which explains a number of perceptual phenomena (D.Hubel, 1996).

The product of the primary visual area, in the form of numerous signals, passes through a very complex path to the secondary and tertiary areas of the sensory cortex and then to the motor cortex. It must be stressed that different areas of the secondary and tertiary cortex are concerned with movement and stereopsis (the ability to judge depth based on the differences between the images produced in the two eyes). The number of neurons involved is of the order of magnitude of 10 giga.

The technique of positron emission tomography (PET) has been used to measure changes in the regional cerebral blood flow of normal subjects, while they were discriminating different attributes (shape, colour, velocity) of the same visual stimuli.

5.4.3 *Visual illusions as gate to cognitive activity*

The human eye is rather poor at detecting the effect of chance, and has the capacity to construct patterns even where none exists.

In the excellent book of D.H. Hubel (1988) 'Eye, brain and vision' there are no keywords 'visual illusion' or 'visual hallucinations'. Actually what our brain 'sees' is often far from the electromagnetic signals reaching our eyes.

It is important to know that the sensors, especially the eyes, are actively influenced by the brain itself. F.Crick (1994) stressed the role of the visual system in the conscious presentation of the real world. He makes the following general remarks: 1)You are easily deceived by your visual system, 2)The visual information provided by our eyes can be ambiguous (any one aspect of the visual information provided by our eyes is usually ambiguous), 3)Seeing is a constructive process: What you see is not what is really there; it is what your brain believes is there. And all these in spite of well-known phrase 'seeing is believing'.

The registration of signals arriving from the real world is corrected and modulated by nervous signals sent from the brain back to the sensors. During each eyeblink we lose sight of visual world for more than 100 milliseconds without usually perceiving the discontinuity. R. Harl (1994) hypothetizes that the observed blink related responses in the human posterior parietal cortex are related to spatial working memory, necessary for maintaining a continuous image of the environment despite the 100 milliseconds loss of input during each blink. It has been measured in the primary visual cortex of macaque on the basis of responses to patterns called 'illusory contours' (D.H. Grosof, 1993).

Blindsight patients are people who 'see' but do not 'understand'. Objects can be detected and visually followed, but not identified and not accompanied by awareness (A.Cowey, P.Stoerig, 1995). Because they are unaware of what they have seen, they have not acquired any knowledge. Their 'vision' is useless (S.Zeki, 1992).

Illusion is a unconscious self-deception aimed to improve the bonus level.

5.4.4 *Complex information is initially decomposed*

How does information processing take place at the stage of intelligence? We know very little about this, but what we know can be described in the following way.

At the beginning, objects in the real world are analysed by our senses, then by some 'analytical meshes', existing on the basis of previous experience. Finally, internal representations of the real objects are generated. The partial internal representations are stored in the neuronal networks, possibly mostly in the tertiary areas of the cerebral cortex. If the information must be retrievable from these dispersed partial internal representations, then it must be synthesized holistically in a form corresponding to the real object. Let us consider an example (Fig. 5-7).

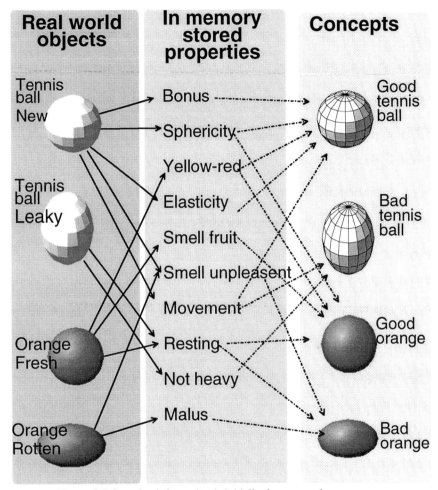

Figure 5-7 Complex information is initially decomposed

Before us are two objects. Both are rather similar, spherical, of the same size, and lie on a table. Both are yellow-red in colour and are elastic to touch. One has a fruity, pleasant smell, while the second has a rather unpleasant odour. The first is an orange, the second a tennis ball. The stored partial internal representations are: sphericity, yellow-red colour, elasticity, fruity smell, unpleasant smell, texture, etc.

Of course, these attributes are directly or indirectly connected with other partial internal representations, such as movement, play, tennis, good taste, and fruit, all of which are labelled with more or less bonus qualities, though they could also have malus qualities if we have recently lost a game.

If, at any time, we must retrieve the concept of a tennis ball, the appropriate partial representations are activated and synchronized into one complex representation of the concept of a tennis ball. By making time available as an additional dimension in which patterns of neural activity could be organized, such a temporal coding mechanism could help solve a number of fundamental problems in understanding how the brain works. A temporal code would also allow multiple representations to exist in the same region of the cortex (A.K.Engel et al, 1992).

5.4.5 *Synchronization of neural networks*

Pattern storage in the memory is ruled by the rather simple method of linking pieces of a pattern (how large and how complex?) into one concept, if the pieces are perceived more or less simultaneously. A whole pattern is formed, stored and retrieved when a group of pieces is associated in time. Because some pieces belong to different parts of different neuronal networks the 'synthesis' of a pattern from pieces evokes other patterns. But how are such links established?

How are distributed representations of the visual features that have to do with a single object in the world put together so that they can create a perception, a concept, and then even influence other mental activities or generate a physical reaction?

How these widely distributed processing areas get bound together to form a given perception, the so-called 'binding problem' is one of the central 'mysteries' of neurosciences. The primary visual cortex is organized into columns of cells that respond separately to different aspects of perceived objects. Even cells in a cat´s brain separated by 7 millimetres, an 'astronomical distance' in the brain, oscillated in near-perfect lockstep at 30 to 60 hertz, that is firing rapid bursts 30 to 60 times a second. Neural impulses travel at up to 1 millimetre per millisecond (that equals to a velocity of 1 m/s; a sprinter is running with a velocity of 10 m/s).

Neurons in the visual cortex, activated by the same object in the world, tend to discharge rhythmically and in unison (CM.Gray, W.Singer, 1989; W.Singer, 1995). The synchronized oscillations occur over less than half a second, and have a frequency of 40 hertz. M.A.Whittington et al (1995) have been able to demonstrate that these 40 hertz oscillations can be evoked in thin slices of brain that were kept alive in a nutrient bath. Only a small subset of the neurotransmitter receptors on hippocampal neurons were needed to generate the 40-Hz oscillatory activity of the network. Similar oscillations have been seen in the olfactory cortex, which is involved in discrimination between different odours. Where does the rhythm come from? Who is the orchestral conductor? How exact is the synchronization of the neuron activity? This phenomenon may be the transient glue that binds together different neural centres ('binding problem').

Synchronisation amplifies signals and creates the concept of the "tennis ball"

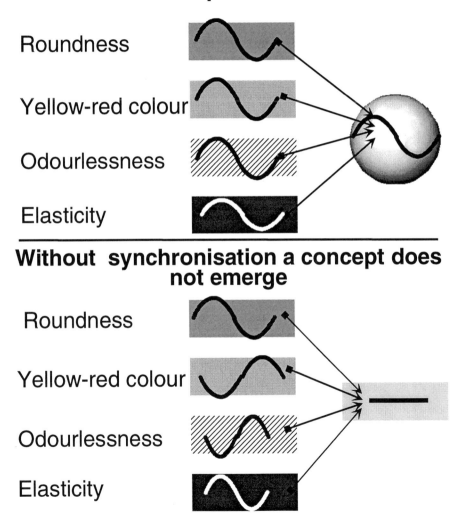

Without synchronisation a concept does not emerge

Figure 5-8 Synchronization of neuronal signals

F.Crick, Ch.Koch (1997): "Rhythmic and synchronized firing (of neurons) may be the neural correlate of awareness and might bind together activity concerning the same object in different cortical areas. ...Then there are representations of the parts of face, as separate from those for the face as a whole. Further, the implications of seeing a face, such as that person's sex, the facial expression, the familiarity or unfamiliarity of the face, and in particular whose face it is, may each be correlated with neurons firing in other places."

P.R. Foelfsema et al (1997): "Information processing in the cerebral cortex invariably involves the activation of millions of neurons that are widely distributed over its various areas. These distributed activity patterns need to be integrated into coherent representational states. A candidate mechanism for the integration and co-ordination of neuronal activity between different brain regions is synchronization on a temporal scale. In the visual cortex, synchronization occurs selectively between the responses of neurons that represent related features and that need to be integrated for the generation of coherent percepts; neurons in other areas of the cerebral cortex also synchronize their discharges."

Experiments show that when two neurons are induced to fire together within a short time window, the functional connection between them is potentiated, and when simultaneous firing is prevented, the connection is depressed (E. Ahissar et al, 1992).

It has been shown that correlated firing between single neurons, recorded simultaneously in the frontal cortex of monkey performing a behavioural task, evolves within a fraction of a second, and in a systematic relation to behavioural events (E.Vaadia et al, 1995).

Exploring the rhythms of the brain, revered by the pioneers of electroencephalography but now mostly dismissed as irrelevant to neural information processing, may even come back into fashion.

When a single sensory stimulus drives many neurons to fire at elevated rates, the spikes of these neurons become tightly synchronized, which could be involved in 'binding' together individual firing-rate feature representations into an unified object percept (R.C. deCharms, M.M. Merzenich, 1996).

Complex information is initially decomposed and if needed composed back into one complex information (Fig. 5-8).

V.G.Hardcastle (1994): "Given what we know about the segregated nature of brain and the relative absence of multi-modal association areas in cortex, how percepts become unified is not clear. However, if we could work out how and where the brain joins together segregated outputs, we would have a start in localizing the neural processes that correlate with conscious perceptual experiences.... In particular... the possibility that 40 Hz oscillatory firing patterns in cortex are important lower-level neuronal events related to perceptual experiences. '...Binding (is) understood as a process of hooking together disparate psychological units. However, this reduction may not reflect the firing rates of individual cells, but instead a 'higher level' order superimposed on individual cell's activity."

5.4.6 The processing of just one word is measurable

The idea that a word consists of separable physical, phonological, and semantic codes and that operations may be performed on them separately, has been basic to many theories. These theories suggest that mental operations take place on the basis of codes related to separate neural systems. However, it is not easy to determine if any operation is fundamental or whether it is based on only a single code.

Significant progress has been realized in the field of word processing. The abstract of the following publication is the best way to present their results.

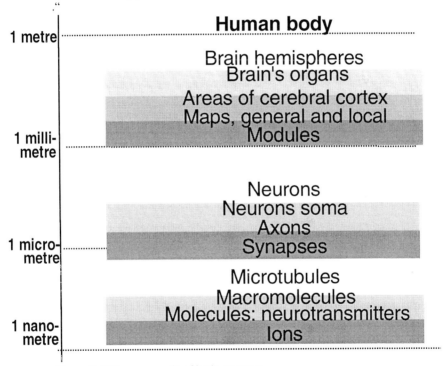

Figure 5-9 The space scale of brain structure

S.E.Petersen et al (1990): "Visual presentation of words activates extrastriate regions of the occipital lobes of the brain. When analyzed by positron emission tomography (PET), certain areas in the left, .. visual cortex were activated by visually presented pseudowords that obey English spelling rules, as well as by actual words. These areas were not activated by nonsense strings of letters or letter-like forms. Thus visual word form computations are based on learned distinctions between words and nonwords. In addition, during passive presentation of words, but not pseudowords, activation occurred in the left frontal area that is related to semantic processing. These findings support distinctions made in cognitive psychology and computational modelling between higher-level visual and semantic computations on single words and describe the anatomy that may underlie these distinctions."

M.E.Raichle (1994): "(This experiment) was of particular interest, because it provided a portrait of pure mental activity (perception and speech—or input and output—having been subtracted away). This image permitted us to view what occurs in our brain as we interpret a meaning of words and, in turn, express meaning through their use. It renders visible conscious function because much of our thinking is carried out by concepts and ideas represented by words.... The apparently simple task of generating a verb for a presented noun is not accomplished by a single part of brain but rather by many areas organized into networks".

Evidence indicates that a word may activate its internal visual, phonological, and even semantic codes without the person having to pay attention to the word. Cognitive studies have often shown the superiority of proper English words over nonwords (nonsense words). A region of the posterior fusiform gyrus responded equally to words and non-words and was unaffected by the semantic context in which words are presented (A.C.Nobre, 1994).

Speech production (that is, simply repeating out loud the presented nouns) predictably involves motor areas of the brain. Regions thought to be Broca's and Wernicke's areas do not appear to be engaged routinely in this type of speech production, an activity that would be viewed by many as quite automatic for most fluent speakers in their native language. We occasionally speak without consciously thinking about the consequences.

The normal process of retrieving words that denote concrete entities depends in part on multiple regions of the left cerebral hemisphere, located outside the classical language areas. Moreover, anatomically separable regions tend to process words for distinct kinds of items (persons, non-manipulable animals, manipulable tools (H.Damasio et al, 1996).

5.4.7 *Attention! Attention is very important*

The number of signals coming from the environment, from the body, and, not in the least, from the mind is very large. The first is assumed to be of the order of magnitude of a giga bits per second, although this number is not easy to confirm. Only a small part of this total can be transformed into nervous signals, into internal sensory information. This fraction is assumed to be of the order of magnitude of a hundred bits per second, though this is not universally accepted. The overabundance of information is overwhelming. It is obvious that selective elimination must be used to reduce the information avalanche to a manageable quantity of information which is sufficient for a given individual in a given situation.

According to our data, the reduction of the input information is by a factor of ten million. This ability to selectively eliminate is called 'attention' by psychologists. Attention is, generally speaking, the application of the mind to any object of sense or thought (M.I.Posner, 1995).

Visual-spatial attention is an essential brain function that enables us to select and preferentially process high priority information in the visual fields. Attended visual stimuli are selected as early as 80-90 milliseconds after stimulus onset (H.J.Heinze, 1994).

Experiments by means of PET show that the right hemisphere may play a special part in human attention. Psychophysical evidence indicates that the sensitivity for discriminating subtle stimulus changes was higher when subjects focused attention on one attribute than when they divided attention among several attributes. Localized increases in blood flow have been found in the prefrontal and superior parietal cortex, primarily in the right hemisphere, regardless of the modality or laterality (left or right) of sensory input. Correspondingly, attention enhanced the activity of different regions of the visual cortex that appear to be specialized for processing information related to the selected attribute, such as shape, colour, or velocity. These phenomena reflect cognitive control of visual processing. In brief, we can say that we see what we want to see.

Experiments of R.Desimone (1995) on macaque monkey suggest that a form of competition is taking place. Nerve cells extending from the regions of the brain where memories are stored—probably the prefrontal cortex—bias the outcome as neurons in the visual pathway vie to become active. The bias operates in such a way that unfamiliar objects and remembered objects of great significance are more likely to win in the competition than the familiar, unimportant ones. From the point of view of our book the impact of the bonus-malus label connected with each memorized information is evident here. The memory system filters what should get into consciousness. Attention is linked to anticipation. Anticipation, an act of mental preparation for expected activity, can be experimentally measured in humans by means of high-resolution electro-encephalogram systems (Section 4.4.4) and recently also by PET method (W.C.Drevets et al, 1995; S.Kinomura et al 1996).

Another question is whether biotic evolution in nature was able to generate programs for intelligence, such as was the case for the lower stages of biotic information processing—reflex and instinct. As we already know, the higher stage of discent is only partially pre-programmed (that means phylogenetically generated), and is essentially influenced by the experience of the individual, through his or her learning ability (that means, is ontogenetically produced). The extreme opinion has been formulated by the British philosopher John Locke, in the 17th century, namely that the mind begins as a 'blank slate' that is inscribed on by experience.

From all this, it can be concluded that human intelligence is not the product of programs, or the blind executor of some form of externally supplied instructions. But, of course, the contrary statement is also false. Human intelligence is not a free, self-determining system. It is not a 'blank slate'. Much more, it is a system containing such well-defined programs and strong-acting subcomponents as reflex and, partially, instinct (Fig. 1-5).

Selection of information is necessary to ease the computational problems induced by the enormous number of signals present on the sensory surfaces, such as in the eye, and to ensure that people respond to stimuli that are relevant to their goals. People can pay attention to, or 'look for', specific types of visual information. Concentrating on one visual attribute, such as colour, might be expected to modulate neuronal activity in brain areas that specialize in processing that attribute. Focusing on basic visual attributes, such as shape, colour, or the velocity of an object, appears to influence behavioural and physiological aspects of visual processing. Behaviourally, sensitivity for making subtle discrimination is increased by attention. Neuronal activity is increased in regions specialized in processing information related to the selected visual attribute.

5.4.8 *Planning and execution*

To achieve a goal, we need to execute multiple movements in a specific temporal order. After repetitive performance of a particular sequence of movement, we are able to memorize and execute the whole sequence without external guidance. Where and how in the brain do we store the information necessary for the orderly performance of multiple movements? A group of cells in the supplementary motor area of cerebral cortex contributes to the planning and coding of several multiple, complex movements.

Most human actions, result from the planning, execution, and prior learning of complex motor actions, and is a mixture of sequencing and concurrency. Motor programs are sets of muscle commands put together before the beginning of a movement sequence. They permit that sequence to be carried out without peripheral feedback. G.M. Edelman assumes that motor plans are essentially carried by the prefrontal cortex. The prefrontal cortex of the frontal lobe plays a most important role in goal-directed behaviour.

5.5 THE HUMAN BRAIN AND ITS FUNCTIONS ARE ASYMMETRIC

5.5.1 *Macro- and micro-asymmetry*

Lower animals which live in water, either free-moving or sitting on the bottom, often have radial symmetry. Self-locomotive animals have mirror symmetry. The advantages of such body symmetry are numerous.

The human body contains a number of asymmetric phenomena, for example the heart, which is mostly on the left side of the body, or asymmetric X and Y sex chromosomes, or right-handedness.

The nervous system of higher animals, including the central nervous system, shows mirror symmetry, left and right halves, and, in the case of the brain, left and right hemispheres. A very simple but significant question then is: Are both halves exactly symmetrical, or only superficially so and, in detail, are differences observable?

From the general point of view, full symmetry, including full mirror symmetry, is possible only in the lowest level of structures; for example, at the level of elementary particles, chemical molecules, and simple crystals. More highly developed structures, such as the central nervous system, and pre-eminently the human brain, as the most complex structure in Nature, have a very low probability of being exactly mirror-symmetric. Therefore, it must be assumed that both brain hemispheres cannot be mirror-symmetric. A further question is: Are the differences between the left and right hemispheres accidental or deliberate?

The probability of the random appearance of different properties on the left or right sides must be small, to prevent difficulties in the social co-operation of individuals. It must also be clear that each accidental difference, sooner or later, will be fixed in the genetic code. It must therefore be expected that differences between the left and right halves of the brain, especially in the human brain, are fixed and can be clearly described.

5.5.2 *Emergence of the asymmetric human brain*

It appears that, except possibly for birdsong, no example of lateralized ability in animals, including primates, has been satisfactorily established. Only a weak right hemispheric advantage for learning, remembering and generalization of facial discrimination was significant for a group of male and female monkeys. The earliest prosimians, 60 megayears ago, clung to tree branches with the right hand while reaching for food with the left—as the bush babies usually do today.

Most chimpanzee, gorilla, and probably also orang-utan mothers have a strong preference for cradling their baby by holding it against the left side of their body. Now, 80 per cent of women also do the same. The reason for this behaviour might be that the mother can monitor the baby with her left ear and visual field, hence using the right hemisphere of the brain, which is best able to interpret emotions. On the other hand, the baby can see the left side of the mother's face, the emotionally most expressive side. This behaviour probably originated in the common ancestor of African apes and humans, between 6 and 8 megayears

In rhesus monkeys, the left side of the face begins to display facial expression earlier than the right side and is more expressive. That the right hemisphere determines facial expression, and the left hemisphere processes species-typical vocal signals, suggests that human and nonhuman primates exhibit the same pattern of brain asymmetry for communication (lateralization) (M.D.Hauser, 1993).

As mentioned before a significant jump in the evolution of the brain is the reduction of redundancy; that is, of the overabundant production of the brain's tertiary areas and the emergence of differentiation between both left and right hemispheres—so-called lateralization. The loss of symmetry of the two halves of the human cerebral cortex is one of the most effective achievements in the evolution of biotic information processing. It is the most effective, the most recent, and perhaps also the last in the whole history of the evolution of terrestrial intelligence (Fig. 5-10).

Reduction in redundancy results in an enormous increase of brain capacity, and therefore allows the jump from the higher levels of discent to the stage of intelligence.

Modern Homo sapiens sapiens is a product of the lateralization of the brain's hemisphere. Observation of some one hundred normal foetuses has shown that around 95 per cent suck the thumb of the right hand, with only 5 per cent favouring the left. This must be considered as a demonstration of the existence of behavioural asymmetry, in the form of right-handedness before birth. The left tertiary motor area has taken over the function of the speech production area the Broca´s area, and the left tertiary sensoric area, the function of speech reception and understanding, the Wernicke´s area.

This differentiation between the left and right areas, which are roughly symmetrically positioned, is the most important expression of the human brain's lateralization. According to our hypothesis, these left-lying areas, are the carriers of the phenomena which is called by N.Chomsky 'universal grammar' (Section 6.5.4).

H.P.Killackey (1995): "(There) is one other aspect of the human brain that should be mentioned, which, although well documented, is poorly understood. This is the functional asymmetry of the human neocortex in the cognitive sphere. This seems to me to be a distinctly human trait...In my opinion, the gap between the functional asymmetries in other species is roughly the same size as the gap between human language and the communication skills of other species.... Perhaps the functional asymmetry was a necessary part of the cognitive system."

Asymmetry of the features of the human brain, and even some functional asymmetry (Fig. 5-11 and Table 5-4), is at least partially controlled by genetic mechanisms. The question arises as to whether part of the functional asymmetry results from epigenetic influences. This is the very old question 'nature or nurture'. However, the most reasonable answer to this question should be 'nature via nurture'.

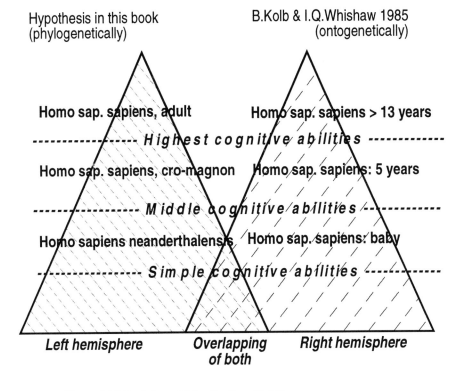

Figure 5-10 Evolution of lateralization of brain

5.5.3 Lateralization—significant or not?

Some anatomical macroscopic differences exist in both human hemispheres. Some of these differences are visible at the first glance. Others require more sophisticated observation and measurement. For example, the temporal lobe has a large flat surface called the planum temporale, which is larger, in most foetuses and neonates, on the left hemisphere. This cannot be the product of epigenesis and evidently has a genetic cause.

Some areas of the temporal lobe which consist of anatomically distinct neuronal layers are larger on the left hemisphere. This area is part of what is called the auditory association cortex, dealing especially with speech sounds. Another area, defined by another neuron category, lies mainly in the angular gyrus, between the temporal and parietal lobes, and has also been found to be larger in the left hemisphere. Because these cellular (microscopic) properties are associated with macroscopically observable asymmetries, the link must have a deeper meaning.

In addition to language, sequential processing or serial ordering is associated with left-hemisphere functioning. There is a cluster of functions that, under normal circumstances, are more strongly represented in the right cerebral hemisphere, and another cluster of functions that are more strongly represented in the left hemisphere.

The most important question is, whether identified microscopic and macroscopic asymmetries are related in a meaningful way to the observable functional asymmetries of both hemispheres, and in what manner these asymmetries are connected with genetic and epigenetic mechanisms. Some difficulty occurs in interpreting data on anatomical asymmetries, as most of it is derived from post-mortem measurements of brains where nothing is known about the kind of functional asymmetries that may have existed before death. In many cases, even the handedness of the individuals is not known. Needless to say, there is no possibility of obtaining information concerning asymmetries present in the pre-natal period of these individual (Section 5.2.4).

G. Schlaug et al (1995) have found in vivo evidence of structural brain asymmetry in musician with perfect pitch which revealed stronger leftward planum temporale (posterior superior temporal gyrus) asymmetry than non-musicians or musicians without perfect pitch. The results indicate that individual variability in cognitive performance can covary with features of external brain morphology. This study does not reveal the mechanism creating this structural asymmetry. Probably the prenatal factors are likely to play a role. The mechanism can still progress by the age of seven, but also could result from later specific stimulation.

Complex interplay exists between the two hemispheres and adaptation of specialization of each hemisphere towards the other one. Depending on which movement and which hemisphere is activated the other side may either benefit or be inhibited in performance of similar motor acts (G.Thut, N.D.Cook, M. Regard,

K.L. Leenders, 1996).

Asymmetry of the hemisphere probably also exists at the level of biochemical properties. In some cases, it has been observed that the right hemisphere is more sensitive or vulnerable to the presence of the hallucinogenic drug LSD. One hypothesis is that LSD affects a certain neurotransmitter that may be used to a greater extent by the right hemisphere.

Table 5-4: Left and right hemisphere differences (for typical right-handedness)

	Left hemisphere	Right hemisphere
Anatomical properties		
Hemisphere mass	smaller	larger
Planum temporale	larger	smaller
Grey matter	more	less
White matter	less	more
Specific density	higher	lower
Biochemical properties		
LSD influence	smaller	larger
Noradrenalin, amount	larger	smaller
Speech controlling hemisphere		
For right-handedness	96%	4%
For left-handedness	70%	30%
Sensors and motor control		
Visual system	reading letters	seeing faces
Acoustical system	hearing speech	hearing music
Musicians with perfect pitch	larger planum temporale	
Body movement control	complex and fine	spatial rough
Cognitive and emotional		
Cognitive processing	verbal analytical	spatial processing
Emotional processing		predominance

Remark: The 'language-dominant' hemisphere is 'motor-dominant' (J. Netz et al, 1995)

It has been discovered that some neurotransmitters, such as norepinephrine, are distributed differently in the right and left halves of the thalamus, a subcortical structure that has, among other things, a role in speech production. An area on the left temporal lobe shows a greater response to the enzyme choline-acetyltransferase than does the corresponding area of the right hemisphere.

Of course, to a first approximation, all these macroscopic, cellular, and biochemical hemisphere asymmetries seem to have a genetic cause, but epigenetic influence cannot be excluded.

The left hemisphere is open to introspection, whereas the right's activities are not subject to a person's direct awareness.

R.W.Sperry (1983): "The left and right hemispheres of the brain are each found to have their own specialized forms of intellect. The left is highly verbal and mathematical, and performs with analytic, symbolic, computer-like, sequential logic. The right, by contrast, is spatial, mute, and performs with a synthetic, spatio-perceptual, and mechanical kind of information processing not yet simulatable in computers."

5.5.4 Female and male peculiarities of the brain

Human genes provide the program that make us what we are; they determine not only what we look like but also our various ways of reacting to the challenges of the world around us.

F.Nottebohm (1989) discovered that in adult male canaries, which sing complex songs, the so-called 'higher vocal centre' in their brain, and another forebrain nucleus, were some three to four times larger than in female canaries, which sing simpler songs. It appeared that the amount of brain volume devoted to a particular skill was considerably greater in the sex excelling at the skill. This instance of so-called sexual dimorphism disproved another long-held view, namely that the brains of vertebrates exhibited no marked anatomical differences between the sexes.

Male and female gorillas and orang-utans differ dramatically in body size and features. It is obvious that brain mass is also significantly smaller in the females, but not the brain/body ratio. The sex difference of body size and brain mass in chimpanzees is insignificant, as in humans.

The genome, the set of genetic information of an organism, consists in humans of 3 giga of nucleotides, resulting in a 7-centimetre-long macromolecule of deoxyribonucleic acid, DNA. DNA consists of two strands wrapped around each other in a double helix, and is contained in chromosomes. Humans have 46 chromosomes: 22 pairs of autosomes and one pair of sex chromosomes, XX in females and XY in males. The Y chromosome consists of 70 million nucleotides and seems to contain no genes necessary for life. Its absence in women simply results in the absence of male characteristics. The Y-genes (gene, an element of genetic material responsible for a specific unit function, mostly for the synthesis of one protein) probably contain a couple of genes. They are the carriers of 'manhood' and are responsible for *'la petite difference'*. The question arises what influence genetic information has on the structure and properties of the central nervous system and especially on the brain. Do differences exist between the male and female brain?

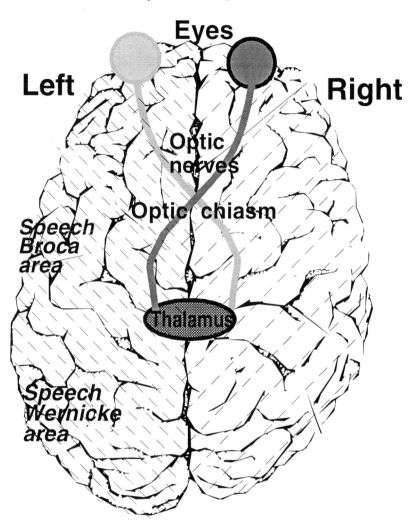

Figure 5-11 Asymmetry of human brain

Sex differences are seen primarily in each neuron. The soma of neurons in females contains the so-called Barr-body, which is a product of the second X chromosome, and is deactivated in each female cell. The male cell has only one X chromosome, the second being a Y chromosome, and therefore does not contain a Barr-body. Of course, the Barr-body is inactive from the point of view of genetics, but it is possible that its presence can have some influence on the functions of the neurons. If this is the case, then the male and female neurons are not similar from the point of view of morphology, and are perhaps also different from the point of view of their functional properties (Table 5-5).

Table 5-5 Differences in the brain structure between gender

	Female	Male
Number of neurons in neocortex *)	19 giga neurons	23 giga neurons
Foetal development period	at 35 weeks, the female brain is fully developed	male brain is not.
Mass of brain (correcting for differences on body size)		about 100 grams heavier
Volume of the preoptic area in the human hypothalamus		2.5 times larger: number of neurons
Percentage of cortical grey matter	smaller	larger
Regional cerebral glucose metabolism during a resting state (by PET)	higher metabolism in cingulate regions	higher metabolism in temporal-limbic regions and in cerebellum
Age-related changes: decrease in brain volume and increase in the cerebrospinal fluid		greater with age in men. In men atrophy of the left hemisphere being more pronounced.
Sexual hormones, such as testosterone		larger concentration
Left hemisphere		slower growth
Right hemisphere		greater development
Difference between the hemispheres		greater in males
Corpus callosum (connection of both hemispheres), size	larger; about 30 mm^2 (mean cross-section area: 680 mm^2)	smaller size **)
Electroencephalographic phenomena	in the eyes-open condition, more engaged in scanning their visual environment	

*)For normal brains of Danes: B, Pakkenberg, H.J. Gundersen (1997);
**) H. Steinmetz et al, (1995)

The average numbers of neocortical neurons was shown to be 19 giga in female brains and 23 giga in male brains, a 16 per cent difference. Ten percent of all neocortical neurons are lost over the life span from 20 to 90 years. Sex and age were the main determinants of the total number of neurons in the human neocortex, whereas body size, per se, had no influence on neuron number (B.Pakkenberg,

H.J.Gundersen, 1997).

D.Kimura (1993): "Women's brains are endowed with better digital control and men's brains are better endowed for targeting external stimuli."

Given that the sex chromosomes, and their deactivation products, produce measurable differences between male and female bodies, do they also produce any difference between male and female brains? They certainly do in many animals, and although anatomical and physiological evidence for human beings is not abundant, significant differences have been found. One obvious influence of sex on the neural system is the difference in hormone activity. It has been shown that neurons are more densely packed in the cerebral cortex of women than of men; the total average number is not detectably different. Women have a mean brain mass about 100 g less than men of the same height.

B.A.Shaywitz et al (1995): "A much debated question is whether sex differences exist in the functional organization of the brain for language. A long-held hypothesis posits that language functions are more likely to be highly lateralized in males and to be represented in both cerebral hemispheres in females, but attempts to demonstrate this have been inconclusive. ...(Using) the echo-planar functional magnetic resonance imaging to study ..(normal) right-handed males and females during orthographic (letter recognition), phonological (rhyme) and semantic (semantic category) tasks. These data provide clear evidence for a sex difference in the functional organization of the brain for language and indicate that these variations exist at the level of phonological processing."

B.A. Shaywitz (1995): "A new model of the reading disorder, called 'dyslexia' emphasizes defects in the language-processing rather than the visual system. It explains why some very smart people have trouble learning to read. Our investigation has already revealed a surprising difference between men and women in the locus of phonological representation for reading. It turns out that in men phonological processing engages the left frontal gyrus, whereas in women it activates not only the left but the right inferior frontal gyrus as well. These differences in lateralization had been suggested by behavioural studies, but they have never before been demonstrated unequivocally. Indeed, our findings constitute the first concrete proof of gender differences in brain organization for any cognitive function. The fact that women's brains tend to have bilateral representation for phonological processing explains several formerly puzzling observations."

Hormones could affect the growth rate of axons and dendrites on neurons sensitive to them. The hormone-sensitive neurons would end up making different connections in male and female brains, because the timing of their meeting up with other growing neurons would be different. Hormones might influence the growth of synapses, and the death of neurons and synapses. Anatomic, physiological and behavioural data indicate that the brains of males and females are not identical and are organized along different lines from very early on in life onwards.

D.Kimura (1992): "I would not expect, for example, that men and women would necessarily be equally represented in activities or professions that emphasize spatial or math skills, such as engineering or physics. But I might expect more women in medical diagnostic fields, where perceptual skills are important. ..Women and men differ not only in physical attributes and reproductive function but also in the way in which they solve intellectual problems. ..The (difference) in brain organization occurs so early in life that from the start the environment is acting on differently wired brains in girls and boys. ..Major sex differences in intellectual function seem to lie in patterns of ability rather than in overall level of intelligence."

It should not be surprising that there are neural structures that help to determine sexual behaviour. Even some anatomic details of human brain structures differ in homosexual and heterosexual males. This may mean that homosexuality is strictly genetically 'hardwired' although there are some serious doubts concerning these results. It has been theorized that the genes implicated in homosexuality are probably those involved in prenatal brain development—specifically in masculization of the hypothalamus during sexual differentiation.

5.5.5 Handedness

Irrespective of culture, about 90 per cent of human beings use the right hand for writing and for the skilful manipulation of objects (M.J. Morgan, 1992). Prehistoric man, Cro-Magnon man, produced on the walls of caves negative handprints, of the left hand in 80 per cent of cases. Thus, people who outlined the hand must have used their right hands to apply the colour. Drawings of hunters show weapons being carried by the right hand. But is right-handedness a genetic or an epigenetic phenomenon? No such overwhelming preference is found in the rest of the animal kingdom. Why, then, are some 10 per cent of human beings left-handed? Why has no single culture ever been found in which a left-hand bias has predominated?

On average, right-handed children are born in 92 per cent of cases where both parents are right-handed, in 80 per cent of cases where one parent is right-handed and the other left-handed, and in 45 per cent of cases where both parents are left-handed. Can these observations be interpreted on the basis of a genetic model?

As anecdote only: In 1992 all three candidates for the presidency of the USA were left-handed: G.Bush, R.Perot and B.Clinton.

5.6 BIBLIOGRAPHY

Adams MD et al (1992) Sequence identification of 2,375 human brain genes. *Nature* **355** 632

Ahissar E at al (1992) Dependence of cortical plasticity on correlated activity of single neurons and behavioral context. *Science* **257** 1412

Baddeley A (1995) Working memory. in Gazzaniga MS (edit) in *'The cognitive neuro sciences'*. MIT Press, Cambridge, Mass.

Bain S (1955) *Les sens et l'intelligence.* (according to Changeux PJ, 1984).

Barondes SH (1994) Thinking about Prozac. *Science* **263** 1102

Bechara A et al (1997) Deciding advantageously before knowing the advantageous strategy. Science, 275,1293

Calvin WH (1991) Islands in the mind: dynamic subdivisions of association cortex and the emergence of a Darwin Machine. *Semin Neurosci* **3** 423

Changeux J-P, Connes A (1992) *Gedanken-Materie.* Springer, Berlin

Churchland PS (1986) *Neurophilosophy. Toward a unified science of the mind-brain.* MIT Press, Cambridge, Mass.

Churchland PS (1989) From Descartes to neural networks. *Sci Am* **261** 1 100

Cowey A, Stoerig P (1995) Blindsight in monkeys. *Nature* **373** 247

Charms. de R.C, Merzenich M.M (1996) Primary cortical representation of sounds by the coordination of action-potential timing. *Nature* **381** 610.

Crick F, Koch Ch (1997) The problem of consciousness. *Sci. Am. Spec.Iss. Mysteries of Mind*, Vol **7**,18

Damasio H et al (1996) A neural basis for lexical retrieval. *Nature* **380** 499

Desimone R (1992) The physiology of memory: recordings of things past. *Science* **258** 245

Drevets WC et al (1995) Blood flow changes in human somatosensory cortex during an ticipated stimulation. *Nature* **373** 249

Dudai Y (1989) *The neurobiology of memory.* Oxford Univ. Press. Oxford

Fischbach GD (1992) Mind and brain. *Sci Am* **267** 3 24

Foelfsema PR et al (1997) Visuomotor integration is associated with zero time-lag synchronization among cortical areas. *Nature* **385** 157

Gawin FH (1991) Cocaine addiction. *Science* 251,1580

Gazzaniga MS (1980) The role of language for conscious experience: from split-brain man, in H.H. Kornhuber, L.Deecke (eds) *'Progress in brain research'*. Elsevier, Amsterdam

Gibbs WW (1996) Jungle medicine. *Sci Am* **275** 6 14

Goldman-Rakic PS (1992) Working memory and the mind. *Sci Am* **267** 3 73

Gur RC et al (1995) Sex differences in cerebral glucose metabolism during a resting state. Science **267** 528

Hari R et al (1994) Visual stability during eyeblinks. *Nature* **367** 121

Heinze HJ et al (1994) Combined spatial and temporal imaging of brain activity. *Nature* **372** 543
Hepper PG et al (1990) Origins of fetal handedness. *Nature* **347** 431
Hinton GE (1992) How neural networks learn from experience. *Sci Am* **267** 3 105
Hopfield JJ, Tank DD (1986) Computing with neural: a model. *Science* **233** 625
Hubel DH (1996) A big step along the visual pathway. *Nature* **380** 197
Hutchison JB (1991) Hormonal control of behaviour: steroid action in the brain. *Curr Opin Neurobiology* **1** 562
Jonides J et al (1993) Spatial working memory in humans as revealed by PET. *Nature* **363** 623
Kalil RE (1989) Synapse formation in the developing brain. *Sci Am* **261** 6 8
Karni A et al (1994) Dependence on REM sleep of overnight improvement of perceptual skill. *Science* **265** 679
Killackey HP (1995) Evolution of the human brain: a neuroanatomical perspective, in Gazzaniga MS (edit) *'The cognitive neurosciences'*. MIT Pr, Cambridge, Mass
Kimura D (1992) Sex differences in the brain. *Sci Am* **267** 3 81
Kinomura S et al (1996) Activation by attention of the human reticular formation and thalamic intralaminar nuclei. *Science* **271** 512
Klopf AH (1982) *The hedonistic neuron.* Hemisphere Pub Co, Washington
König M et al (1996) Pain responses, anxiety and aggression in mice deficient in pre-proenkephalin. *Nature*, **383** 535.
LeDoux JE (1992) Brains mechanisms of emotional learning. *Curr Opin.Neur* **2** 191
LeDoux JE (1994) Emotion, memory and the brain. *Sci Am* **270** 32
Livingston MS, Hubel DH (1987) Psychophysical evidence for separate channels for the perception of form, color, movement, and depth. *Jour Neuroscience* **34** 16
Livingston MS (1988) Art, illusion and the visual system. *Sci Am* **258** 1 68
Luria AR (1981) *Lectures on language and cognition.* Wiley, N.York
McGehee DS, Role LW (1996) Memories of nicotine. *Nature*, **383** 670
Morgan MJ (1992) On the evolutionary origin of right-handedness. *Curr Biology* **2** 1 15
Netz J et al (1995) Hemispheric asymmetry of callosal inhibition in man. *Exp Brain Res* **104** 527
Nobre AC et al (1994) Word recognition in the human temporal lobe. *Nature* **372** 260
Pakkenberg B, Gundersen HJ (1997) Neocortical neuron number in humans: effect of sex and age. *J.Comp.Neurobio.* **384** 312
Penrose R (1994) *Shadows of mind.* Oxford Univ Press, Oxford, UK
Petersen SE et al (1990) Activation of extrastriate and frontal cortical areas by visual words and word-like stimuli. *Science* **249** 1041
Posner MI, Raichle ME (1994) *Images of mind.* Sci Am Library, N.York
Posner MI (1995) Attention in cognitive neuroscience, in Gazzaniga MS (edit) *'The cognitive neurosciences'*. MIT Press, Cambridge, Mass
Schacter DL (1995) Implicit memory: a new frontier for cognitive neurosciences. in Gazzaniga MS (edit) *'The cognitive neurosciences'*. MIT Press, Cambridge, Mass.

Schlaug G et al (1995) In vivo evidence of structural brain asymmetry in musicians. *Science* **267** 699

Shaywitz BA et al (1995) Sex differences in the functional organization of the brain for language. *Nature* **373** 607

Singer W (1995) Time as coding space in neurocortical processing: a hypothesis, in Gazzaniga MS (edit) *'The cognitive neurosciences'*. MIT Press, Cambridge, Mass.

Smith RE, Jonides J (1995) Working memory in humans: neuropsychological evidence, in Gazzaniga MS (edit) *'The cognitive neurosciences'*. MIT Pres, Cambridge, Mass.

Snyder LH et al (1997) Coding of intention in posterior parietal cortex. *Nature*, **386** 167

Steinmetz H et al (1995) Corpus callosum and brain volume in women and men. *Neuro Report* **6** 1002

Steriade M, et al (1993) Thalamocortical oscillations in the sleeping and aroused brain. *Science* **262** 679

Talbot JD et al (1991) Multiple representations of pain in human cerebral cortex. *Science* **251** 1355

Tanaka K (1992) Inferotemporal cortex and higher visual functions. *Curr Opin Neurobio* **5** 202

Tanji J, Shima K (1994) Role for supplementary motor area cells in planning movements ahead. *Nature* **371** 413

Thorpe S et al (1996) Speed of processing in the human visual system. *Nature* **381** 520

Thut G, Cook ND, Regard M, Leenders KL et al (1996) Intermanual transfer of proximal and distal motor engrams in humans. *Exp Brain Research* **108** 321

Thut G, Schultz W, Leenders KL et al (1997) Activation of the human brain by monetary reward. *NeuroReport* **8** 1225

Young MP, Yamane S (1992) Sparse population coding of faces in the inferotemporal cortex. *Science* **256** 1327

Vaadia E et al (1995) Dynamics of neuronal interactions in monkey cortex. *Nature* **373** 515

Watanabe M (1996) Reward expectancy in primate prefrontal neurons. *Nature* **382** 629

Whittington MA et al (1995) Synchronized oscillations in interneuron networks driven by metabotropic glutamate receptor activation. *Nature* **373** 612

Wickelgreen I (1997) Getting grasp on working memory. *Science*, **275** 1580

Wiesel TN (1982) Postnatal development of the visual cortex and the influence of environment. *Nature* **299** 583

Wise R (1980) The dopamine synapse and the notion of 'pleasure- center' in the brain. *Trends Neurosci* **91** 4

Zeki S (1992) The visual image in mind and brain. *Sci Am* **267** 3 4

Chapter 6
Tools of Intelligence

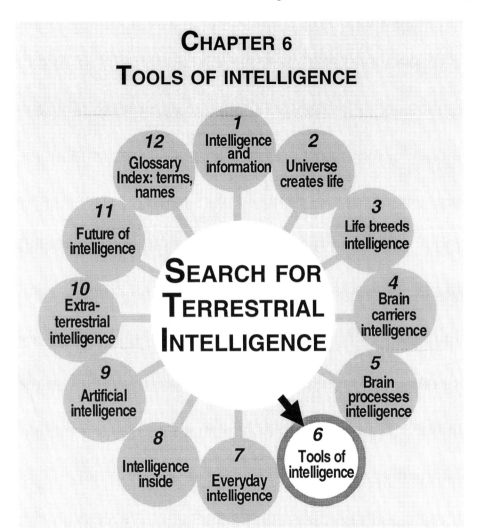

Does a theory of human intelligence exist? Universal and terrestrial definition of intelligence. Evolution from reflex, instinct, discent to intelligence. Emotions; important component of intelligence. Animal intelligence: yes or no? Lost stages of evolution: sub-intelligence.

Learning ability and bonus-malus system. Four stages of animal communication. Development of human speech. Universal grammar? How a child learns to speak. Percepts and concepts. Qualia.

How important are questions? Brain damage and speech. Deaf-and-dumb language. Writing and reading: human uniqueness.

6 TOOLS OF INTELLIGENCE 205
6.1 DEFINITION OF HUMAN INTELLIGENCE 205
6.1.1 Common sense, intellect, reason, wisdom 205
6.1.2 Does a theory of human intelligence exist? 205
6.1.3 Terrestrial or universal definition 207
6.1.4 Intelligence and different branches of science 209
6.2 INTELLIGENCE AS THE HIGHEST STAGE OF INFOR MATION PROCESSING 210
6.2.1 From reflex to intelligence; evolution 210
6.2.2 Intelligence influences reflex, instinct and discent 212
6.3 EMOTIONS, IMPORTANT COMPONENTS OF INTELLIGENCE 212
6.3.1 Emotions, the human predominance 212
6.3.2 Animal intelligence: yes or no? 215
6.3.3 Lost stage of evolution: Sub-intelligence or superdiscent? 217
6.4 LEARNING ABILITY AND UNDERSTANDING 217
6.4.1 Learning ability 217
6.4.2 Semantics; what do we understand as 'understanding'? 219
6.4.3 Learning and the bonus-malus system 220
6.4.4 Highest complexity of the bonus-malus system 221
6.5 SPEECH, THE MOST IMPORTANT MEDIUM OF INTEL-LIGENCE 222
6.5.1 Four stages in the evolution of communication 222
6.5.2 Animal communication; some remarks 225
6.5.3 Levels of speech development 226
6.5.4 Universal grammar: some doubts 232
6.5.5 How a child learns to speak 234
6.5.6 Continuity of human speech 235
6.6 SPEECH AND ABSTRACTS 237
6.6.1 Percepts, concepts, etc. 237
6.6.2 Heuristic and algorithmic problem solutions 239
6.6.3 Qualia 240
6.6.4 Mutual basis of speech: world knowledge 240
6.6.5 How important are questions? 241
6.7 SPEECH, SOME ABNORMALITIES 243
6.7.1 Brain damage and speech 243
6.7.2 Deaf-and-dumb language 244
6.8 WRITING AND READING: INTELLECTUAL ABILITIES 246
6.8.1 Writing and reading; human uniqueness 246
6.8.2 Pictorial representations 248
6.9 BIBLIOGRAPHY 250

6 TOOLS OF INTELLIGENCE

6.1 DEFINITION OF HUMAN INTELLIGENCE

6.1.1 *Common sense, intellect, reason, wisdom*

Human beings have the inherent characteristic of thinking of themselves as belonging to the highest creatures, and, of course, the wisest and most (if not exclusively) intelligent. There are at present, and were in the past, numerous methods and different points of view for classifying the intelligence of Homo sapiens sapiens.

Because of the extreme complex and multi-dimensional nature of intelligence and other related characteristics, there exists a large spectrum of rather inexact terms, from different scientific areas and from everyday language, which are used to describe them. In spite of this wealth of terms, there are encyclopaedias and dictionaries in which terms such as 'intelligence' are even not mentioned. Also, the use of the term intelligence is far from being standardized. Different authors, different scientists, not to mention different journalists, use this term in a more or less arbitrary fashion. Some of them, use the term 'intelligence' not only for man, chimpanzees, doves, and fish, but even for amoeba.

6.1.2 *Does a theory of human intelligence exist?*

It seems necessary that one of the most important aspects of human development—intelligence—should have an adequate theoretical background. The question is, whether such a theory does exist? There is currently little chance to provide a generally accepted theory, but what about more or less accepted working hypotheses?

How can a theory of human intelligence satisfy the criterion of being correct? A good theory, especially one concerning intelligence, must fulfil the criteria formulated before in the Table 1-1.

Present understanding of the nature of human intelligence makes it impossible to formulate a definition in one sentence. The only reasonable course is to create a longer definition which includes all necessary aspects. From what has been said before, it will be clear that a definition must take into account the pathway of the evolution of biotic information processing. In this evolution, intelligence has evolved from discent and, in a deeper sense, from instinct and reflex (Fig. 1-5).

This means that a definition of intelligence must consist of four more or less clearly separated stages. The best way of presenting such a complex system is using a diagram (Fig. 6-1).

Each of these stages can be described by the most important behavioural parameters, such as type of memory, ability to learn, and communication with

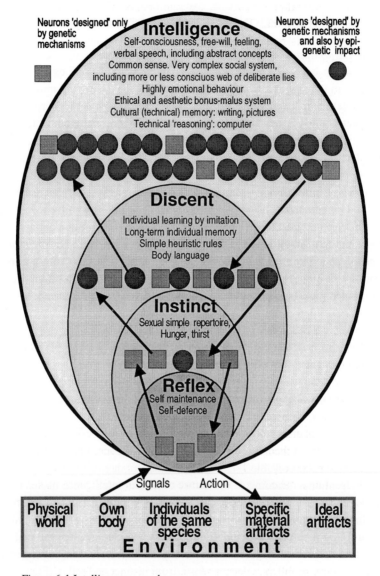

Figure 6-1 Intelligence—a scheme

other individuals. The environment of animals with the appropriate stage of information processing can be divided as is shown on the scheme. Even if these components of the environment do not belong directly to the definition of every level of biotic information processing, the significance of the relationship between them cannot be underestimated.

Figure 6-1 is perhaps not fully comprehensive, but it contains the most significant properties of intelligence, in a condensed form. It is obvious that a definition of intelligence must include the lower stages of the biotic information processing, discent, instinct and reflex. Only in few publications an analogous differentiation of biotic information processing is found. The hypothesis of Luria, Gould and Singer has been mentioned in Section 3.4.4 and illustrated in Fig. 3-8. The definition of Pfenninger (1989) and his scheme is displayed in Fig. 6-2 (see also Table 6-1 prepared according to J-P Changeux, A. Connes, 1989).

6.1.3 Terrestrial or universal definition

It should be clear that the above description of intelligence deals primarily with terrestrial intelligence. However, our task also covers the general case of universal intelligence. Is there a significant difference in the definition of terrestrial and universal intelligence? Why would extraterrestrial intelligence be bothered by terrestrial intelligence? In spite of this one conclusion seems to be obvious. Assuming that terrestrial (or extraterrestrial) intelligence evolves further and achieves a higher stage, called in this book superintelligence (Section 11.5), a definition of universal validity will be essential.

The number of published definitions of human intelligence tends to become rapidly uncountable! Here are some examples.

H. Gardner (1984) proposed a theory in which human intelligence is based on the following skills: verbal, logic- mathematical, spatial, musical, bodily, and personal. This classification seems to be rather arbitrary.

R.J. Sternberg (1984) developed under the name of a 'triarchic theory of human intelligence', a theory consisting of three sub-theories: 1) context subtheory, 2) two-faceted subtheory, and 3) components subtheory. However, the same author, in his book 'Handbook of human intelligence' (1982), including 1030 pages, writes: "The term intelligence is used in many different ways by the many different authors of the chapters of this book."

H. Eysenck (1987) ascertains that each theory of human intelligence must include the three following components: 1) biological intelligence, including learning, memory, problem-solving and other cognitive processes, 2) social intelligence, and 3) psychometric intelligence, which is the object of intelligence-quotient measurements.

One of the best-known theories of intelligence has been proposed by the English psychologist J.P.Guilford (1967), who claimed that 'intelligence' is multi-dimensional, embracing 5 types of mental operations, 6 types of relations, and 4 types of representations. All these result in 5 x 6 x 4 = 120 highly specific abilities. Later on, the number of abilities was increased by Guilford himself to 150. Evidence for these numerous abilities did not derive primarily from an inductive program of original experiments; they were formulated on logical (in many ways a priori) grounds.

Table 6-1 Stages of thinking machine

Properties	Level according to Changeux and Connes, 1992	Biotic information processing in this book
Machines with given target, with bonus-malus, but without gain-will, and without pain	**Zero** level is not mentioned (M.T., K.L.)	Reflex
Machines with given target, that is with valuation function, (with gain-will) without emotionality	**First** level	Instinct
Target is given. Valuation function over valuation function; emotionality, reflection (thinking about thought). No feeling of harmony	**Second** level	Discent
Reflection over reflection. Creativity, even if the target is not sufficiently clear (creation of new unforeseeable target). Feeling of harmony of the creation, aesthetics (bonus-malus)	**Third** level	Intelligence

D.R. Hofstadter (1982) claimed that perfect intelligence cannot exist—though the definition of intelligence, and especially of perfect, was not made.

K. Lorenz (1982) claimed that animals, such as his dog, at least have the same emotions and feelings as humans, and express many attributes of 'Self'.

G.G. Simpson (1950) believed that the emergence of intelligence has such a low probability that perhaps, even on the Earth, it does not exist at all!

Continuing our practice of quoting standard references, here are some citations, without comment:

Encyclopaedia Britannica (1979): "Definitions (of intelligence) vary, but common elements clearly may be discerned. Intelligence is defined as a cognitive disposition (knowing) distinct from the affective (emotional) or motivational. It is thought of as exceedingly general and, in contrast to more specific abilities, as influencing a broad range of human performance. Defined primarily as capacity or potentiality, rather than fully developed attainment, it is almost universally accepted as having a biological basis. Cognition manifestly is a function of the central nervous system, and individual differences are related to biological-genetic endowment.. The word (intelligence) provides a convenient portmanteau description of the highest level of co-ordinated thinking by an individual."

R.C.Schank, P.G.Childers (1984): "The simplest and perhaps safest definition of intelligence is the ability to react to something new in a nonprogrammed way. The ability to be surprised or to think for oneself is really what we mean by intelligence. What are the signs of intelligence? These are very difficult questions to answer. They are among the central questions being wrestled with in psychology, philosophy and anthropology. Many people have found their own answers for these questions and are ready to be supreme arbiters of what or who is really intelligent."

T.Stonier (1990): "In a subsequent work ('Beyond Information') the present author intends to examine the relationship between intelligence (all biological systems exhibit some measure of intelligence) and the negation of entropy."

R.Penrose (1994): "Of course I have not defined any of the terms 'intelligence', 'understanding', or 'awareness'. I think that it would be most unwise to give full definitions here. We shall need to rely, to some extent, on our intuitive perceptions as to what these words actually mean."

The Penrose's book 'Shadows of the mind' includes 450 pages and there is enough place to try to define the crucial term 'intelligence'.

6.1.4 *Intelligence and different branches of science*

One of the difficulties in the formation of a general theory of human intelligence is the number of branches of science which claim to have the correct approach. Some of these branches are:

– Philosophy (Greek *philein* to love + *sophos* skilled, clever, wise)
– Psychology (Greek *psyche* life, spirit, soul)
– Psychiatry (Greek *iatros* physician)
– Anthropology (Greek *anthropos* man)
– Neurobiology (Greek *neuron* string)
– Cognitive science (Latin *cognoscere* to know)
– Informatics (Latin *informare* to form)
– Theology (Greek *theos* god).

In different times, in different cultures, and in different social spheres, one scientific branch or another has claimed to be the correct tool and to offer the best methodology for research into what is one of the most complex subjects in Nature.

6.2 INTELLIGENCE AS THE HIGHEST STAGE OF INFORMATION PROCESSING

6.2.1 *From reflex to intelligence; evolution*

It was some 800 megayears ago when the first multicellular marine animals succeeded in reaching the higher stage of evolution, and generated the simplest nervous system (the lowest stage of biotic information processing—reflex—has been described in Section 3.4.1).

Reflex itself is very successful and a very large number of animal species have survived up to the present day, on land and in the sea, whose evolution stopped at the reflex stage. If assumptions concerning the number of worms and other simple animals are true, then the number of species, as well as the number of individuals, is the largest of all terrestrial animals. It is probable that reflex is the most abundant type of information processing at present, and there are many arguments why this will continue to be so in the future.

The higher stage of biotic information processing, instinct, probably emerged some 570 megayears ago, when a hard exoskeleton was so useful in the evolution of more complex animals, moving on the ocean floors. Later, when insects began to emerge and conquer the continents, and the first flying animals conquered the atmosphere, the number of species and individuals possessing the complex and more highly developed nervous system corresponding to instinct (Section 3.4.2) became so large that they began to rival the number of lower animals, which operated on the level of reflex.

It is of the greatest importance in our search to see that the emergence and development of instinct does not reduce the chances of the lower (reflex) information processing stage to exist also. Both stages can coexist. Moreover, according to our understanding of the architecture of biotic information processing, instinct includes reflex as a lower stage.

There is now evidence that the emergence, around 150 megayears ago, of the next higher stage of biotic information processing, descent (Section 3.4.3), as exhibited by amphibians, reptiles and birds, did in any case reduce the potential for survival and development of animals which were at the lower level of evolution of the nervous system.

It is now evident that the number of species and individuals at the stages of discent is smaller than the number at the stages of instinct and reflex. The pyramidal structure of the different stages of biotic information processing is strongly expressed.

The emergence of discent cannot be considered as evidence of higher success or of better species adaptability. It is only evidence that, in some ecological niches and under some environmental conditions, animals with a more highly developed nervous system could emerge side by side with lower animals. It is also evidence that the pyramidal hierarchy enriches the realm of animals and the whole terrestrial biosphere, and does not act destructively towards the lower species. Snails, which as a species are more than 400 megayears old, have very primitive nervous systems but have survived to the present day, while the mastodon, with a rather highly evolved and large brain, has disappeared (Fig. 3-7).

The emergence of the highest stage of information processing—intelligence, as displayed by Homo sapiens sapiens—was, in the beginning, a similar link in the evolutionary chain to previous links. The emergence of intelligence enriched the realm of animals and the terrestrial biosphere. We should remember that the development of biotic information processing doubtless proceeded in a continuous manner. However, the discontinuity between the present highest level of animal discent and human intelligence must be explained (Fig. 6-2).

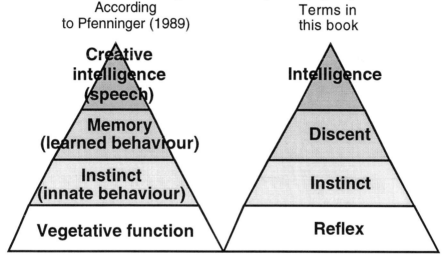

Figure 6-2 Four stages of intelligence

As we very well know, the current global situation seems to show that the highest stage of the biotic information processing (intelligence) threatens the pyramidal structure of the terrestrial biosphere and tends to destroy its equilibrium, due to the explosive increase of world population, by direct holocaust of a number of species, and because of its technical activity. We can, and must, hope that it is not too late to restore the equilibrium of the terrestrial biosphere, including human intelligence, perhaps even to a new and different, but stable, level.

J.Khalfa (1994): "The emergence of intelligence is an evolutionary problem: why is instinct suddenly insufficient at a particular stage of history of adaptation? ...An anyway, is the difference between instinct an intelligence such a clear one?... (Later).. compares attempts at explaining animal behaviour on the basis of associative learning with intentional approaches (which in this book corresponds to the stage 'discent')... At the next level, that of the emergence of intelligence in humans.....There is no doubt that .. language is the essential criterion of human specificity."

6.2.2 *Intelligence influences reflex, instinct and discent*

Biotic evolution not only generates, step by step, higher stages of information processing, but also makes the best use of the previously developed stages. The emergence of instinct consisted of two parallel steps: the generation of a higher stage, and the adaptation of the lower stage of biotic information processing, reflex, to the requirements of the higher stage which at the same time meant improved cooperation between both stages. However, it would be an inexcusable error to neglect the strong influence of the lower stage on the activity of the higher. Reflex influences the activity of instinct (Fig. 6-3).

The exploitation of the former efficiently working structure and its adaptation to the new emerging higher structure is a principal strategy of the whole of biotic evolution. Discent, as the third stage in the development of biotic information processing, follows the same strategy and, after some adaptation, makes use of the lower stages of instinct and reflex. Reflex and instinct have a significant impact on the activities of discent. Discent is therefore, in our terminology, a two-way pyramid, with influence from top to bottom, as well as from bottom to top. Intelligence follows the same path. It includes discent and, therefore, also instinct and reflex. The mechanisms of instinct, also called the driving force, or impulse, or urge, not only adapted to the new hierarchical structure, but also played an appropriate role in the formation of discent and intelligence. In spite of this, it seems to be erroneous to call this phenomenon 'animal instinct', and say that it is this which influence 'human intelligence'. It must be clear that instinct, as a stage of biotic information processing in Homo sapiens sapiens is 'human instinct', and it is this which plays its role—an important role—in the whole phenomenon called 'intelligence'.

A.R.Damasio (1994): "Both 'high-level' and 'low-level' brain regions,... cooperate in the making reason. Yet the dependence of high reason on low brain does not turn high reason into low reason... Current research .. generally supports the idea of asymmetry in process of emotion, but also indicates that the asymmetries to not pertain to all emotions equally"

6.3 Emotions, important components of intelligence

6.3.1 *Emotions, the human predominance*

In common sense pictures of 'intelligent robots' almost always the lack of emotions is the most significant difference to human beings. And it seems that this is a right judgement. Emotions are an indivisible component of human intelligence.

Tools of intelligence 213

One of 1995 world best-sellers was D.Goleman's 'Emotional Intelligence: Why it can matter more than IQ?'.

There is evidence that in the 'lower' stages of biotic information processing, such as reflex, instinct and even discent emotions are nonexisting or al least very limited. Only the human being is the bearer of this phenomenon.

The human face, expresses emotions, conveys information about a person's identity, and often blends of different emotions at the same time, all of which are elements critical to social behaviour. Facial expressions can convey both basic emotions whose expression may be partly innate, as well as subtler emotions whose meaning is partially determined by culture (R.Adolphs, 1994).

Emotions and fear are the most significant components of the bonus-malus system, and allow to make fast decisions (mental or muscular).

S.Pinker (1997): "We instinctively fear snakes, but we appear not to be afraid of fast cars, which are a real danger now. This suggests our emotions were shaped by our evolutionary environment, not the one we grew up in."

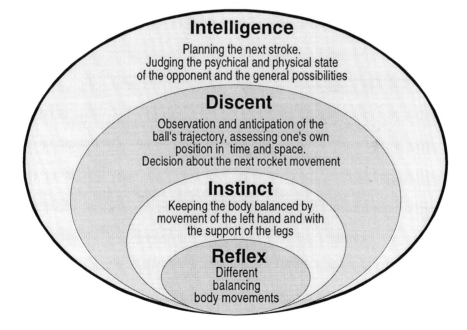

Figure 6-3 Tennis player

Not seldom contrapuntal to intelligence are mentioned emotions. This is not correct. Emotions are inseparable components of discent and especially of intelligence. The highest level of the bonus-malus system (reward-punishment) is the emotional level. Emotional behaviour is not only the means for one animal to communicate its motivational state to another of its own species. It is also a tool for mobilizing its own body for expected action, especially its hormonal system. For example, an expression of aggression is generated in the case of non-maintenance of a safe distance from another individual. Emotions are in some way connected to action in the future; that is with the motivation for this future action. There is no universally agreed upon definition or theory of emotion that delimits emotional phenomena in a way that is useful for relating those phenomena to neural systems (J.E.Ledoux, 1995).

A.R.Damasio (1994): "I began writing this book to propose that reason is not as pure as most of us think it is or wish it were, that emotions and feelings may not be intruders in the bastion of reason at all; they may be enmeshed in its network, for worse and for better. ... It is thus even more surprising and novel that the absence of emotion and feeling is no less damaging, no less capable of compromising the rationality that make us distinctively human and allows us to decide in consonance with a sense of personal future, social convention, and moral principle. I suggest only that certain aspects of the process of emotion and feeling are indispensable for rationality."

J.E. LeDoux (1997): "Emotional and declarative memories are joined seamlessly in our conscious experience. That does not mean we have direct conscious access to our emotional memory."

In humans, the emotions cause changes in heart rate, breathing, and stomach and intestine contractions. However, the dominant role of the brain, both consciously and unconsciously, cannot be questioned. Of course, in spite of the significant role of the thalamus and other components of the limbic system, the areas controlling the emergence of emotions are the various circuits, zones, and pathways of the whole neuronal network.

One of the most significant emotions is fear. Some studies are addressing this problem by identifying specific brain processes that regulate fear and its associated behaviours. Such information is still extremely difficult to obtain in humans. Researchers are teasing apart the neuro-chemical processes that give rise to different fears in monkeys. The results may lead to new ways to treat anxiety in humans Opiates (morphinelike substances) and anti-anxiety drugs (Valium) act on neurons in the prefrontal cortex, amygdala and the hypothalamus. But how they might cooperate is unclear (N.H.Kalin, 1997). Fear-conditioning circuitry is essentially connected with amygdala.

Some studies implicate the amygdala in both fear conditioning and face perception. In humans the neural response in the left amygdala was significantly greater to fearful as opposed to happy expressions. Furthermore, this response showed a significant interaction with the intensity of emotion (increasing with increasing fearfulness, decreasing with increasing happiness). (J.S Morris et al,1996).

In fact, the higher one goes up the scale of biotic information processing, the more emotion is displayed. On the level of intelligence, the areas of the cerebral cortex (which controls intellectual activities) which are most active in generating emotions are the frontal lobes. The cognitive factor—the sense that one has some control of a situation—turns out to be significant, not only in emotional arousal, but also in the experience of pain, stress, aggression, etc.

The amygdala is a temporal lobe structure which is considered to be involved in the neural substrates of emotion. Bilateral damage results in impairment of social perception, including defective recognition of facial expressions of conveying negative emotions, especially of fear and anger. The amygdala's role in the recognition of certain emotions is not confined to vision (S.K. Scott et al, 1997).

Human beings are the most emotional creatures of all, with many highly differentiated emotional expressions and a wide variety of emotional experiences. The facial expression is common for all people in the world, independent of cultural and social differences, from the Oxford alumnus to the Bushman.

6.3.2 *Animal intelligence: yes or no?*

Having emphasized the mutual relationship between the lower stages of information processing (reflex, instinct and discent) and intelligence, two questions must be answered. First, can one apply the term 'intelligence', to animals which are at the lower stages of evolution of the nervous system? This applies primarily to primates. The second question concerns the problem of the members of genus Homo.

Each of us knows very well how 'intelligent' his or her favourite dog, cat or horse is, or from time to time can be. At the beginning of this century a number of scientific papers were devoted to 'der kluge Hans'—'the clever Hans', a circus horse able to perform some arithmetical operations, such as addition. Even seventy years later, a scientific conference in Zürich was devoted to the phenomenon of 'der kluge Hans'. This time the real problem was, how was it possible to lead many scientists astray at the beginning of the century, in spite of the existence of clearly formulated scientific methods? And how far can the co-operation between an animal and its master extend? Independent of this, is such a co-operation a conscious forgery, or unconscious and uncontrolled?

The principal differences between humans (intelligence) and primates (discent) is in the co-ordination of muscle movement, which is also needed for speaking abilities:

W.H.Calvin (1994): "Ballistic movement are extremely rapid actions of the limbs that once initiated, cannot be modified. Ape have only elementary forms of ballistic arm movements, at which humans are experts—hammering, clubbing and throwing. Ballistic movements require a surprising amount of planning. For sudden limb movements lasting less than one fifth of a second,... the brain has to determine every detail of the movement in advance."

D.Premack (1983), one of the exponents of the existence of 'speaking apes', formulated the following hypothesis. He proposed that reason, being the highest biotic information processing stage, contains the following components: 1) simple imagination 2) abstract representation 3) syntactic speech. All species attain the first level, primates the second, and only humans the third.

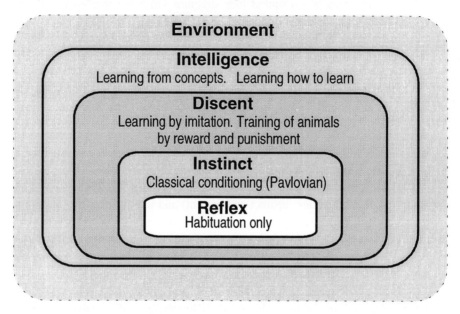

Figure 6-4 Levels of learning

N.Chomsky is of the opinion that a young chimpanzee can learn sign language as well as a human child can learn to fly, by straightening his arms. The ability to speak is decided at the genetic level (including so-called 'universal grammar', Section 6.5.4), and is only further developed at the epigenetic level. This genetically controlled ability exists at present only in Homo sapiens sapiens. Even if this viewpoint is extreme, the criticism of experiments in the field of chimpanzees' ability to speak is justified. However, even in our day, discussion concerning the existence of the ability of primates to learn sign language has not ended.

J. Lehtmate (1994) speaks about the 'intelligence' of orang-utans, making a remark that because of very limited social contact between adult individuals the emergence of 'intelligence' is handicapped. There are reports of monkeys (in South America) eating hallucinogenic plants (W.W.Gibbs, 1996).

In this book, we assume that even the highest level of primate information processing does not exceed the stage of discent. Animals do not have 'intellect' and therefore are not intelligent, in the strict sense of the term.

6.3.3 *Lost stage of evolution: Sub-intelligence or superdiscent?*

All the above-mentioned arguments cannot be used without deeper insight into the case of other members of the genus Homo, such as Homo erectus and, especially, Homo sapiens neanderthalensis. The reason for such caution can be seen from a simple analysis of the evolution of the higher stages of biotic information processing. Figure 6-8 shows a simplified picture of this evolution. It is evident that there exists a gap between discent, even at its highest level, and intelligence, at its lowest level. Is such a break in evolution possible, or is it caused by the elimination of some links? It may be that the extinct species of Homo erectus and the Neanderthals were carriers of the missing links between primate discent and present-day human intelligence. It is so easy to assume that Homo habilis had a 'superdiscent' and Homo erectus a 'sub-intelligence'. Neanderthals may have been representatives of 'proto-intelligence', or even of 'intelligence'. Why not? However, it is obvious that such a purely verbal classification, without hard facts, is worth very little.

6.4 LEARNING ABILITY AND UNDERSTANDING

6.4.1 *Learning ability*

Intelligence is the highest stage of biotic information processing and includes all the lower stages—discent, instinct, and reflex. The two lowest stages are almost entirely without the ability to learn. Discent is the stage which is characterized by the ability to learn, and to acquire and memorize new information and new skills. In particular, skills in the performance of body movements are very typical of discent. There is enormous experience in, and theoretical study of, the training of animals, especially domesticated ones, in motor skills using operand conditioning techniques (Table 6-2).

According to our definition, discent also acts in humans, but unconsciously. Some of our learning processes, for example those concerned with motor activity, are fully subconscious throughout. It is essential, when learning human skills, to have mental concentration, with a planned strategy of action and subsequent evaluation and correction of errors in successive attempts. When well learned, skills can be accomplished without voluntary attention, as in walking and swimming (by children), cycling, skiing, and other similar activities where the performance of the skilled action has become automatic. Today, the list of learned motor skills is enormous, and includes almost all kinds of sports, as well as such artistic activities as ballet and music and, last but not least, the driving of automobiles.

It is generally agreed that the cerebral cortex is the main organ involved in human learning, especially in the first phases, encompassing motivation, attention, evaluation, and correction. However, the performance of automatic motor activities is controlled mainly by the cerebellum, and the basal ganglia. Both belong to the oldest components of the human brain (Section 4.1.3).

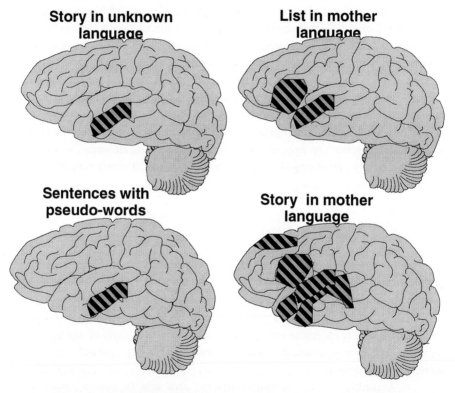

Figure 6-5 Reading known and unknown words
(accord. to M.Posner, M.Raichle, 1994)

Learning changes the pathways for performing the task of remembering. A clear example is presented by M.Posner and M.Raichle (1994) in the case of reading a word (Fig. 6.5). During the early development, the speed and accuracy with which an organism extracts environmental information can be extremely important for its survival. Many aspects of a particular natural language must be acquired from listening experience. Long before infants begin to produce their native language, they acquire information about its sound properties.

Young humans are generally viewed as poor learners, suggesting that innate factors are primary responsible for the acquisition of language. The results of experiments with 8-month-old infants raise the possibility that infants posses experience-dependent mechanisms that may be powerful enough to support not only word segmentation but also the acquisition of other aspects of language. If this is the case, then the massive amount of experience gathered by infants during the first postnatal year may play a far greater role in development than has previously been recognized (Saffran JR et al, 1996).

According to G.M.Edelman (1989), 'primary consciousness', which in this book is called 'preconsciousness', requires the following: 1) the formation of percepts, mostly by selective systems in secondary areas of the cerebral cortex, 2) memory, connected with the developed hedonic system, 3) learning abilities, and 4) the distinction between 'own body' and 'world outside'. All these functions, in this book, belong to the third stage of biotic information processing: discent. 'Higher-order consciousness', called here 'consciousness', contains the following components: 1) the formation of concepts, which is linked to the linguistic system, 2) memory of semantic and syntactic properties, connected with a very highly developed hedonistic system, 3) the ability to learn how to learn, and 4) the distinction between 'self' and the rest of the world—the 'non- self'.

6.4.2   Semantics; what do we understand as 'understanding'?

One of the results of the learning process, especially of the learning of knowledge, is the emerging feeling of 'understanding', which doubtless has a relatively high bonus label.

One of the most significant, but probably not very exact, terms used in every discussion concerning human intelligence is the term 'understanding'. What do we understand when we say, 'I have understood'? Is this equivalent to saying that we know, consider or accept as a fact, truth or principle, without further mention or explanation, and have achieved a mental grasp of the nature of something, in connection with and in relation to information acquired beforehand?

The last condition expresses the fact that the process of understanding depends very strongly upon the current state of knowledge of the individual, from her or his cultural background. If the information is presented by another individual, then the state of understanding depends upon the way it is communicated.

A.Einstein (1937): "The most difficult thing is to understand why we understand at all."

Table 6-2 Semantics-learning about the real world

Stage of biotic information processing	Object of learning	Relation of the central nervous system to reality
Reflex	Nothing new can be learned	All vital information must be stored genetically
Instinct	Very little new can be learned	Some defined things. Successful also in changing
Discent	Able to be learned by the individual-trained (dressage)	Very much is learnable. Successful in a fast changing
Intelligence	Able to create new information. Limitation is Gödel´s incompleteness	Ability to create scientific systems falsified by experiments

In one of the novels of Thomas Mann, a magician in a strolling circus takes a rabbit out of his top hat. To a question as to how he did this, his answer was, 'It is black magic'. One young spectator then said, 'Oh, now I understand!' Perhaps human understanding is in some sense related to black magic.

R.Penrose (1987): "Understanding is not just a question of carrying out some appropriate algorithm —and still less is it merely a matter of the output of an algorithm. 'Understanding' is not just a question of software (that is, of programs and algorithms), but the 'hardware' should also be important. By 'hardware' is meant the actual physical construction of the object, either computer or brain. I do believe that the nature of the hardware is likely to be vitally important."

6.4.3 *Learning and the bonus-malus system*

Learning, memorizing and the bonus-malus system are intimately linked (Section 5.3.2). Learning is the experience-dependent generation or modification of enduring internal representations of the external real world, or in some cases of an ideal, imaginary realm.

Learning—that is, the technique for building some new information into long-term, even lifelong, memory—is common to all animals at the stage of discent, and of course for animals at the stage of intelligence.

G.M.Edelman (1989): "Learning must be related to evolved species-specific hedonic, consummatory, appetitive, and pain- avoiding behaviours that reflect ethologically determined values."

We should not neglect the fact that our sensors, the primary receptors of signals from the real world and from our own bodies, which send corresponding signals to the brain, are continuously controlled, influenced, and guided by the same brain. Incoming information from he real world is transformed and censored by the brain. Our internal representation of the real world is clearly extensively transformed, but in spite of this the correspondence between the internal representation and the real world is adequate enough to give us the chance of survival

6.4.4 Highest complexity of the bonus-malus system

The highest stage of biotic information requires the most complex and most extended bonus-malus system. Our hypothesis of the bonus-malus system must include a number of behaviours, emotions, feelings, concepts etc. which are characteristic of intelligence. Some of the phenomena associated with intelligence are listed below, but discussed in detail in following Sections: –Self (Section 8.1.1) –Selfconsciousness (Section 8.1.2) –Taboo (Section 7.3.4) –Euphoria (Section 7.5.2) –Feeling of free will (Section 8.3) –Ethical beliefs (Section 8.4) –Moral considerations, altruism (Section 8.4.3) –Feeling of happiness.

The lower stages of biotic information processing (reflex, instinct and discent), with their specific behaviour repertoires, have relatively high corresponding bonus-malus values. The behaviour characteristic for intelligence shows a lower absolute value. However, this statement is extremely simplified. It is in the nature of intelligence that, in some situations, and in some individuals, some feelings and emotions, or ideas and hopes may receive an extremely high bonus-malus label value, higher than the value currently covered upon more primitive behaviour states, such as hunger or even fear of death.

S. Kosslyn (1997): "Emotion apparently is not something that necesserily clouds reasoning, but rather seems to provide an essential foundation for at least some kinds of reasoning."

Of course, one could ask, why we should discuss the question of absolute values in the bonus-malus system at all, if in most cases this value changes so significantly that the whole scale breaks down. However, on the other hand, the search for a better understanding of this quality scale is one possible way for humans to obtain greater understanding of themselves.

Feeling of happiness is recently also an object of scientific research (D.G. Myers, E.Diener, 1997). Happiness and satisfaction with life—or what psychologists call 'subjective well-being'. Researches use various methods to survey people's subjective sense of well-being. Some formulate following sentences: "In most ways my life is close to my ideal", "The conditions of my life are excellent", "I am satisfied with my life", "If I could live my life over, I would change almost nothing" and ask for answer in following form: "Do you strongly disagree, disagree, slightly disagree, neither agree or disagree, slightly agree, agree, strongly agree". In some researches in USA, 1989 around 80 per cent of women and men have been 'satisfied' and about 20 per cent 'very happy'. And this was valid for all age groups between 15 and over 65 years. Mostly the 'happy' people are optimistic.

There is little doubt that the moment state of the individual bonus-malus system, which so significantly influences the information processing, could be described by only one 'quotient' called 'happiness'. Even the most abstract reasoning is labelled by the subjective feeling of happiness or unhappiness.

6.5 SPEECH, THE MOST IMPORTANT MEDIUM OF INTELLIGENCE

6.5.1 *Four stages in the evolution of communication*

The four stages of biotic information processing—reflex, instinct, discent and intelligence—differ in the means of communication between individuals. On the one hand we can differentiate between various media and 'messengers', such as mechanical, tactile stimuli (direct touch), chemical stimuli (body smell, and special chemical substances called 'pheromones'), visual stimuli (body movement, gesticulation, facial expression) and acoustic stimuli (grasshoppers´ 'music', birdsong), and the highest: the human speech. Vagueness in the classification of human communication is mirrored in the terms used to describe it (Fig 8-6).

On the other hand, it is of interest to investigate the evolution of each of the various ways of communication during the emergence of the higher stage of biotic information processing. Along the evolutionary path, the amount of information processed in the mechanical, chemical, visual, and acoustic media increases dramatically. Human speech seems to be so rich that it can be treated as an open-ended information system.

Our presentation of the communication media is obviously very imprecise. The origins of these communication media are different. The following are exclusively of genetic origin: body expressions, some acoustic signals, including cries of pain and a baby's weeping and babbling, and some chemical signals, such as bodily odours. Other media are of cultural origin: facial expressions, at least in some more sophisticated cases, perfume, as a chemical signal, and of course, primarily human speech. The differentiation between genetically and culturally derived means of communication is not specifically represented in Figure 6-7.

D.C. Dennett 1994): "We are without any doubt at all the most intelligent (species on the planet). We are also the only species with language. What is the relation between those two obvious facts?"

S.Pinker (1994): "Language is not is not a cultural artifact....is a complex, specialized skill, which develops in the child spontaneously, without conscious effort or formal instruction... I prefer the admittedly quaint term 'instinct'. It conveys the idea that people know how to talk in more or less the sense that spiders know how to spin webs. Spiders spin spider webs because they have spider brains. Although there are differences between webs and words. I think it is fruitful to consider language as an evolutionary adaptation, like the eye. If language is an instinct, it should have an identifiable seat in the brain, and perhaps even a special set of genes that help wire it into place."

Tools of intelligence 223

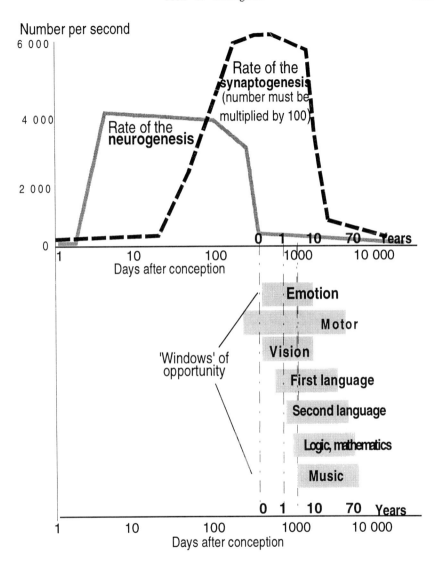

Figure 6-6 Windows of opportunity

Table 6-3 Four stages of language

	First stage	Second stage	Third stage	Fourth stage
Language (accord. to J.Diamond, 1995)	**Velvet** language includes at least ten different grunts, including separate 'words'	**Pidgin** language consists of strings of words, with little grammar or phrase construction and composed mainly just of nouns, verbs, and adjectives. Remark: pidgin emerges due to contact of two languages	**Creoles** are fully expressive languages with newly invented grammars, and with the modular hierarchical organization, but lack verb conjugations for person and tense, noun declinations for case and number	**Modern human speech**, including technologically primitive peoples
Human evolution (J.Diamond)	**Monkey**	**Homo erectus, Neanderthals** (proto-language)	**Homo sapiens**	**Homo sapiens sapiens**
Human children	Children: **7-9 months**	Children: **1 year** (Table 7-2)	Children **>1.5 year**	Children **3 years**
Speech in this book (Fig. 7-4)	**Rudimentary speech:** Homo habilis	**Sub-speech:** Homo erectus	**Proto-speech:** Neanderthals	**Modern speech:** Homo sap. sapiens

Proto-language is a collection of words that may be juxtaposed, but which does not create propositions in their juxtaposition. It is like the utterance of a child in the earliest stages of language development. It is like a 'pidgin', which occurs when two groups which do not share a common tongue come in contact. Sub-language (in every-day called 'Creole') develops in the mind of a child (Homo sapiens sapiens) learning a language for the first time who is exposed to a proto-language (Pidgin). The child will hypothesise a syntax, then use it to produce novel constructions.

Language thus constitutes a form of extrasomatic memory, a medium of information storage that exists outside any individual´s brain and which survives any individual´s death. Language makes it possible, at any time, for human cognition to be collective. Language allows us to transcend our individual cognitive weakness and to conjoin our individual strengths (P.M. Churchland, 1995).

Why the human speech plays so important role in the emergence of Homo? The relatively fast reaching of every corner of the planet based sending scouting parties ahead to report on the best routes and a sophisticated information about the new territories, its climate, plants, animals, possible danger. Certainly, most of reports of the individuals possessing a strong spirit of adventure included at least some phantasmagoria (stories about the lost paradise). Speech is not only a excellent medium for transportation of information between individuals but because is the best tool for deception, for lying, for transportation of intentional false information (Section 7.4.4)

The Italian proverb *Se non e' vero, e' ben trovato*, (If it is not true, it has been found well) explains this property of human speech in an excellent way.

6.5.2 Animal communication; some remarks

Who does not know the astonishing details of the 'language' of honeybees? Who has not heard the fine poetry of a singing nightingale? And how often have we been informed about 'speaking chimpanzees'?

The terms used to describe the communication between animals of the same species, and especially the communication between chimpanzees and humans, are far from being well defined, and are often deliberately falsely used, to create a more spectacular impression on the layman.

F.Nottebohm (1989) claimed that in his favourite object of investigation, the canary, instinct is responsible only for the primitive form of song, which is innate (that is, on the stage of instinct) to young canaries, who have not heard the full song of an adult bird. The centre of this skill is in the left hemisphere of the brain. However, the full canary's song can be learned only by hearing the songs of an adult bird (that is, on the stage of discent). In round 4 thousand species of songbirds (total number of living birds species: 9,000) young birds learn vocalizations by imitating the signals of adult conspecifics. In absence of adult vocal models, these birds never develop normal vocalizations.

In the 1960s and 1970s, an impressive number of experiments, supported by government funding, were performed in first-class scientific centres to investigate 'animal intelligence', including the speaking ability of apes and especially of chimpanzees.

Some young chimpanzees, were the objects of experiments to learn the deaf-and-dumb sign language. Numerous papers, published by excellent scientific journals, claimed to have achieved positive results in teaching the human sign language to the young chimpanzee and achieving a state of mutual communication between teacher and animal. Critics stated that apes were doing little more than mimicking their instructors, a skill that, in different ways, rats and pigeons acquire with ease. The apes displayed no sense of grammatical structure in their use of symbols. The ape-language researchers were victims of self-delusion.

W.H.Calvin (1994): "Our closest animal cousins, the common chimpanzee and the bonobo (pygmy chimpanzee) can achieve surprising levels of language comprehension, when motivated by skilled teachers. Kanzi, the most accomplished bonobo, can interpret sentences he has never heard before... Neither Kanzi nor the child constructs such (complex) sentences independently, but each can demonstrate understanding."

From our viewpoint, all these facts can be summarized in one short and unequivocal statement: Chimpanzees, not to mention other primates, are principally not able to learn to speak, or even to learn the deaf-and-dumb sign language. Therefore we must classify primates as belonging to discent—the third stage of information processing—and not to intelligence—the fourth and highest stage.

Animals, including primates, and even the famous chimpanzee 'Neam Chimpsky', also cannot learn sign language, because of the lack of the appropriate cerebral area, and the corresponding genes.

The vocal calls of primates are controlled not by their cerebral cortex but by phylogenetically older neural structures in the brain stem and limbic system, structures that are heavily involved in emotion. In monkey brains exist areas that correspond to the human speech areas. But these regions corresponding to human Wernicke´s area are not involved in producing monkey´s calls, nor are they involved in producing their gestures. The monkey uses the regions to recognize sound sequences and to discriminate the calls of other monkeys from its own call. The areas corresponding to the human Broca´s area are involved in controls over the muscle of the face, mouth, tongue, and larynx.

Human vocalizations other than language, like sobbing laughing, moaning, and shouting in pain, are also controlled subcortically. Human speech is controlled by cortex cerebri.

6.5.3 Levels of speech development

Human speech, ultimately the most significant and unique property of human intelligence, is based on some processes which are doubtless present in reflex, instinct and discent (Fig. 6-7, Fig 6-8).

The origin of human speech is a topic that has a bad reputation among linguists. After the publication of Darwin´s 'Origin of species' (1858), many uncritical ideas about the evolution of language were proposed, to such an extent that, in 1866, the French Academy of Linguistics announced that its journal did not accept papers on the origin of language. In any event, the problem of origin and nature of human speech is an indivisible part of the problem of intelligence.

It is possible that all modern languages develop from a single ancestral language 'mother tongue', which have been spoken 100 thousand years ago (L.L.Ca-valli-Sforza, 1995). It must be remembered that 100 thousand years ago the population of Homo sapiens sapiens was probably only 10,000 individuals ('population bottleneck') (Section 3.5.6).

Tools of intelligence 227

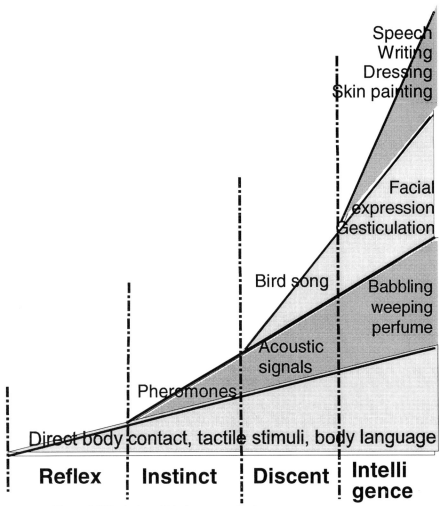

Figure 6-7 Evolution of biotic communication

The daily working vocabulary of the average American and Englishman includes 600 words only. The average high school graduate (seventeen years old) uses around 60,000 words, including proper names, numbers, foreign words, acronyms. Is 60,000 words a lot or a little? In the Bible they are only 7,706 words, which covered an enormous bandwidth of meanings. Fuzziness of these texts comes also from the lack of vocals and words-spaces in old Hebraic and Old-Greek.

According to Chomsky a visiting Martian scientist would surely conclude that aside from their unintelligible vocabularies Earthlings speak a single language.

In some linguistic hypotheses the around 5,000 terrestrial languages could be merged to only six groups: 1) Scan (including 'Nostratic' and 'Amerind') in Eurasia, the Americas and northern Africa, 2) 'Khoisan' and 3) Congo-Sahan in sub-Saharan Africa, 4) Austric in South-east Asia and the Indian and Pacific Oceans, 5) Australian and 6) New Guinean. Possibly as many as 90% of the world's languages are threatened with extinction in the next century. Not only due to bombardment by electronic media ('cultural nerve gas'). The few remaining languages may only reflect historical accident rather than revealing anything of importance about the linguistic potential of the human brain.

A.R.Damasio and H.Damasio (1992) believe the brain processes language by means of three interacting sets of structures. First, a large collection of neural systems in both the right and the left cerebral hemispheres represents non-language interactions between the body and its environment, as mediated by varied sensory and motor systems—that is to say, anything that a person does, perceives, thinks or feels while acting in the world. The brain not only categorizes these non-language representations, it also creates a higher level of representation. Successive layers of categories and symbolic representations form the basis for abstraction and metaphor.

Second, a smaller number of neural systems, generally located in the left cerebral hemisphere, represents phonemes, phoneme combinations and syntactic rules for combining words. When stimulated from within the brain, these systems assemble word-forms and generate sentences to be spoken or written. When stimulated externally by speech or text, they perform the initial processing of auditory or visual language signals.

A third set of structures, also located largely in the left hemisphere, mediates between the first two. It can take a concept and stimulate the production of word-forms, or it can receive words and cause the brain to evoke the corresponding concepts.

In spite of the risk of being wrong, we will try to classify different components of the human communication system according to the four stages of information processing:

–Human reflex: control of vocal apparatus at the lowest stage, cry,

–Human instinct: baby's weeping and babbling,

–Human discent: learning words and sentences of the mother tongue by imitation; learning of semantics through contact with the environment,

–Intelligence: conscious formulation of concepts and sentences, emergence of 'Self'', conscious production of lies, deceptions and false information, learning of another language, learning of synthetic languages, such as a computer language, or the language of mathematics.

Some 'futurists' claim the possibility or even the need for the direct exchange of thoughts between the brains of partners. This is evidently a contra-social idea because of the notorious use of social deception by man.

F.Crick (1994): "The understanding of the evolution of language will not come only from what the linguists are doing, but from finding how language develops in the brain and then finding the genes for it and trying to work out when those genes come on evolution."

If, as many molecular biologists believe, modern Homo sapiens arose as a discrete evolutionary event some 120 to 200 thousand years ago (the so-called hypothesis of 'mitochondrial Eve', the mother of us all who lived in Africa), monogenesis of language is likely. The role of the older sub-species, the Homo sapiens neanderthalensis, in the evolution of speech remains an open question.

Table 6-4 Components of human speech

Phoneme	The individual sound units
Morpheme	The smallest meaningful units of words
Word	Combination of morphemes
Syntax (Grammar)	The rule of admissible combination of words
Phrase (Sentence)	The admissible combinations of words into sentences
Lexicon	The collection of all words in given language
Semantics	The meanings that corresponds to words and sentences
Prosody	The vocal intonation that modify the meaning of words

Some fundamental principles of language are already imbedded in the human brain genetically. All humans are born with essentially the same neuronal network, but the switches flip over into different positions—corresponding to different rules of grammar—depending on whether a child learns Japanese, or the !Kung-bushmen's language, or English.

Some students of the fossil endocranial casts claim that Homo habilis may have been capable of rudimentary speech. P.Lieberman (1991) claims that the extinction of Homo neanderthalensis may be based on missing 'human speech'. Neanderthals had a nonhuman supra-laryngal vocal tract, although they walked perfectly upright and had brains that were as large as or larger than ours. At least they would have had less efficient vocal communication—more confusable speech, and perhaps a very slow rate (and, last but not least, with less intentional false information).

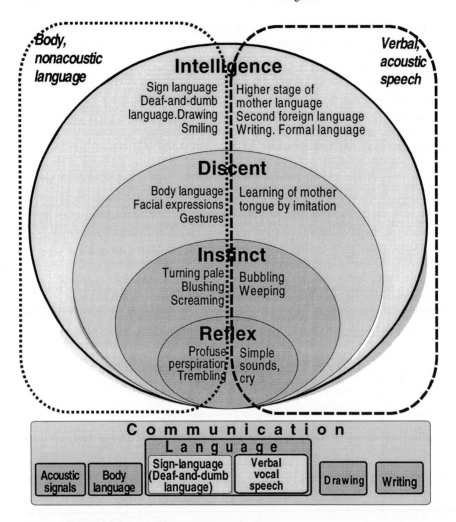

Figure 6-8 Stages of human communication

S.Pinker (1994): "Broca's aphasia and Specific Language Impairment are cases where language is impaired and the rest of intelligence seems more or less intact. But this does not show that language is separate from intelligence. Perhaps language imposes greater demands on the brain than any other problem the mind does to solve. .. For language, all systems have to be one hundred per cent. (On the other hand) fluent grammatical language can in fact appear in many kinds of people with severe intellectual impairments, like schizophrenics, Alzheimer's patients, some autistic children, and some aphasics. Or are our thoughts couched in some silent medium of the brain—a language of thought, or 'mentalese'—and merely clothed in words whenever we need to communicate them to a listener? No question could be more central to understanding the language instinct. Many creative people insist that in their most inspired moments they think not in words but in mental images. People do not think in English or Chinese or Apache; they think in a language of thought. This language of thought probably looks like all these languages; presumably it has symbols for concepts, and arrangements of symbols.... But compared with any given language, mentalese must be richer in some way and simpler in others. On other hand, mentalese must be simpler than spoken language; conversation —specific words and constructions (like *a* and *the*) are absent, and information about pronouncing words, or even ordering them, is unnecessary. Knowing a language, then, is knowing how to translate mentalese into strings of words and vice versa. People without a language would still have mentalese, and babies and many nonhuman animals presumably have simpler dialects."

According to Lieberman the presence of a functionally modern human vocal tract 125,000 years ago and its subsequent retention and elaboration are consistent with the presence in this period of brain mechanisms allowing automatized speech motor activity, vocal tract normalization, and the decoding of encoded speech.

Characteristic of fully developed human language is its symbolic nature, consisting of arbitrary symbols that designate other objects or actions. Animal communication lacks the symbolic potential that underlies human language.

As it is only about 40,000 years ago, that forms of repeated iconic structures (engravings, sculptures and ultimately cave painters) appear, it is the only true evidence for existence of language. Before that, all artifacts can only be described as having forms closely related to their functions, and therefore lacking the key criterion of symbolic content. But it does not imply that early hominids (Homo erectus) were not capable of flexibility, cognitive sophistication or abstract and symbolic system of thought, or indeed that they did not have some kind of sub-speech. It just means they kept their thoughts to themselves.

N.Chomsky (1983): "Language depends on a set of genes that are on a par with the ones that specify the structure of our eyes or circulatory system, or determine that we have arms instead of wings. The gene-control problem is conceptually similar to the problem of accounting for language growth. In fact, language development really ought to be called language growth, because the language organ grows like any other body organ."

Here it is postulated that Homo sapiens neanderthalensis reached the level of so-called 'proto-speech' (Fig. 6-9). It must be stressed that this hypothesis is far from being established, but seems to be useful in our considerations. The emergence of fully developed speech is, according to this hypothesis, linked to the emergence of lateralization of the brain's hemispheres.

K.Popper (1980) believes that humans are able to create themselves, through creation of speech. Speech is the first product of the human mind. It is the first 'ideal artifact'. Human language is like other tools, in that it develops outside the human body.

S.Pinker (1994): "Why does the sequence have to take three years? Could it be any faster? Perhaps not. Complicated machines take time to assemble, and human infants may be expelled from the womb before their brains are complete. A human, after all, is an animal with a ludicrously large head, and a woman pelvis, through which it must pass, can be only so big. If human beings stayed in the womb for the proportion of their life cycle that we would expect based on extrapolation from other primates, they would be born at the age of eighteen months. That is the age at which babies in fact begin to put words together. In one sense, then babies are born talking!"

Using speech in a most efficient and most rational way must be only in knowledge of the significant weakness of human speech: the genuine semantic uncertainty of most words, the vagueness of syntax of all natural languages, the ambivalence and ambiguity of sentences, the intentional deception of most social partners (including children, parents, husband, etc.).

6.5.4 Universal grammar: some doubts

The concept of the existence of a special brain organ devoted to the uniquely human ability of managing speech, primarily controlled by genetics, was formulated by N.Chomsky, and in the fifties and sixties initiated very fruitful discussions and research.

N.Chomsky (1969): "What is universal grammar? It is the sum total of all the immutable principles that heredity builds into the language organ. These principles cover: grammar, speech sounds, and meaning. Put differently, universal grammar is the inherited genetic endowment that makes it possible for us to speak and learn human language. If a Martian landed from outer space and spoke a language that violated universal grammar (Author's note: terrestrial 'universal grammar'), we simply would not be able to learn that language the way that we learn a human language like English or Swahili. We would have to approach the alien's language slowly and laboriously, the way that scientists study physics."

The idea of a 'universal grammar' has recently been strongly critisized.

G.M.Edelman (1989): "A basic distinction must be made between the present project (of G.M.Edelman) and proposals by Chomsky of a universal grammar embodied as a set of rules in a brain module whose functioning is genetically predetermined. Broca's and Wernicke's areas, which are evolutionary adaptations unique to language, are not themselves sufficient for the realization of meaningful speech. Language 'centres' considered as little microcosms containing structures for a universal grammar do not exist in the brain."

The question can be raised of why are the evolutionary adaptations which Edelman requires—both in the left hemisphere, when almost all other specific areas are symmetrically distributed? The answer has been given by the emergence of lateralization in the human brain (Section 5.5.2).

L.L.Cavalli-Sforza (1991): "It is perhaps surprising that so much of expected correlation between languages and genes remains, despite the blurring caused by gene or language replacements. Genes, always transmitted from parents to children, describe a vertical path through the generations. Culture can also pass vertically from generation to generation, but unlike genes, it can also be transmitted horizontally, between unrelated individuals."

M.S.Seidenberg, (1997): "Modern thinking about language has been dominated by the views of Noam Chomsky, who created the generative paradigm. ....Linguists equate knowing a language with knowing a grammar.. and that acquisition (of language) is possible only because children posses innate knowledge of grammatical structure. An alternative view... emphasizes continuity between how language is acquired and how it is used. It retains the idea that innate capacities constrain language learning, but calls into question whether they include knowledge of grammatical structure. ... This approach does not deny that children are born with capacities that make language learning possible; rather, it questions whether these capacities include knowledge of linguistic universals per se. Innate capacities may take the form of biases or sensitivities toward particular types of information inherent in environment events such as language, rather than a priori knowledge of grammar itself."

Table 6-5 Development of children's language (partially according to S.Pinker, 1994)

Age of children	Vocal activity
0-2 months	Producing cries, grunts, sighs, licks and pops
2-5 months	Additional: coos and laughs
5-7 months	Babies begin to play with sounds, rather than using them to express their physical and emotional states and their sequences of clicks, hums, glides, trills, and smacks begin to sound like consonants and vowels
7-9 months	Baby suddenly begin to babble in real syllables like *ba-ba-ba, neh-neh-neh,* and *dee-dee-dee.* The sounds are the same in all languages
9-12 months	Babies vary their syllables, like *neh-nee,* and *meh-neh,* and produce that really cute sentence-like gibberish
1 year	Babies begin to understand words, and around that birthday, they start to produce them. Words are produced in isolation
1,5 year	Language take-off. Vocabulary growth jumps to new-word-every-two hours minimum rate. Syntax begins, with strings of the minimum length: two
3 years	Child is a 'grammatical genius'—master of most constructions, obeying rules far more than flouting them. Child, parallel to 'grammatical genius' is 'liar genius', often with intention to win a desirable object or parents action
3-6 years	In some areas exist: the phenomena of 'Wunderkind' (infant prodigious especially in music , mathematics, and chess. Why only in these fields of human intellectual activity? (Remark: maybe all these areas belong to a 'jigsaw puzzle world' that is a world with well defined limits and rules)

6.5.5 *How a child learns to speak*

An infant's babbling is a clear example of the strategy of overabundant production and subsequent selective elimination. The number of sounds a baby makes is very large and is, it seems, very similar in different nations and cultures, being controlled by a genetically formed vocal mechanism. An infant which imitates the word 'mama' without understanding its meaning is not engaging in true speech. Only the influence of the mother ('mother tongue') enables the selective elimination of many of the overabundant, sounds and the reinforcing of those positively selected. Most children say their first words at about 6 to 8 months, children continue to acquire single words in a fairly slow fashion until they have acquired approximately 50 words, then their vocabularies often increase rapidly in size as they acquire between 7 and 9 words a day from ages two to six years.

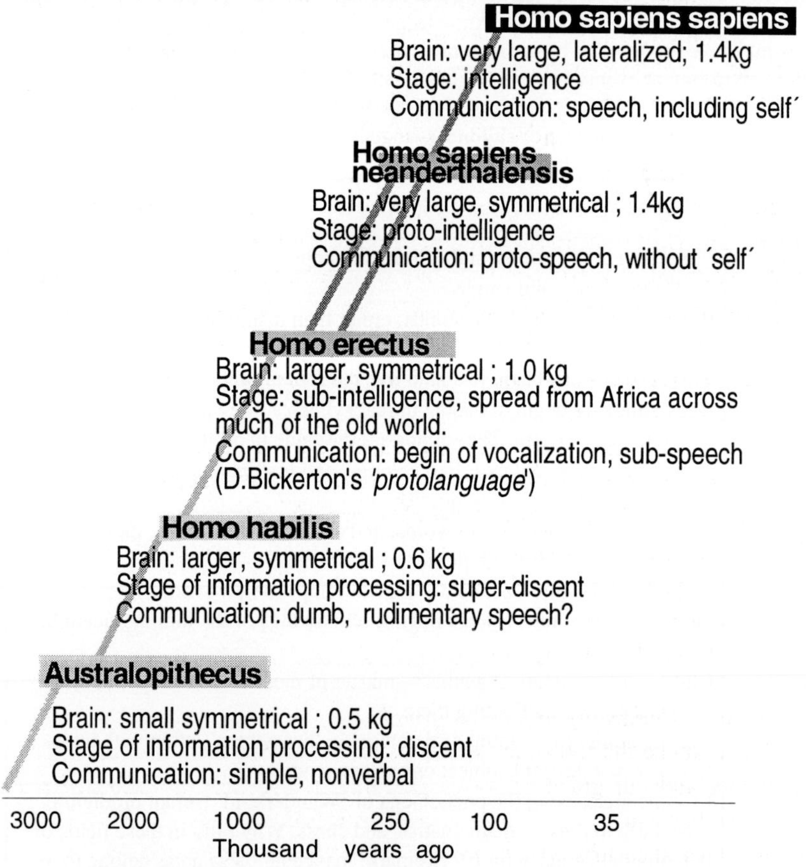

Homo sapiens sapiens
Brain: very large, lateralized; 1.4kg
Stage: intelligence
Communication: speech, including 'self'

Homo sapiens neanderthalensis
Brain: very large, symmetrical ; 1.4kg
Stage: proto-intelligence
Communication: proto-speech, without 'self'

Homo erectus
Brain: larger, symmetrical ; 1.0 kg
Stage: sub-intelligence, spread from Africa across much of the old world.
Communication: begin of vocalization, sub-speech (D.Bickerton's *protolanguage*')

Homo habilis
Brain: larger, symmetrical ; 0.6 kg
Stage of information processing: super-discent
Communication: dumb, rudimentary speech?

Australopithecus
Brain: small symmetrical ; 0.5 kg
Stage of information processing: discent
Communication: simple, nonverbal

| 3000 | 2000 | 1000 | 250 | 100 | 35 |

Thousand years ago

Figure 6-9 Human speech—evolution

In order to learn the sounds in a linguistic system, the child must inhibit many of the sounds found in babbling, as they are later not required in the child's speech. Later, the child learns many words and, by the age of between two and six, oral speech is a major task, involving both expression and comprehension. By about the age of four he or she has mastered the fundamentals of the systematic grammar of his language (K. Stromswold, 1995).

Young children talk to themselves—sometimes as much or even more than they talk to other people. This behaviour is called 'private speech' (L.E.Berk, 1994). In fact, private speech is an essential part of cognitive development for all children and is a healthy, adaptive and essential behaviour.

By the age of six, the average child has increased his vocabulary to about 2500 words, learning 2 to 3 words per day throughout the preceding three years. A 19-year old college student knows about 40,000 words. Including all idioms, names, etc. this number must be almost doubled. This means that for 16 years, beginning at the age of three around ten words per day must be learned.

During the whole life around 50 million of spoken words are processed by a typical human: 70 years with 365 days per year with 2-3 hours per day and 3600 seconds per hour with rate 1 word per 3 seconds.

Manual babbling has now been reported to occur in deaf children exposed to sign languages from birth. The similarities between manual and vocal babbling suggest that babbling is a product of brain-based language capacity, in which phonetic and syllabic units are produced by the infant as a first step toward building a linguistic system. Babbling is tied to the abstract linguistic structure of language and to an expressive capacity capable of processing both types of signals, signed and spoken (L.A.Petitto, 1991).

It seems that astonishingly little attention has been dedicated to the problem of the children's storyteller, to the phenomena of the 'spontaneous ability' to lie, to 'fantasy' which is coupled with the clear intention to achieve a desired thing or action by the child. This phenomenon of lie is obviously an integral component of human language and occurs relatively early in the evolution of personality. How far this is connected with later 'creativity' and 'fantasy' or with social activity should be an object of further investigations

6.5.6 *Continuity of human speech*

Let us make the following mental experiment. We invite for lunch our father or mother and our grandmother or grandfather. Mentally we also invite our great-grandmother and so on. The number of generations increases. After 10 generations we have ancestors who lived approximately 250 years ago. We keep on going further back, until we get to our thousandth ancestor, who lived around 25,000 years ago.

We all take our seats along one very long table and begin our meal. Don't forget: our great-great-great-grandparents were much smaller than we are. We joke with one of our parents; the joke must be simple and concerning the simplest things, for example something about our son. This joke is repeated to the next partner, to the grandparent. He, or she, tells this joke to his or her neighbour and so on. There are no linguistic problems in doing this. The premise is only that the joke is simple enough. After 1000 steps the joke arrives at the oldest of our ancestors and probably all around the table are laughing.

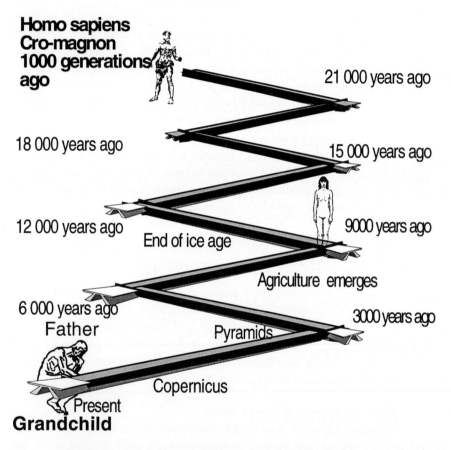

Figure 6-10 A thousand past generations at one table

It is worth-while to remember that this 'language proximity' corresponds to sexual proximity, sexual breeding, between modern humans and Cro-Magnon's. It remains an embarrassing problem of possible slavery and other forms of discrimination when 'races' have the chance to live in a common neighbourhood. The other question is the reason for the full extinction of Neanderthals 35 000 years ago.

This simplified picture of the thousand generations around one table does not obscure the reality that, at the present time, there would exist some thousand different languages and dialects. At the beginning of this table would exist a language which, although totally different from any known to us, could be analysed by linguists on the same basis as other human languages, and translated into all known languages.

Many words in a natural language have significantly different meanings for different speakers, and for the same speaker on different occasions. And this imprecision (or richness) of natural languages is ineradicable. However, the enormous difference in the cultural and social environments of our ancestors and ourselves must influence the semantic meaning of words, and therefore cause difficulty during translation.

A similar 'experiment' was realized in this century when linguists and anthropologists, had for the first time, contact with some African tribes. Anthropologists believe that the culture of the !Kung San in the Kalahari Desert, has evolved relatively little during the past 25,000–40,000 years.

From the point of view expressed in this book, we postulate that our ancestors, at least over the past 25,000 years, have had an intelligence level similar to our own. No qualitative jump is evident. However, if we go back in time and invite our ancestors from not just 25,000 years ago (1000 generations) but, let us say, 100,000 years ago (individuals of Homo sapiens neanderthalensis), or even 500,000 years (individuals of Homo erectus), then the situation is different.

We are forced to the conclusion that such distant ancestors are at a lower stage of information processing—called 'subintelligence' for Neanderthals and 'proto-intelligence' for Homo erectus.

6.6 SPEECH AND ABSTRACTS

6.6.1 *Percepts, concepts, etc.*

J-P. Changeux (1985): "The mental object is identified as the physical state created by correlated, transient activity, both electrical and chemical, in a large population or 'assembly' of neurons in specific cortical areas. ..The primary percept is a labile mental object, whose graph and activity are determined by and dependent on interaction with the outside world. Percept is experienced in the presence of the object in the outside world. The formation of a global percept that gathers all these (different) features together is thus the result of activity in several secondary areas. Various maps contribute to a single percept. The mental image is an autonomous and transient memory object, not requiring direct interaction with the environment. Its autonomy can be conceived only if there is a temporally stable coupling between neurons of the graph, which exists as a material trace before the image is evoked. Mental images arise spontaneously and voluntarily in the physical absence of the original object. The concept is, like the image, a memory object, but it contains only a small sensory component or even none at all, because it is the result of neuronal activity in association areas such as the frontal lobe or in a large number of areas in different regions of the brain. The concept is a simplified, formalized 'skeletal' representation of the object in question, reduced to its essentials. Even if, in certain cases, concepts may seem totally abstract and universal, we nonetheless regard them as 'representations' and classify them in the category of mental objects. The particular mental objects, or pre-representations, exist before the interaction with the outside world. They are labile and transient. Only a few of them are stored. This storage results from a selection. Selection follows from a test of reality. The test of reality consists of the comparison of a percept with a pre-representation."

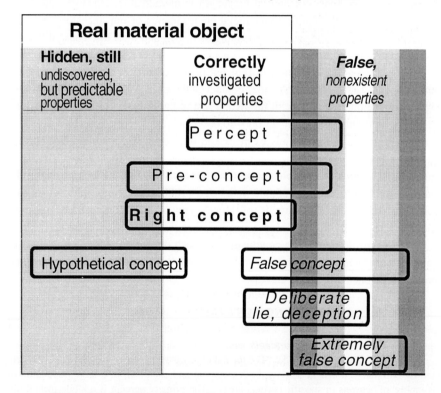

Figure 6-11 Material objects and true and false concepts

Intelligence, and especially human speech, are the most important carriers of social communication. There is no place here to discuss the extent of these phenomena, but it is worth noting that, if linguistic abilities are associated with the left hemisphere (Section 4.1.8) the fundamental aspects of such behaviour as social communication are in some way related to the right hemisphere. This is evidently caused by the role that the right cerebral hemisphere plays in perceiving emotions expressed on the face of the social partner.

S Freud (1930): "Normal forgetting takes place by way of condensation. In this way it becomes the basis for the formation of concepts."

One of the most important properties of language is the ability to report the state of the 'internal subjective life' of an individual.

G.M.Edelman (1989): "Concepts are difficult to define, and their existence and neural bases must be inferred indirectly from experimental evidence. Moreover, one is constantly tempted to consider concepts as properties of language. But this is not so—a good case can be made that animals without true linguistic abilities, such as chimpanzees, have concepts and that concepts are acquired prior to language. It is important to resist the temptation to think that concepts are merely mental images or (even worse) that they themselves are the 'language of thought'. Moreover, because the word 'concept' is ordinarily used in connection with language and with thought, there is a terminological difficulty at the outset. .. I feel that it is preferable to introducing a neologism like 'pre-concept' for the present non-linguistic usage. Unlike language (or at least speech), concepts are not conventional or arbitrary, are not linked to a speech community, and do not depend upon sequential presentation. ..There is no 'language of thought', for concepts have none of the properties of language. There is thought and there is language. One can reasonably suppose that the frontal (and possibly parietal and temporal) cortex and the basal ganglia are good candidates to serve as bases for the ..having of concepts. If this is true, lobsters and perhaps even birds do not have concepts, but dogs might. .. Pre-linguistic infants appear to be able to construct concepts... This proposed model for concepts... avoids the postulation of an entirely new kind of brain microstructure."

There must exist an objective process which allows the differentiation to be made between true and false concepts and percepts, and other mental objects. Such a test of reality must occur at each level of biotic information processing. At the stage of intelligence, the test of reality corresponds to a semantic proof. (Fig. 6-11).

6.6.2 *Heuristic and algorithmic problem solutions*

One way of classifying mental activity is to follow the path from the formulation of a problem to the finding of its solution. A typical task at the level of intelligence is the performance of a calculation. For any significant mental activity of a human brain, the algorithm would have to be very complicated but, according to the hypothesis called 'Strong Artificial Intelligence' (Section 9.4.2), an algorithm nevertheless. All mental qualities—thinking, feeling, intelligence, consciousness—are to be regarded, according to this hypothesis, merely as aspects of this complicated functioning; that is to say, they are features merely of the algorithm being carried out by the brain.

There is no doubt that a large proportion of mental activity is performed, not by means of algorithms, but by other 'mental techniques'. Very often this non-algorithmic method is called 'heuristic' (Greek *heuriskein* to invent, to discover). It uses rules of thumb, empirically derived, from the previous experience of the individual or from rules learned from a teacher. These rules of thumb give no explanation as to how to find a good solution step by step, but they propose a solution without giving any details of the complete path between the formulation of the problem and its solution.

Heuristics and intuition can be defined as follows: heuristics uses known 'rules of thumb', while intuition (Section 8.1.4) unconsciously processes memorized concepts in working memory.

R.Penrose (1989): "There must be an essentially non-algorithmic ingredient in the action of consciousness. Algorithms, in themselves, never ascertain truth! It would be as easy to make an algorithm produce nothing but falsehoods as it would be to make it produce truths."

6.6.3 *Qualia*

What is the substance of qualia: "The phenomenal qualities of the things of which we are conscious, the raw feels and sensations that make up much of our conscious life" (Korb, 1993, according to S.R.Hameroff, 1994). Consider the qualia of redness; red is 'merely' a specific frequency of the electromagnetic spectrum, a brand of photon. Red of the electromagnetic spectrum can correlate with a specific excitation in brain. Then the sensation of redness in two people could be compared. The redness of red that I perceive so vividly cannot be precisely communicated to another human being, at least in the ordinary course of events.

More complex qualia are: love, painfulness of pain , etc. Qualia, as private things, are an emergent property of the neural network arrangement. One property of qualia is that they are private. The problem of qualia might be within our grasp before the end of the century (F.Crick, 1994).

Ch.Koch (1996): "According to (D.Chalmers) experience (or qualia) cannot be dealt within a reductionist way. It might well be possible that ....we understand the entire cascade of events leading from light of certain wavelength impinging on my retina to activation of neurons that correlate with my subjective feeling of 'redness' the ...problem will have disappeared (by analogy to the disappearance of the problem of vitalism)."

6.6.4 *Mutual basis of speech: world knowledge*

Speech is not just the ability to produce and receive sounds, even in a well-defined form and sequence. Speech is primarily a medium for mutual communication. This means that both the sender and receiver are well enough acquainted with the same environment, that they both have knowledge about the same world.

Human speech is based on an enormous wealth of knowledge. The world knowledge of a 3-year-old child is so large that, at present, there is no computer which can compete. The amount and type of knowledge varies from one individual to another, because it results from his own, personal history. For a given individual, this knowledge changes continuously, and sometimes dramatically.

It is difficult to describe the content of the world knowledge of an average individual. However, in written form this knowledge would fill some tens of thick volumes. Here, (Table 6-6) in brief, are some examples, each probably corresponding to a full book. Don't forget all your lies, deceptions, false information and mistakes.

Of course, a professional writer, especially one of crime novels, can write many more volumes, but the amount of information concerning the real world will be much larger in our volumes and of much greater significance. Even the author of scientific books is limited to a few volumes, especially if he remembers not to repeat himself!

An average person on this planet, based on his own knowledge of the real world and of his own internal world, would be able to write some dozen volumes. Therefore, from this point of view, the difference between a renowned author and the man on the street is not so large. But the difference between an average human being and the most highly developed primate is enormous. It cannot be repeated often enough how significant for all communication, especially by means of speech, is the similarity or even equality of world knowledge of both speaker and listener. Of course, one important parameter is that the roles of speaker and listener often reverse, in a more or less ritualistic pattern (Fig. 6-12). The movement of conductor's baton contains information corresponding only some tens of bits, but is able to influence the action of some tens of orchestra members and finally change millions of bits in the symphony performance. This is because it influences the enormous amount of information in the brains of orchestra members.

6.6.5 *How important are questions?*

The question is the central concern of problem posing. The answer is the central concern of problem solving. The amount of information which is retained at the stages of reflex and instinct is limited.

The quantity of information processed at the stage of discent (that is, acquired by means of simple, rather passive, learning) is significantly greater, but is far from the enormous capabilities of the brain at the stage of intelligence. An active means of acquiring information is by using questions. Intelligence is indivisibly related to the asking of questions. Inquisitiveness is the human means of acquiring knowledge.

Table 6-6 Volumes of average individual knowledge

Volume	Content
1	All knowledge about my own mother tongue, which I use for speech, writing, thoughts, dreams, fears, etc.
2	Knowledge of my own body, in the past and at present; including all my biotic functions such as: breathing, eating, drinking, moving, seeing, hearing, and state of health
3	Descriptions of my family members, my friends, casual and intimate, including physical and psychological characteristics, their histories, their influences on myself, etc.
4	My direct physical environment: house, city, country with geographic, climatic, economic details, etc., including even the simplest and most common facts, which are sometimes of greatest importance for my survival
5	All my school knowledge, even if seldom having any close connection with my real life, but which is the basis of all other knowledge. Just the history of my own and other nations seems to need an enormous effort to remember
6	Information about my hobbies (including chess, with all its combinations), my own sport experiences, all my knowledge about sports records, Olympic Games, etc.
7	Reminiscences from books which have been read, films seen, concerts and other music listened to, paintings studied, assuming that a suitable description of the latter can be recalled at all
8	Everyday activities, such as cooking and shopping, with all the associated problems, such as prices and time-tables
9	Moving within the environment, including driving situations, streets scenes, etc.
10	My experiences in the use of technical devices in my direct environment
11	Knowledge concerning my own profession, including all more or less important details.
12	Moral principles, my religious experiences and those of my friends, and the social and political environment
13	Social activities and details of my intimate, personal life, with various large or small secrets and lies

G.Cantor (1867) in his doctoral thesis: "In mathematics the art of asking questions is more valuable than solving problems".

Human children surpass an adult chimpanzee, not only by using speech but particularly by their verbal questioning. Even the most involved investigators of 'speaking chimpanzees' have not claimed that their pupils have tried to formulate questions. Of course, the answer to a question must not always be verbal. A very effective method of finding an answer is by the use of physical action, trying to find the real answer in the environment. This is often one of the most important motivations for action. One can differentiate between the following three types of questions

–Those which can be answered with a simple yes or no, e.g. Must I do something?

–Those requiring a description or definition, e.g. What is this? Where is that?

–Questions which expect an instruction or algorithm, e.g. How can I do this?

Questions can be right or wrong, but even wrong questions are the product of highly developed intelligence. It may be that most questions are initially wrong and only by later selective elimination, as the result of reaction from the real world, are they corrected.

We are limited in our ability to ask questions as well.

6.7   SPEECH, SOME ABNORMALITIES

6.7.1   *Brain damage and speech*

Normal production of vocal sounds is carried out by respiratory mechanisms, our laryngeal anatomy. However, the most important component of speech generation is the activity of the higher nervous system. Modern knowledge concerning the biotic basis of human speech began with the investigation of brain injuries resulting in speech abnormalities. Speech disorders, such as lisping and stuttering also belong to these abnormalities, but the most important are such disorders as aphasia and dysphasia. Aphasia is essentially the inability to remember the meaning of language and how it is produced. A sufferer can no longer remember the intricate patterns required for articulation or form a word in speech or writing, even though he or she may know what they want to express. Dysphasia means the partial or total loss of language as a result of lesions in those part of the brain that are directly related to the language function.

S.Pinker (1994): "Among anomic patients (those who have trouble using nouns) different patients have problems with different kinds of nouns. Some can use concrete nouns but not abstract nouns. Some can use abstract nouns but not concrete nouns. Some can use nouns for nonliving things but have trouble with nouns for living things; others can use nouns for living things but have trouble with nouns for nonliving things. Some can name animals and vegetables,... foods, body parts, clothing, vehicles, or furniture. There are patients who have trouble with nouns for anything but animals, patients who cannot name objects typically found indoors, patients who cannot name colours, and patients who have trouble with proper names. One patient could not name fruits or vegetable: he could name an abacus and a sphinx but not an apple or a peach."

Injuries and other damage to the human brain can cause substantial, and often irreversible, loss of memory. In some cases the loss of memory—amnesia—is linked to difficulties in recalling events prior to the injury or damage, so-called retrograde amnesia. In other cases, the inability to remember events subsequent to the accident occurs, known as anterograde amnesia. In such severe cases of amnesia it is clear what a prominent role memory plays in the phenomenon called personality.

A novel (and quite unexpected) form of writing deficit has been observed in two Italian-speaking patients who had suffered damage to the left hemispheres of their brains and who were selectively impaired on writing vowels. For example, requested to write the name of the town in which he lived (Bologna), the patient wrote B L G N. There is no obvious reason why vowels should be more difficult to 'remember' than consonants or more difficult to execute motorically.

There are some impairments in human speech ability which are caused by mutation in a single gene and the possibility exists that the gene could be located and characterized. The affected people have some difficulties with plurals and expression of possession, and past tense. The affected children have an otherwise normal mental development (M.Gopnik, 1990).

L.L.Cavalli-Sforza (1995): "And if there are genes controlling hair colour, why shouldn't there be genes controlling self-discipline or sense of humour?"

6.7.2 *Deaf-and-dumb language*

Deaf people use a sign language instead of normal speech. This type of communication between intelligent beings is just as much a language as English or Chinese. It has no written form, but neither do two-thirds of the world's languages. It is a language, just as rich in vocabulary, syntax, and grammar as the spoken and written varieties, and it is learnt in the same way. Signs consist of arrangements of hand shapes, as well as the hands' location, orientation, and movement. Hearing people tend to think of spoken language as linear, with the words following one another in one dimension. In sign language, things or ideas can exist simultaneously in space, in three dimensions. American Sign Language (ASL) has a very different syntax from that of English. There is no doubt that sign language can be learnt only by intelligent beings—by humans.

The same cortical tertiary areas which are responsible for the processing of spoken language, such as Broca's and Wernicke's areas, process sign language, too. Study of sign language users who have had strokes has revealed some parallels between spoken and sign languages. Those with damage to the left side of the brain showed impairments in sign language, but they remained able to process non-language visuospatial relations. Those with damage to the right hemisphere showed the reverse pattern.

Tools of intelligence 245

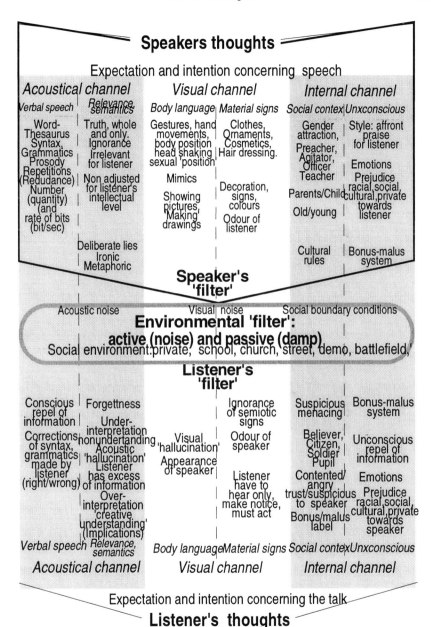

Figure 6-12 Speaker-listener communication

6.8 WRITING AND READING: INTELLECTUAL ABILITIES

6.8.1 *Writing and reading; human uniqueness*

The earliest artifacts connected with written counting are notched bones found in caves in present-day Israel and Jordan, dating about 15,000—10,000 BC The bones seem to have been used as Lunar calendars.

Pre-writing in the form of tokens may date back to Paleolithic times as the 25,000 years old cave drawings themselves. The western writing systems may have evolved from the token system (spheres, disks, cones, and so on) developed for accounting purposes in Mesopotamia beginning in the ninth millenium BC. The system developed by the ancient Sumerians, consisted of a set of clay tokens of distinctive shapes and markings, used to keep records of sheep, cattle and other animals, and goods of various kinds such as oil and grain. In the fourth millenium BC, roughly at the time of the growth of cities, the variety of tokens increased greatly. Shortly thereafter they were placed in envelopes, which, like the string, could mark off a single transaction. This possibly constitutes the first writing. Around 3,200 BC Uruk scripts are fully developed. When literary Sumerian texts were written, beginning about 2,600 BC, they were written in that 'administrative script' and were, therefore, not complete renderings of a text but rather 'an aid for someone who was given an oral performance' (D.R. Olson, 1994).

S.Pinker (1994): "Writing is clearly an optional accessory. Writing was invented a small number of times in history, and alphabetic writing, where one character corresponds to one sound, seems to have been invented only once. All modern phonemic alphabets appear to be descended from a system invented by the Canaanites around 1,700 BC."

From Etruscan to Sanskrit, all alphabetic writing systems display sign lists that include 20 to 40 distinct symbols; the syllabic systems from Persian to Cherokee or Japanese count some 40 to 85 signs; beyond that, 'logographic' systems enter, with symbols for words—not at all picture writing—where the count runs from 500 for Hittite up to above 5,000 for Chinese.

It has been mentioned earlier that there exists an astonishing parallel between the number of selected amino acids—which is 20—and the number of letters in modern language—around 25 (Section 3.1.5). Among the possible millions of proteins, only two to four thousand have been selected as bricks for the construction of living beings, from micro-organisms up to humans. Among the possible millions of words, and some hundreds of thousands in the English language, an average human being needs to know only two to four thousand words.

G.A.Miller (1991): "The available evidence strongly suggests that writing did not originate as a representation of speech, but instead grew out of an interest in pictures and pictorial representation. The phonetization of writing began about 5,000 years ago, which began with logographs for objects, numbers, and personal names."

Table 6-7 Written and drawn representations; quality

Quality of representation	One-dimensional:	Two-dimensional	Three-dimensional
Highest	Mathematical and physical formula, computer program		
Very high		Geographic chart obliges to fulfil all areas	3D-globe hologram.
High		Tomogram. Photography. Scheme. (Design obliges to fulfil all areas)	
Good		Table: obliges to fulfil all areas. Graphic diagram, curve obliges to fulfil all connections	Quasi-3D graphics
Mean	Prose, prosaic description: large freedom parameter		
Low		Photo of body language	
Lowest		Symiotic tables. Traffic lights	

The Arabic numerals are now internationally recognized ideographic (logographic) characters—most people understand 1-2-3, even if they pronounce it differently.

The influence of the differences between phonetic and ideographic writing can be observed in modern Japan, where both types are used in parallel by the same people. Modern Japanese writing uses two systems of signs. 'Kana' is phonetic and, like an alphabet, is made up of sixty-nine symbols, each corresponding to a distinct sound. Kana (Japanese script) contains symbols that correspond to particular sounds. Kana letters always sound the same, by contrast to English with a lot of different pronunciation in 'difficult' words, such as 'nausea' and 'aisle'.

'Kanji', on the other hand, is ideographic. Each of its signs has a specific meaning, but not a defined sound. There are some thousands of Kanji signs and around three thousand must be known in order to read a newspaper. When Kana or Kanji characters are presented to the right or left hemisphere, through one visual field or the other, the results suggest that the left hemisphere is more specialized for Kana recognition and the right for Kanji. This can be explained because the left hemisphere is better adapted to the processing of sequences of phonetic (alphabetic) writing—Kana. The right hemisphere, would prefer the processing of picture-like ideograms—Kanji.

At the present time one can investigate the influence on mental activity and respective brain functions in both 'left-to-right' and 'right-to-left' writing in the same culture, and even by the same persons. For example, American Jews, have both English and Jewish heritage within the same culture, and have been studied from this point of view.

It is worth underlining the enormous influence on intelligence development which occurs with the emergence of the ability to draw. Drawing, in the form of wall painting, is a relatively old type of extrasomatic memory and learning tool. Drawing is primarily related to the right hemisphere. Drawings probably appear parallel to the emergence of more highly developed speech, which includes such abstract terms as 'Self'.

Let us turn to the question of the link between intelligence and writing. One can argue about the speaking ability of chimpanzees, but writing/reading ability is without doubt the unique property of humans. The value of the handwriting in the development of culture, including science, especially mathematics, cannot be overestimated. This almost 'magic' value of writing and of personal signature traces the whole history of civilization. The jump to the print technique was one of the elements of the revolution in cultural development. Even in this situation the confidence to printed words remains unbroken. The open question is if in the future the further development of electronic techniques for storage information in all forms, acoustic, typed and even direct transformation of spoken to printed texts will influence the value of handwriting.

6.8.2 *Pictorial representations*

As we have mentioned in Table 6-7 the quality of written and drawn representations depends strongly on the form. Pictures, even when only two-dimensional representations, are rich in information and are relative very fast 'read' by humans at all civilization levels.

Representation in material form influenced very strong the intelligence itself. The role of written 'holy books' of 'holy pictures', of material 'symbols' such as talismans, crosses, relics, flags, etc. was, is and probably will be, significant.

D.R. Olson (1994): "Just outside my office door is a map.... of the floor plan of the building ... Near the centre of this map is a conspicuous arrow to which is attached the caption 'You are here'. Like all successful maps it orients the viewer in the building...... I don't need a map to tell me where I am; I know where I am, 'I am here, right where I stand'. The map, so to speak, contradicts me, insisting that I am at the point indicated by the arrow. The map undertakes to lift me from my firm stance on the floor and transpose me into this geometry of lines and angles. We have not paid sufficient attention to the fact that our representations have a way of telling us, dictating to us, what and where we are. We are nowhere until our location is identified on the map."

D.R Olson (1994): "There are six deeply held and widely shared beliefs or assumptions about literacy on which current scholarship has cast considerable doubt. 1) Writing is the transcription of speech. 2) The superiority of writing to speech. 3) The technological superiority of the alphabetic writing system. 4) Literacy as the organ of social progress. 5) Cultural development. 6) Literacy as an instrument of cognitive development..... What is required is a theory or set of theories of just how literacy relates to language, mind, and culture. No such theory currently exists perhaps because the concepts of both literacy and thinking are too general and too vague to bear such theoretical burdens."

The believe that the representation of things carried some of the properties of the things presented is called 'metonymy'. To this day we are tempted to believe in efficacy of charms, curses, blessings and well-wishing. Even the most civilized among us would still feel a twinge if we were to poke a pin through the eye of a photograph of a friend and we do not allow our children to mutilate their dolls (and not only because of financial reasons).

D.R. Olson (1994): "Sworn testimony was replaced by written documents, private study replaced public performance, silent reading replaced oral reading, ... a rhetorical persuasive style gave way to a modern prose style, and so on. McLuhan explained these changes by appeal to a general principle that each medium altered one's perception of the content; in the McLuhan argot 'The medium is the message'. The principle applies equally to the alphabet, to, printing, to television and to the computer."

The relationship between the semantic processing of words and of pictures is a matter of debate among cognitive scientists. R.Vandenberghe et al (1996) claim that semantic tasks activate a distributed semantic processing system shared by both words and pictures, with a few specific areas differentially active for either words or pictures. A related claim, also based on the performance of visual experience is that verbal semantics is represented only in the left hemisphere, but visual semantics is represented in both, the right and the left hemisphere.

6.9 BIBLIOGRAPHY

Barlow HB (1983) Intelligence, guesswork, language. *Nature* **304** 207
Bickerton D (1984) The language bioprogram hypothesis. *Behav Brain Sci* **7** 173
Calvin WH (1994) The emergence of intelligence. *Sci Am* **271** 4 78
Cattel RB (1987) *Intelligence: its structure, growth and action.* N-Holland, Amsterdam
Changeux J-P, Connes A (1989) *Matiere a` pense'e.* Odile Jacob, Paris. Deutsche Über setzung; K.Hepp (1992) Springer, Berlin
Chomsky N (1983) Interview. *Omni* **11** 10
Chomsky N (1980) *Rules and representation.* Columbia Univ Press, N.York
Corbetta M, and all (1990) Attentional modulation of neural processing of shape, color, and velocity in humans. *Science* **248** 1556
Damasio AR (1986) Sign language aphasia during left-hemisphere Amytal injection. *Nature* **322** 363
Damasio AR (1994) *Descartes' error. Emotion,reason, and the human brain.* Putnam, N.York
Diamond J (1995) The evolution of human inventiveness. in MP Murphy, LAJ O'Neill (edit) *'What is life? The next fifty years'.* Cambridge Univ Press, N.York
Frey S, Kempter G, Frenz HG, (1996) Multimedia-Gesellschaft. *Spekt Wiss* **8** 32
Gardner H (1984) *Frames of mind: the theory of multiple intelligence.* Heineman, London
Gibbs WW (1996) Jungle medicine. *Sci Am* **275** 6 14
Goleman D (1995) *Emotional Intelligence: why it can matter more than IQ.* Bantam,N.Y
Griffin RD (1991) Animal thinking. *Sci Am* **266** 11 104
Guilford JP (1967) *The nature of human intelligence.* McGraw-Hill, N.York
Hardcastle VG (1994) Psychology's binding problem. *J Conscious Study* **1** 1 66
Jensen AB (1983) The definition of intelligence. *Behav Brain Sci* **2** 313
Kalin NH (1997) The neurobiology of fear. *Sci. Am. Spec.Iss. Mysteries of Mind*, Vol **7**,76
Kandel ER, Hawkins RD (1992) The biological basis of learning and individuality. *Sci Am* **267** 3 53
Khalfa J (edit) (1994) *What is intelligence?* Cambridge Univ Press, Cambridge, Mass
Koch Ch, Poggio T (1987) Information processing in nerve cells, in Adelman G (edit) *'Encyclopaedia of Neuroscience'.* Birkhäuser, Boston
Koch Ch (1996) Hard-headed dualism. *Nature* **381** 123
Kuipers B (1984) Commonsense reasoning about causality. *Artif Intell* **24** 169
LeDoux JE (1995) In search of an emotional system in the brain, in MS Gazzaniga (edit) *'The cognitive neurosciences'*, MIT Press, Cambridge, Mass
LeDoux JE (1997) Emotion, memory and the brain. *Sci. Am. Spec.Iss. Mysteries of Mind*, Vol **7**,68
Lehmate J (1994) Intelligenz von Orang-Utangs. *Spekt Wiss* **11** 78
Lenneberg EH (1975) *Foundations of language development.* Acad Press, Cambridge, Mss
Morris JS et al (1996) A differential neural response in the human amygdala to fearful and happy facial expressions. *Nature*, **383** 812.

Myers DG, Diener E (1997) The pursuit of happiness *Sci. Am. Spec.Iss. Mysteries of Mind*, Vol **7**,40
Neisser U (1979) The concept of intelligence. *Intelligence* **3** 217
Olson DR (1994) *The world on paper. The conceptual and cognitive implications of writing and reading.* Cambridge Univ Press, N.York
Pardo JV, Fox PT, Raichle ME (1991) Localization of a human system for sustained attention by PET. *Nature* **349** 61
Pinker S (1994) *The language instinct. The new science of language and mind.* Penguin Books, London
Posner MI, Raichle ME (1994) *Images of mind.* Sci Am Library, N.York
Premack D (1983) The codes of man and beasts. *Behav Brain Sci* **6** 124
Premack D and AJ (1983) *The mind of an ape.* Norton, London
Saffran JR et al (1996) Statistical learning by 8-month-old infants. *Science*, **274** 1926
Schabarov-Kuscharenko JP (1987) *Teoryia intellekta.* Wyssch Schkola, Charkow
Schank RC, Childers PG (1984) *The cognitive computer.* Addison-Wesley, Reading, Mass
Schull J (1990) Are species intelligent? *Behav Brain Sci* **13** 63
Scott SK et al (1997) Impaired auditory recognition of fear and anger following bilateral amygdala lesions. *Nature,* **385** 123
Searle JR (1980) Minds, brains, and programs. *Behav Brain Sci* **3** 417
Seidenberg MS (1997) Language acquisition and use: learning and applying probabilistic constraints. *Science* **275** 1599
Simon HA (1981) Studying human intelligence by creating artificial intelligence. *Am. Sci* **269** 301
Simon HA (1983) Search and reasoning in problem solving. *Art Intelligence* **21** 7
Skoyles JR (1984) Alphabet and the Western mind. *Nature* **309** 409
Sternberg RJ (1985) Human Intelligence: The model is the message. *Science* **230** 1111
Taube M (1982) *Human intelligence, the ultimate resource.* (Script) Swiss Federal Institute of Technology, Zürich
Vandenberghe R et al (1996) Functional anatomy of a common semantic system for words and pictures. *Nature,* **383** 254.
Viaud G (1960) *Intelligence, its evolution and form.* Hutchinson, London (cited by V. Reynolds 1980).
Winograd T (1983) *Language as a cognitive process.* Addison-Wesley, Reading, Mass.

Chapter 7
Everyday Intelligence

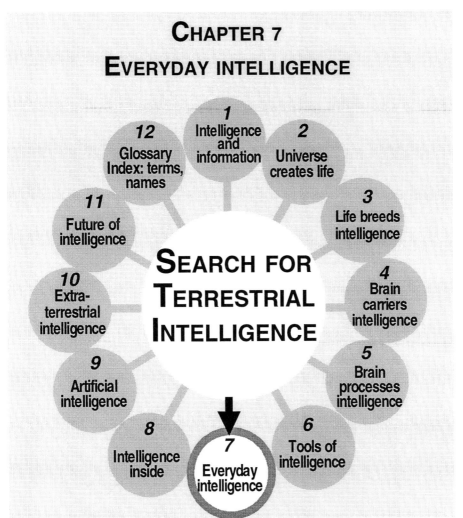

Common sense, humanity's greatest attribute. Common sense and bonus-malus system. What is wisdom?. Science and common sense. Is human intelligence measurable? IQ problem. Creative thinking and fantasy. Genius, a summit of intelligence? Artistic skills. Psychic disorder. Personality, how does it emerge. Intelligence: nurture or/and nature. Taboo, sexuality, placebo etc. What do we think about thinking? Some paradoxes. Myths and rationality. Deception, cheating, lie and humour. Culture, product and source of intelligence. Psychoactive agents and culture. Mental games.

7 EVERYDAY INTELLIGENCE 254
7.1 COMMON SENSE, HUMANITY'S GREATEST ATTRI-BUTE 254
7.1.1 Common sense: what is it? *254*
7.1.2 Common sense and heuristic problem solution *255*
7.1.3 The quality of biotic information: the bonus-malus system *256*
7.1.4 Wisdom *258*
7.1.5 Science and common sense *260*
7.2 DIFFERENTIATION BETWEEN INDIVIDUALS 263
7.2.1 The 'IQ' question—is human intelligence measurable? *263*
7.2.2 Lateralization as one of the causes of personal differentiation *265*
7.2.3 Creative thinking, fantasy *266*
7.2.4 Genius, the summit of intelligence? *267*
7.2.5 Idiot savants, remarkable individuals *268*
7.2.6 Artistic skills *270*
7.2.7 Mental disorders and brain diseases *271*
7.3 PERSONALITY: THE PRODUCT OF GENES OR OF CULTURE 274
7.3.1 How does a personality emerge? *274*
7.3.2 Intelligence: nature or nurture *278*
7.3.3 School and intelligence *280*
7.3.4 Taboos, sexuality, placebos, hypnosis and psychosomatics *280*
7.4 THINKING: THE QUINTESSENCE OF INTELLIGENCE 281
7.4.1 What do we think about thinking *281*
7.4.2 Paradoxes and the principle of incompleteness *282*
7.4.3 Myths and rationality *286*
7.4.4 Deception, fraudulence, cheating *287*
7.4.5 Humour is unique *289*
7.5 CULTURE: IDEAL ARTIFACT 289
7.5.1 Culture: product and source of intelligence *289*
7.5.2 Psychoactive agents and culture *290*
7.5.3 Mental game *295*
7.5.4 Aesthetics, some remarks *296*
7.6 BIBLIOGRAPHY 299

7   EVERYDAY INTELLIGENCE

7.1   COMMON SENSE, HUMANITY'S GREATEST ATTRIBUTE

7.1.1   *Common sense: what is it?*

'Common sense' generally refers to practical attitudes (cultural, moral, social, economical, environmental) and widely accepted beliefs which are hard to justify, but which are generally assumed to be reliable. Extreme deviation from common sense beliefs may be evidence of psychological disturbance, but may, on the other hand, be the products of genius, sometimes becoming accepted later as common sense. In this book, the term 'common sense' is used as the state of information processing at the level of the average human being at the present time, more or less independent of his cultural environment. Of course, such a definition is somewhat imprecise and requires further elaboration.

'Common sense' in reality differs in:
–Different cultures, even in geographical neighbourhoods,
–Different age groups-different social groups,
–Both genders,
–Different religious groups especially in fundamentalists,
–Different political ideologies, especially in extremity

These differences show that the common sense is not so 'common' as we think.

P. Feyerabend (1995): "There is not one common sense, there are many. Nor is there one way of knowing; there are many such ways. Science itself has conflicting parts with different strategies, results, metaphysical embroideries. It is a collage, not a system. ..Scientific institutions are not 'objective'... They often merge with other traditions, are affected by them, affect them in turn. The material benefits of science are not at all obvious. There are great benefits, true. But there are also great disadvantages."

7.1.2 *Common sense and heuristic problem solution*

In Section 6.6.2 the significance of the two methods of problem solutions was discussed: algorithmic and heuristic. Common sense is based on both, but the heuristic methodology is its most characteristic property. A large part of the learning process present within the framework of family and society is the learning of heuristic rules.

Heuristics allows us to solve even a complex problem in an extremely short time, without a deeper understanding of the nature of the problem being necessary. Everyday problems are the best example.

For some hundreds of years, the idea that the Earth is spherical was doubtless the expression of highly sophisticated thinking. At present, when almost all people have seen pictures of the Earth taken from cosmic space and are confronted with this idea at every step, the concept belongs to the realm of common sense. Thirty years ago, contact with and the use of computers was an example of highly sophisticated procedures, but nowadays many of the concepts and even some of the operations of computers could be classified as common sense. The continuous increase of the content of common sense is good evidence of the enormous, and still inexhaustible, resources of the human mind (Fig. 7-1).

The common sense includes the strategy to omit the shadow-side of human activity, in every-day activity but also in communication with other individuals.

All our thoughts express only our wishful thinking, often coloured with belief (even if very weak) that they have a self-fulfilling power.

7.1.3 *The quality of biotic information: the bonus-malus system*

Common sense is based on the ability to interpret how good or bad each problem is. The subjective quality of information is for common sense of highest importance. The simplest way to begin considering the quality of information is with the following quotation, taken from the book of one of the fathers of the information theory:

L.N. Brillouin (1956): "The present theory of information completely ignores the value of the information handled, transmitted, or processed. Many other writers seem not to have realized the importance of this restriction. How to introduce into the theory the element of value. Ignoring the human value of the information is just the way to discuss it scientifically, without being influenced by prejudices and emotional considerations. Whether this information is valuable or worthless does not concern us. The idea of 'value' refers to the possible use by a living observer. This is beyond the reach of our theory, and we are unable to discuss the process of thinking or any other problem about the use of the information by living creatures... The present theory considers information as always positive and never negative. The value of the information can, and must in certain cases, be regarded as negative. At any rate, one point is immediately obvious: any criterion for 'value' will result in an evaluation of the information received. This is equivalent to selecting the information according to a certain figure of merit."

The situation in the information theory remains unchanged until today. In this book, we have already introduced such a figure of merit, a scale for measuring the subjective, actual value of information for a given individual. We have called this biotic information quality system, the 'bonus-malus system'. Let us try to sum up the most important features:

–The bonus-malus system (also called the 'reward-punishment', 'pleasure-pain', or 'hedonic' system) exists at all four stages of biotic information processing—reflex, instinct, discent and intelligence. It controls the quality label of information (good or bad), the quality intensity (important or unimportant), and the quality of recall of memorized information, that is, on all phases of information processing.

–A bonus-malus label changes continuously, influenced by new experiences and by each act of information reprocessing. The bonus-malus system includes among other attributes, a sophisticated and very complex system. In spite of rudimentary knowledge of these mechanisms, a very general picture seems to be appearing (Fig. 7-2).

Some natural or synthetic substances influence the bonus-malus labelling so strongly that they are even able to change the sign from positive (bonus, reward, pleasure) to negative (malus, punishment, pain), or vice versa. This affects both freshly acquired and old, memorized information. The partial permeability of the brain-blood barrier has enabled Homo sapiens sapiens to use some natural substances, such as nicotine and alcohol, and later synthetic ones, as drugs for thousands of years (Section 7.5.2).

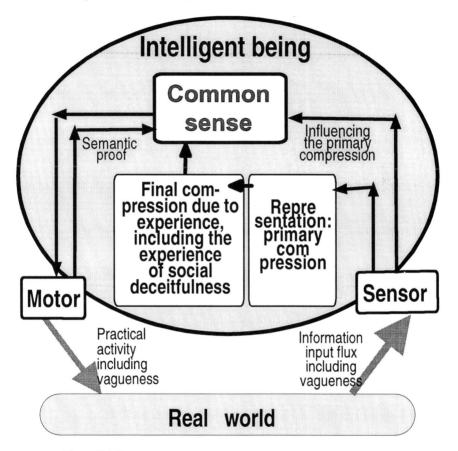

Figure 7-1 Common sense

These facts demonstrate the possibilities for humans to influence their own internal information processing system in a very dramatic way. Because the latter is directly connected with information processing in the brain, we must take into account the fact that humans directly and intentionally influence their own free wills (Section 8.3).

Some of the known neurotransmitters, such as norepinephrine and dopamine (Section 4.5.2), which are both catecholamines, play the role of the bearer of pleasure in rats, monkeys, and probably, humans. At the present time, it is evident that reward pathways play an important role in learning and memory, both of which are the most significant operations in information processing.

J.S. Morris et al (1996) believe that the human brain measures fear and happiness as opposite ends of a single scale.

7.1.4 Wisdom

R.J.Sternberg (1990): "To understand wisdom fully and correctly probably requires more wisdom than any of us have."

H.Simon (1985): "Human beings know a lot of things, some of which are true, and apply them. When we like the results, we call it wisdom."

Wisdom is often considered as a peak performance, perhaps even as a possible end state of human knowledge and its development. An everyday definition could be: good judgement and advice about important but uncertain matters of life. Some theoretical definitions claim that wisdom is an expert knowledge system in fundamental life pragmatics (life planning, life management, life review). On the other hand, it must be mentioned that even massive handbooks of general psychology do not index the subject of wisdom.

D.A. Kramer (1990) writes that there are five possible functions of wisdom in adult life:

–Life planning, involving both intra- and interpersonal knowledge and skills

–Advising others

–Management and guidance of society

–Self-reflection in order to provide evaluation and continuity of own life

–Questioning the meaning of life.

According to R.J. Sternberg (1990), wise people know:

–What they know

–What they do not know

–That they can know, given the limitations of understanding

–What they cannot know

From our viewpoint wisdom is based mostly on heuristic rules, particularly those coming from areas of social life. There do not exist analytical methods or algorithmic rules to manage complex social behaviour. If the idea of chaotic phenomena has a significance, then this is the case in this field. If a small cause could really result in 'unexpected' and unforeseeable, but significant, consequences then this is the case in human social relations.

We think that the clever is satisfied with his own cleverness. The intelligent knows that he is not intelligent enough. The wise hopes that greater and superior wisdom than his exists. The intellectual doubts the existence of such a phenomenon as intellect.

Everyday intelligence 259

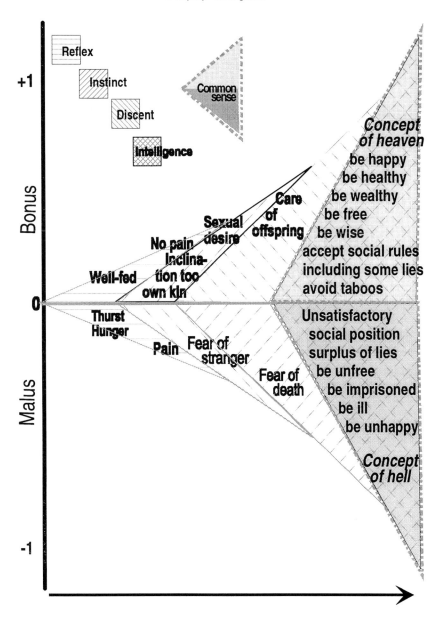

Figure 7-2 Common sense and bonus-malus system

In the term 'wisdom' or 'common sense' should be included also the notion of 'happiness'. Happiness is evidently very high on the scale bonus-malus system, and therefore significantly influences all human information processing. So hat is this 'thing' called happiness? (Section 6.4.4). Now a pioneering band of researchers tried to explain this elusive concept—or at least take its measure. They have proposed a five-item 'Life Satisfaction Scale' and a seven-point 'Delighted-Terrible Scale'. Because all these problems are deeply involved in their cultural and social environment, dependent on the personality, on momentary state of their bonus-malus system, it seems that this 'research' is not worth any further consideration.

7.1.5 *Science and common sense*

Science uses the following method: it separates the real world in portions, cutting more-or-less consciously, even some significant connections between the portions, and then tries to investigate a representation of the separated portion. This 'artificially' isolated portion of the real world we call 'puzzle' (piece of jigsaw puzzle). The success of science in investigation of the 'puzzles' ensues from the use of idiosyncratic methods and tools for each puzzle. On the other hand the differentiation of methods, tools and 'languages' makes it difficult to achieve co-operation of different branches, and prevents the unification of Science. The total Truth may lie in the infinitely complex mosaic of sciences, assuming that the world is basically intelligible. All these lead to the impossibility to construct a formal representation of science, and to determine whether science has limits.

D. Hilbert (1930): "Wir müssen wissen, wir werden wissen" ("We must know, we shall know").

This exhortation is engraved on Hilbert's gravestone. Those six words are cited from Hilbert's address entitled 'Naturerkennen und Logik' in 1930 ('The Understanding of Nature and Logic'). Hilbert's address sums up his enthusiasm for mathematics and the devoted life he spent raising it to a new level. But in 1931 the Austrian-U.S. mathematician Kurt Gödel showed this goal to be unattainable because within any rigidly logical mathematical system there are propositions (or questions) that cannot be proved or disproved on the basis of the axioms within that system and that, therefore, it is uncertain that the basic axioms of arithmetic will not give rise to contradictions. E.P. Wigner, (student of Hilbert and Nobel laureate, in physics, 1963) wrote in 1960 about 'The unreasonable effectiveness of mathematics in the physical sciences' which presented a very impressive list of successes of the mathematical approach in the modern world-picture. And the way in which atoms, acting within the human brain, investigate how the atoms originated in the Big

Bang and how they are acting now in the intelligent brain, is the only one available for us.

The scientific activity cannot be free from influence of the bonus-malus system. The general tendency towards the 'true' science is dominated by the individual bonus-malus label. It is much easier to formulate such a general sentence in this book, than in real own scientific activity (for example in formulation of a proposal for a research grant). In the depths of our mind we discover own weaknesses and doubts.

In a reductionist view science is the way in which atoms investigate the atoms. Truth is what the atoms 'know' about all other atoms. This extreme reductionist point of view seems to be inconsequent. May be the following formulation is more adequate: Atoms in mutual interaction create consciousness which investigate the nature of atoms.

S.Hawking (1997): "To start with, I should say I'm unashamed reductionist. I believe that the laws of biology can be reduced to those of chemistry...And I further believe that the laws of chemistry can be reduced to those of physics. We (Hawking and Penrose) now have very different approaches to the world, physical and mental. Basically, he's (Penrose) a Platonist believing that there's a unique world of ideas that describes a unique physical reality. I, on the other hand, am a positivist who believes that physical theories are just mathematical models we construct, and that it is meaningless to ask if they correspond to reality, just whether they predict observations...Rather, his (Penrose's) argument seemed to be that consciousness is a mystery and quantum gravity is another mystery so they must be related" (S.Hawking in R.Penrose, 1997).

R.Penrose (1997): "I agree that this (process relevant to the brain (M.T., K.L.)) would seem to be 'very unlikely—were it not for the fact something very strange is indeed going on in the conscious brain which appears to me (..) to be beyond what we can understand in terms of our present-day physical world-picture".

The deep relation between the concepts of intelligence, common sense, wisdom and science seems to be self-evident. In spite of this, science has no direct biotic, immediate, survival value.

There is in our present state of self-knowledge a particular reason why we need to be able to achieve this sort of 'deep knowledge' about the Universe and ourselves (that is the phenomenon of intelligence). In fact, many communities, for many thousands of years, have made a perfectly satisfactory living on this planet without having such underlying 'theoretical' knowledge called science (P.Davies, 1995). Paraphrasing the above mentioned title of the Eugene Wigner's paper 'Unreasonable effectiveness of mathematics in the physical sciences' we could express our admiration about the unexplainable effectiveness of human brain (product of the self-evolving Universe) in the understanding of almost all aspects of the real world, even without an explicitly structured 'science'.

A.Cromer (1993): "Science is the heretical belief that the truth about the real nature of things is to be found by studying the things themselves."

According to Cromer (1993) Homo sapiens would never have 'invented' science, except for an unlikely concatenation of historical events. The scientific thinking has been 'invented' by Greeks as a result of different rather unique factors. The democratic political tradition where men first learned to persuade one another by means of rational debate, was based on a widespread Greek-speaking world around which scholars could wander and poets produced masterpieces. A highly developed maritime economy that prevented isolation and parochialism, and which allowed an independent merchant class that could hire its own teachers, was the cause of this development. Cromer claims, that all these factors came together in one great civilisation quite fortuitous; it didn't happen twice.

A. Cromer (1993) in his book 'Uncommon sense. The heretical nature of science': "Most people believe that science—whatever it is—arose from humankind's innate intelligence and curiosity, that it is a natural part of human development.....(I) will argue, scientific thinking, which is analytic and objective, goes against the brain of traditional human thinking."

It seems to us that a more exact formulation should be: The human brain and its emerging product, the mind, is an astonishing phenomenon, finding an excellent correspondence between itself and the Universe.

P.S. Churchland (1986): "So it is that the brain investigates the brain, theorizing about what brains do when they theorize, finding out what brains do when they find out, and being changed forever by the knowledge."

M.Planck (1910): "Science cannot solve the ultimate mystery of Nature. And it is because in the last analysis we ourselves are part of the mystery we are trying to solve."

The last two citations represent exactly our opinion in this matter.

P. Davies (1995): "These remarkably ingenious laws (of physics) are able to permit matter to self-organize to the point where consciousness emerges in the cosmos—mind from matter. We have a closed circle of consistency here: the laws of physics produce complex systems, and these complex systems lead to consciousness, which then produces mathematics... The world hasn't been created for our benefits; we're not at the centre of creation. We are not the most significant thing. I think we do have a place in the universe—not a central place but a significant place nevertheless."

F.Dyson (1980): "I do not feel like an alien in this Universe. The more I examine the Universe and study the details of its architecture the more evidence I find that the Universe in some sense must have known we were coming."

Parallel to this direction in our society there exists an intellectual current which claims that science has flourished for a few hundred years (in the beginning mostly in Europe), but there's no reason to expect it to go on forever (Section 11.6.2). The end is in sight!

7.2 DIFFERENTIATION BETWEEN INDIVIDUALS

7.2.1 *The 'IQ' question—is human intelligence measurable?*

What is intelligence? What does it mean to say that one person is more or less intelligent than another? Are differences in intelligence more or less important than differences in, say, ambition, hard work, ruthlessness, or sheer good fortune in accounting for the success of some and the failure of others? In emotions, in feelings, in believing? As already stated, human intelligence is a more-or-less well-defined phenomenon, which can be described by means of scientific methods. Thus there seems to be a good basis for making measurements of intelligence, in a similar way to that in which measurements are made in other branches of science. Since the beginning of this century, a number of methods for measuring human intelligence have been proposed, and intensively and extensively applied.

The intelligence test consists of a series of tasks designed to measure the capacity to make abstractions, to learn, and to deal with novel situations. The intelligent quotient (IQ) is a number representing the level of the intelligence of an individual. It is obtained by dividing an individual's mental age (determined by the performance on a standardized intelligence test) by his or her chronological age and multiplying the result by 100. An IQ of 100 is considered average, 130-140 or above is gifted, and 70 and below is mentally deficient or retarded. About 3 per cent of the population has an IQ of less than 70 but for about half of these people a genetic abnormality may be to blame. The IQ tends to rise until a person reaches the middle or late 20s, then declines gradually (Fig. 7-3). IQ values have been used in predicting success in school and even in some professions.

Critics of the IQ method, however, charge that there is a cultural bias in test construction and that the IQ may easily be overemphasized or misinterpreted in trying to judge a person's ability. Some critics have been more severe.

S.E. Luria, S.J. Gould, S Singer (1981): "Social scientists have stressed the finding of a 15-point difference in average IQ scores for whites and blacks in the US, a difference that is probably due much more to environmental background than to hereditary factors. In summary we can say with confidence that today there is no evidence for the existence of specific intelligence genes nor of racial differences in genetic determinants of intelligence."

S. Rose, L.J. Kamin, R.C. Lewontin (1984): "How do we know that IQ tests measure 'intelligence'? Somehow, when the tests are created, there must exist a prior criterion of intelligence against which the results of the tests can be compared. People who are generally considered 'intelligent' must rate highly and those who are obviously 'stupid' must do badly or the test will be rejected. IQ tests, then, have not been designed from the principles of some general theory of intelligence and subsequently shown to be independently a predicator of social success. On the contrary, they have been empirically adjusted and standardized to correlate well with school performance, while the notion that they measure 'intelligence' is added on, with no independent justification to validate."

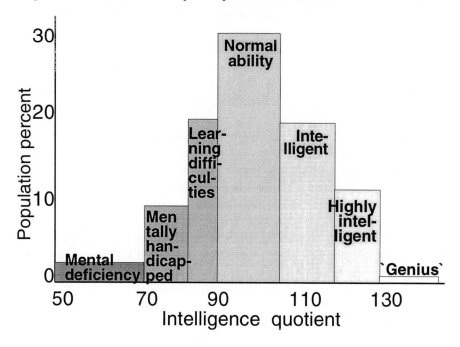

Figure 7-3 Intelligence quotient

K.Popper (1977) is not convinced that such a complex phenomenon as human intelligence, including creativity and so on can be expressed by a one-dimensional function such as an 'intelligence quotient'. Probably even such a genius as Albert Einstein would not have rated a high enough IQ. Popper states that he himself has had contact with IQ geniuses who were far from being brilliant.

In 1994 a book of R.J. Herrnstein and Ch. Murray 'The Bell's curve' could be summarized in the following way:-intelligence could be defined as a general mental capability that includes reasoning and learning abilities;-intelligence can be measured by IQ tests; –IQ tests are not culturally biased; –IQ is of great practical and social importance; –there exists in the USA a persistent black-white IQ gap; –IQ is substantially heritable (neurogenetic determinism); –IQ is also affected by environment; –we do not yet know how to manipulate it to raise a low IQ permanently.

Most reviews of the book have been negative. Here one citation:

L.J. Kamin (1995): "Herrnstein and Murray (in their book 'The Bell curve') labour mighty to show the low IQ is the cause of socio-economic status, and not vice versa... On close examination, this scientific emperor is wearing no clothes. The book has nothing to do with science."

Another opinion (1996): "The work (of Herrnstein and Murray) is a string of half-truths".

From what has been written in this book so far, it can be seen that the intelligence quotient cannot be taken too seriously, because of the multi-dimensional character of biotic information processing (Ciba Found.Symposium, 1993).

L.L.Cavalli-Sforza (1995): "The IQ test does not measure intelligence itself, which is too elusive and includes numerous facets of different capabilities. It measures the ability to carry out certain numerical, geometric, linguistic, and abstract shape operations....Your IQ is also, and perhaps above all, determined by how you are prepared to sweat over intellectual and practical questions that require concentration and analysis."

But it seems to be too simple to negate the impact of genetic components on the intellectual abilities of the central nervous system. On the one hand it is widely believed that the human prefrontal cortex controls the highest and most distinctively human forms of thinking, yet at the same time, extensive damage to this area of the brain is apt to yield only modest decrements in traditional measures of intelligence, by means of standard intelligence tests. Despite the fact that patients with some cortex damage perform well in the IQ tests, they are clearly impaired on a variety of cognitive functions, including planning, monitoring and modifying behaviour, learning complex tasks, and temporal sequencing (N.Robin, L.J.Holyoak, 1995).

To all this criticism one additional question. Why in a 'normal' IQ-set is there no appropriate place for testing the capability to lie? Real social activity is often, may be too often, coupled with the necessity to cheat, swindle and lie; human intelligence bears the appropriate methods and techniques.

7.2.2 *Lateralization as one of the causes of personal differentiation*

The overabundant production and selective elimination as a general strategy of Nature—is expressed in the form of the great diversity of human individuality. The enormous spectrum of human behaviour, abilities, skills, creativity, adaptability, etc. is the best guarantee for the unlimited ability to adapt to future changes in the environment, both social and natural.

It is a generic property of intelligence that each individual is in some way different from any other. The probability that two humans are identical is practically zero. They differ not because of different genes, or different epigenetic somatic development, but because of different experiences during life, and therefore differentiated bonus-malus systems, that is emotional mechanisms which strongly influence the information which each processes.

Now the question arises whether an intelligent computer could be constructed, which does not need to have the characteristics of individuality. Would the impossibility of the existence of two identical intelligent beings be then cancelled? It seems that even with the most sophisticated technology, or perhaps because of it, small unavoidable differences will exist ('genetic mutations'). The computers would be different, and during their operation they would learn different things ('epigenetic cultural influence') which would result in two different 'intelligent devices' at the end (see Chapter 9, Artificial intelligence).

The human brain being asymmetrical (lateralization) may have different properties due to differences in both hemispheres. In a simplified way one could say that some characteristic features of behaviour depend on differentiation in the left and right hemispheres. An anecdotal example is taken from the book of B. Kolb and I.Q. Whishaw (1985) (Table 7-1).

Table 7-1   Laterality and personality

Behaviour	**Professor Alpha**	**Professor Beta**
Interested in:	details	generalities
Style of working	only one thing	more things
Verbal abilities	very good	troublesome
In discussions	wins	loses
Speed of thinking	fast	slow
Kind of writing	clear	unclear
Dominant hemisphere	left	right

A rather interesting book, 'The Science of Mind' (1989), includes a chapter entitled 'Hemisphere dominance in Japan and the West', in which the author (Japanese) maintains that the cerebral dominance for sexual functioning, which is extremely laden with emotion, seems to be located in the left hemisphere for Japanese and in the right hemisphere for occidentals. Even the layouts of the brains are enumerated in two different tables—one for the 'Japanese model' and the other for the 'Western model' .

7.2.3   *Creative thinking, fantasy*

Common sense cannot be opposed to creative and sophisticated thinking, but, on the other hand, creativity is something other than common sense. What is creativity, in the broad sense of the term? How does creativity emerge? And what about the differences between genders?

The human mind is able to produce 'mental objects' which have no connection with reality: so-called fantasy. It is not so easy to define what is fantasy and what is not. The famous science-fiction author, S. Lem (1984) asserts, for example, that a quadratic rainbow is a fantastic object: but is the summation 'two plus two equals seven' also the product of fantasy? If it is assumed that fantasy is a product of a highly evolved system of information processing, and is an indispensable property of all intelligence, will it also be present in an 'intelligent computer', when the time comes for such a device to be constructed?

Several studies now show that creativity and mood disorders (manic-depressive illness) are linked (K.R.Jamison, 1995). During episodes of mania or mild mania patients experience symptoms that are in many ways opposite to those associated with depression. Their mood and self-esteem are elevated. They sleep less and have abundant energy; their productivity increases. Manics frequently become paranoid and irritable. Moreover, their speech is often rapid, excitable and intrusive, and their thoughts move quickly and fluidly from one topic to another.

Some obvious questions arise: the coupling between creativity and the inclination to production of false information; the connection between creativity and an unconscious lie. Is phantasmagoria a pathological phenomenon or a creative process?

Some students are of the opinion that the 'most human' property is the inventiveness (J.Diamond, 1995). In spite of difficulties to define 'inventiveness' it cannot be denied that it is more-or-less directly connected with intelligence.

In the Table 7-3 some arbitrarily selected examples are listed. Continuous managing of fire was an invention of high significance for the evolution of Homo. But is seems that invention of how to use special tools and materials for making fire required a more sophisticated stage of development.

7.2.4 *Genius, the summit of intelligence?*

The Swiss astro-physicist F.Zwicky wrote in 1971 a book with the title 'Everybody is a genius'. In one sense he is right. From our previous definition of intelligence and common sense, it must be clear that the quality of information processing in humans is very high, even extremely high. However, the width of this 'quality of intelligence' is large and encompasses a wide spectrum of different facets, levels, specific features, etc. If this is so, then the emergence of a genius—an intelligent being with, let us say, 'higher intelligence'—is of a low probability. This is a trivial conclusion.

However, less trivial is the conclusion that the very high quality of 'normal' or 'average' intelligence (if such a classification is at all allowed and reasonable) connotes that the difference between the levels of the genius and the normal average person cannot be large. Recently an international news magazine asked: Where do great minds come from? And why are there no Einsteins, Freuds or Picassos today?

Genius is only a little 'higher' than the average, and even this superiority is expressed over a very limited range of human activity such as music, physics, or finance. Genius is only a rocky outcrop on a very high more-or-less flat mountain of intelligence. However, the existence of geniuses is the guarantee for the continual development of mankind. It is also remarkable that sooner or later a brilliant idea or artistic creation is accepted by a large number of people as common sense. This must be evidence that eventually the 'normal' human mind is more or less capable of understanding those concepts, ideas, and artifacts which have been produced by geniuses, who at a first glance were so far above the man on the street.

Some scientists have tried to demystify the term 'genius':

M.Minsky (1986): "I suspect that genius needs one thing more: in order to accumulate outstanding qualities, one needs unusually effective ways to learn. It's not enough to learn a lot; one also has to manage what one learns... better ways to learn to learn."

How far does a genius depend on his cultural environment? G.Vollmer (1986) is of the opinion that Archimedes or Pythagoras would today be a Gauss, or an Einstein.

Despite intellectual acknowledgement of the essential duality of the origins of high ability, most individual researchers are emotionally—sometimes passionately—attached to the defence of one extreme.

Another question concerns the genetic inheritance of genius. Of course, the famous family of musicians, the Bach´s, or the Pleiads of Bernoulli´s are well-known. However, there are many more examples in which the inheritance of genes of 'genius' has not occurred. Assuming that 'genius' is inheritable, and is socially preferable, one can speculate on the creation of a new sub-species—Homo sapiens genialis. What a terrible prospect for the normal individual!

7.2.5   Idiot savants, remarkable individuals

Human intelligence consists of many different, sometimes remarkable, sometimes shocking, but mostly unexpected and incomprehensible, phenomena. The existence of so-called 'idiot savants' is an example of the latter. These are people who are, in general, mentally defective, but who display from their earliest days an unusual aptitude or brilliance in some special field, such as the ability to draw, play the piano, or perform rapid mental calculations (more appropriate name is 'mono-savants') (Ciba Found. Symposium, 1993).

Table 7-2 Human inventiveness

Period, years ago	Species	Invention	Remarks
2,000,000	Homo habilis	Tools for preparation of tools	
1 500,000	Homo erectus	Fire as heat and light source	
500,000	Homo erectus	Fire for fire making	
200,000	Homo sapiens neanderthalen.	Sub-speech as communication tool	Lack of inventiveness in other fields of activity
40,000	Homo sapiens sapiens	Speech for speaking about speech. Invention of 'adventures' (territorial expansion) as agent of 'better life'	Global expansion
35,000		Use of drug as stimulator. Artistic ornaments and wall paintings	Wall painting as carrier of information
11,000		Grains for grains cultivation	Beginning of agriculture
8,000	Modern Homo	Invention of state organization and of religious system	
6,000		Writing: extrasomatical information storage and transportation	Fixed, objective carrier of information
2400	Greeks	Writing about writing. Thinking about thinking Learning how to learn Science about science	Literature criticism Beginning of philosophy Emergence of schools Scientific method
500	Columbus	Discovery that geographic discovery is of practical importance	Beginning of expansion of European (colonization)
450	Copernicus	Discovery that pure intellectual discovery is of general importance	Beginning of new era of scientific development
170	Babbage	Machine for information processing: computer	
50	Shannon, von Neumann, Turing	Information about information Knowledge about knowledge	So-called 'theory of information'
30	Sperry, Eccles, Crick	Intelligence investigates intelligence	Including the investigation of consciousness

Many of us have met a 'living computer', a person with the extraordinary ability to manipulate large numbers or strings of numbers in his or her head. Some of these people are even mentally disabled, but one must remember that among such remarkable calculator-geniuses names as Euler, Ampere and Gauss can be found.

Many people probably have some extraordinary ability for processing information, though generally only in one rather narrow field, such as knowledge of all details of the football league over the past twenty years in many European countries. But the ability to perform enormously complex arithmetical operations seems to be an abnormal property of the human cerebral cortex, and is probably the result of an unconventional way of managing numbers; for example not by the left hemisphere in the form of numerical strings, but perhaps as spatial pictures or even sounds, processed by the right hemisphere and without direct conscious control (S.B.Smith, 1983).

The existence of geniuses and 'idiot savants' on the one hand, and mentally disabled and ill persons on the other, is the price to be paid for the existence of normal people—humans with ordinary, average common sense. It may be that what we call wisdom is the result of the orchestration of numerous small stupidities.

One of the extreme rare abilities is the blind chess play in more spectacular cases as Alechin and Najdorf simultaneously blind playing on 45 chessboards (during more than 17 hours) and winning 39 of them (Section 9.3.3).

7.2.6 Artistic skills

Music, like other forms of expression, requires specific skills for its production, and the organization and representation of these skills in the human brain are not well understood. With the use of PET and MRI, the functional neuroanatomy of musical sight-reading and keyboard performance was studied in ten professional pianists.

Reading musical notations and translating these notations into movement patterns on a keyboard resulted in activation of cortical areas distinct from, but adjacent to, those underlying similar verbal operations. These explain why brain damage in musicians may or may not effect both verbal and musical functions depending on the size and location of the damaged area.

Some recent observation by means of MRI are of such significance for the here discussed problem that it seems to be worth-while to repeat them (Section 5.5.3). G. Schlaug et al (1995) have found in vivo evidence of structural brain asymmetry in musicians with perfect pitch who revealed stronger leftward planum temporale asymmetry than non-musicians or musicians without perfect pitch. The results indicate that outstanding musical ability is associated with increased leftward asymmetry of cortex subverting music-related functions.

However, sight-reading and playing are only a fraction of musical experience, and we are still far from understanding the pleasure and emotions elicited by music, as well as the composer's mind (J.Sergent, 1992). Maurice Ravel, who composed Bolero and other classics, suffered from an unidentified brain disease which destroyed his speech and writing abilities at least a year before his musical skills were affected.

7.2.7 Mental disorders and brain diseases

There exist many human conditions resulting in apparent behaviour or subjective sensations which are commonly considered as psychic disorders. The brain plays an important role in these conditions. Mental illness may occur as an consequence of direct damage to brain tissue which in turn influences brain functions. There are many causes which could be listed here. On the other hand derailed mental function may occur when the brain itself is apparently normal, but psychological, social or other factors have temporarily or continuously altered normal brain functions such that the psychic disorder in question comes to the surface.

Disorders of the human nervous system may influence the output processes, that is motor activities, or input processes, that is sensoric activities and the association processes. The symptoms may range from slight disturbances in personality to tragic crippling, or fatal diseases. The causes are some times genetically determined (Down syndrome), some times have an infectious background (bacteria as in syphilis, viruses such as in rabies) or may be the consequence of injuries.

Of the brain diseases, which directly alter brain tissue and lead to loss of nerve cells or at least to loss of their function, Alzheimer's disease is one of the most well known. Under this name probably several similar brain disease conditions are hidden. Some of them are definitely determined by genetic abnormalities, but of other forms no cause is known. In any event a characteristic pathological feature of these diseases is that neurons degenerate at a slowly progressive pace in typical locations of the brain, gradually die in a special way and finally are lost. With their loss the functions which these cells subserved are also lost or impaired or at least altered. Typically parts of the temporal and frontal lobe in the brain are affected, but also the parietal lobe comes into play. The hippocampus with its important connections responsible for many memory functions, is particularly damaged. Dementia characterised by memory loss is a dominant and generally feared clinical syndrome of Alzheimer's disease, but other phenomena of general intellectual decline may be equally or more disturbing to the patient and the patient's environment. Loss of mental faculties may be obvious to the patient himself in the beginning of the disease, but after some years he may become less aware of this or not at all.

Another well-known but rarer brain disease in which an 'organic' cause leads to localised brain cell loss is Huntington´s disease. Here the cause is clearly an inherited genetic abnormality. Although during mid-life the brain damage becomes evident particularly in the basal ganglia of the brain and thus leads to a typical abnormal movement pattern (called 'chorea'), often the most impressing features constitute the changes in personality accompanied by irritability, aggression, professional and social derailment. Suicide is common. The changes in personality are usually thought to be due to pathological changes which not only occur in the basal ganglia, but also appear in the frontal lobe.

Finally, at the other end of spectrum are brain diseases which affect mainly brain regions responsible for body functions and less so mental capacities. A well-known example is Parkinson´s disease. The damage here occurs at a relatively small site in the brain stem namely the neurotransmitter dopamine producing cells in the substantia nigra (K.L.Leenders at al, 1984, 1986, 1990), which deranges the organization of the body movements by disturbing widely the large hemisphere's functions in this respect. Typical clinical symptoms are slowness of movement, resting tremors and muscle rigidity, but also balance problems, vegetative and other signs may be present. It is astonishing that under certain conditions it can be shown that part of the destroyed neuronal network can be reinstalled by implanting embryonic dopaminergic cells into human brain accompanied by clinical improvements (O. Lindvall et al, 1990).

In a fairly large part of the parkinsonian patients depression can occur as an expression of impaired brain function, but also cognitive alterations associated with frontal lobe involvement can be detected.

Some people predict that within a few years they will be routinely mapping and sequencing genes that influence intelligence, personality and important psychiatric disorders like schizophrenia.

The price which must be paid for having a highly developed and complex brain is far from low or insignificant. The 'normal' functioning of human intelligence is from time to time disturbed in healthy individuals, but in so-called 'mentally ill' people these functions are often and for more prolonged periods deranged. About 2 per cent of the population is affected. The World Health Organization report 1996 predicts that by the year 2020 mental illness will be the most debilitating affliction in the developing world.

Table 7-3 Hallucinations and others

Influence on brain areas	Kind of hallucinations
Electrical impact on the primary visual area	Simple hallucinations—points of lights-lighting, stars
Electrical impact on the secondary visual area	Patient believes he sees a butterfly and tries to catch it Patient believes to see a dog, calls it and wonders why the doctors do not notice it
Chemical impact of mescaline or LSD (probably the tertiary area of cortex)	Very colourful pictures

Schizophrenics do not, as a popular myth holds, have split personalities. Rather, the disease causes a wider fragmentation of their intellect and social selves, attacking the very qualities that make us human. Schizophrenia is still diagnosed today as it was at the turn of the century: by its psychopathology; that is by its abnormal pattern of thought and perception as inferred from language and behaviour. There exists no blood, urine or cerebrospinal fluid test, or X-ray CT, or PET, or MRI or EEG procedure which may prove the diagnosis of schizophrenia. We know that something goes wrong chemically or physically in the brain of a schizophrenic patient, but we do not know what (I.I.Gottesmann, 1991). Because schizophrenia mainly affects thought and emotion, it has been proposed that some abnormality in thalamus could be responsible for this (N.C.Andreasen et al, 1994).

There are a number of studies asking if madness plague great artists. Several studies now show that creativity and mood disorders are linked

K.R.Jamison (1997): "Bouts of depression and manic energy are unusually common among gifted artists, musicians and writers. The painful roller coaster of their emotions may deepen their creative appreciation of the ambiguities of every day life."

Biographical studies of earlier generations of artists and writers show high rates of suicide, depression and manic-depression—up to 18 times the rate of suicide seen in the general population, eight to 10 times that of depression and 10 to 20 times that seen in manic-depressive illness and its milder variants. The manic-depressive temperament is, in a biological sense, an alert, sensitive system that reacts strongly and swiftly. And here it must be remembered that there are suspicions that manic-depressive illness is at least partially genetically influenced.

The prefrontal cortex is particularly relevant to schizophrenia because it contains circuits that are active during manipulation of symbolic information and in a type of working memory. Positive results have been obtained by pharmacological treatment, such as the phenothiazines and the butyrophenones, which ameliorate hallucinations (Table 7-3), delusions, disorganized thinking and inappropriate affects—'positive' symptoms of schizophrenia that are most evident during acute psychotic episodes. It seems that relatives of schizophrenics have an eight times greater risk of becoming schizophrenic than the average person. The data suggests that inheritance plays a role in schizophrenia, but also indicates that nongenetic factors are of critical importance. In spite of extensive and intensive research in this field, the connection between psychiatric disease and a single locus on the human genome has not been found. The risk of illness rises with increasing genetic similarity, but even a good identity of genes does not produce a perfect correspondence. Although schizophrenia and manic-depression can devastate the patient's life, the disorders do not preclude the performance of highly creative work. Manic-depressive illness often occurs in connection with extraordinary talent, even genius, in politics and arts.

One of the main criteria for the functioning of a 'good' theory of intelligence (or better, of the human brain?) is whether this theory can deal with both real and illusory stimuli. Explanation of the mechanism of hallucinations (false perception in the absence of external stimuli) could also be very helpful. Using PET (positron emission tomography) brain states associated with the occurrence of auditory verbal hallucinations (voices talking to or about them) in six schizophrenic patients have been studied (D.A. Silbersweig et al, 1995).

D. Goldman (1996): "Much of the variation in personality traits and in diseases such as schizophrenia, bipolar affective illness, and alcoholism is genetic in origin, at least when examined in one population. ..The amount of neuroticism, a personality trait that can be quantified by testing, is influenced by two alleles of a gene encoding a transporter for the neurotransmitter serotonin. One allele results in more protein—and more neuroticism—and other, less protein and less neuroticism. The number of potential candidate genes—genes that can in any way alter brain function—is formidable:.. more than 200."

Schizophrenia—one of the most debilitating of mental illnesses, in which thoughts and emotions are sometimes disconnected or distorted by delusions—is expected to afflict around 25 million people in low-income societies by the year 2000 a 45 percent increase over the number afflicted in 1985, mostly among women (A. Kleinnman, 1997).

U.Frith (1997): "Autistic individuals seem lost in their own inner world. The isolation stems from biological abnormalities that may interfere with the ability to imagine other's people's mental states."

7.3 PERSONALITY: THE PRODUCT OF GENES OR OF CULTURE

7.3.1 *How does a personality emerge?*

Once more it is worthwhile to refresh the content of the terms 'genetic' and 'epigenetic'. Epigenesis: developmental pattern arising by gradual unfolding of the genetic program under the influence of environmental, cultural elements.

Probably the supreme (highly abstract, most complex and far-reaching) human concept is that of Self (Section 8.1.1). Directly associated with the concept of Self is that of personality. Here the term personality is used in the sense of a complex set of characteristics that distinguish a particular selfconscious human individual from all other individuals.

Descartes (1637): "What then am I? A thinking being. What is a thinking being? It is a being which doubts, which understands, which conceives, which affirms, which denies, which wills, which rejects, which imagines also, and which perceives."

Developmental psychology is a branch of science which investigates the emergence and development of personality, from birth to adulthood, towards a fully developed form.

J.Piaget and his followers are convinced that children advance through four rather regular stages of intellectual development (see Fig. 7-4, where Piaget's own nomenclature has been used): 1) First stage: 'sensorimotor intelligence', from birth up to two years. This stage is divided into three periods: a) from birth until one month, called 'reflex'; b) from one to three months, called the 'primary circular reaction' stage; c) from three months to two years, called the 'secondary co-ordination' stage. 2) Second stage: 'preoperative intelligence', from two to seven or eight years. 3) Third stage: 'concrete operative', up to eleven years. 4) Fourth stage: 'formal operative', older than eleven years.

The significance of the theoretical approach of Piaget cannot be overestimated, but it seems that some elements of intellectual development have been omitted. To these belong all the emotional components, and some specific human behaviour, such as smiling. According to Piaget, in the first stage a child learns to use his or her muscles and senses to deal with external objects and events, while language begins to form. The child also begins to deal with objects and know that they exist, even if they are beyond sight and touch.

Psychologists believe that most people's personalities can be considered in terms of five basic attributes: energy, friendliness, conscientiousness, emotional stability and intellect.

He or she also begins to 'symbolize'; that is, to represent things by words or gestures. In the second stage the child experiences the greatest language growth; words and other symbols become a way to represent both the outside world ('Non-self') and inner feelings ('Self'). He or she begins to gain a sense of symmetry, to learn by trial and error, and to integrate symbolization and elementary types of relationships, such as logical and mathematical, and spatial and temporal.

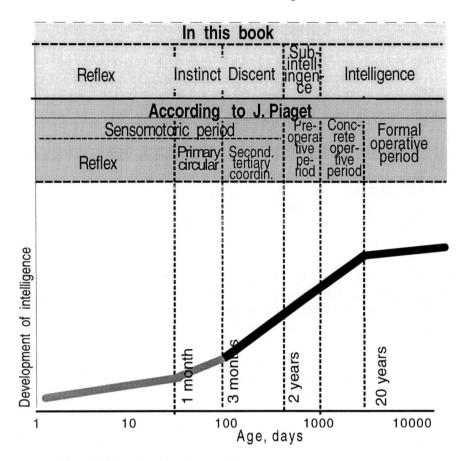

Figure 7-4 Development of personality

M.Minsky (1985) stressed the significance of the explanation, first proposed by S. Papert in the 1960s, for the differences between older and younger children. Most previous theories had tried to explain Piaget's experiments by suggesting that children develop different kinds of reasoning as time goes by. That certainly is true, but the importance of Papert´s conception is in emphasizing not merely the ingredients of reasoning, but how they are organized. A mind cannot rely very much merely on accumulated knowledge; it must also develop better ways of using what it already knows. Minsky called this 'Papert's principle', and formulated it in the following way: "Some of the most crucial steps in mental growth are based not simply on acquiring new skills, but on acquiring new ..... ways to use what one already knows."

Table 7-4 Five main determinants of personality (according to T.J. Bouchard Jr, 1994)

Extroversion, Introversion-Extroversion (-) Dominance, *Positive emotionality*	
Is outgoing, decisive, persuasive and enjoys leadership roles	Is retiring, reserved withdrawn, and does not enjoy being in centre of attention
Neuroticism: Anxiety, Emotional Stability (-), Stress Reactivity, *Negative Emotionality*	
Is emotionally unstable, nervous, irritable, and prone to worry	Quickly gets over upsetting experiences, stable and not prone to worries and fears
Conscientiousness: Conformity, Dependability, Authoritarianism (-), *Constraint*	
Is planful, organized, responsible, practical, and dependable	Is impulsive, careless, irresponsible, and cannot be depended upon
Agreeableness Likability, Friendliness, Pleasant, *Aggression* (-)	
Is sympathetic, warm kind, good-natured, and will not take advantage	Is quarrelsome, aggressive, unfriendly, cold and vindictive
Openness: Culture, Intellect, Sophistication, Imagination, *Absorption*	
Is insightful, curious, original, imaginative, and open to novel experiences and stimuli	Has narrow interests, is unintelligent, unreflective, and shallow
Negative signs indicate trait names that characterized the opposite end of the dimension. The italic trait terms indicate the 'Multidimensional Personality Questionnaire' factors or scales used to measure these five characteristics in the Minnesota study of twins reared apart.	

V.B. Mountcastle (1990): "Each of us lives in the centre of his own perceptual space, and from that central position each experiences the functioning of his own brain. Thus each of us constructs, stores, and recalls his own uniquely private image of the world and events within it. Your images and my images may at times be identical and veridical as regards physical reality. Not often they are constructions that differ between individuals... which contribute to the uniqueness of each human personality."

The process of the spontaneous education of a child through contact with its mother is of greatest importance. As we have mentioned, consciousness requires a fully developed connection between both hemispheres of the brain. The corpus callosum and other commissures begin to evolve in a new-born child's brain, and reaches full maturity only after some years. Selfconsciousness achieves its full form as contralateral connections gain in strength.

Some humans with a severe form of epilepsy, have been operated on by means of commissurotomy; that is, cutting the corpus callosum and other connections between the left and right brain hemispheres. M.S.Gazzaniga, claims that these patients after the operation, do not have a concerted, unified personality, are no longer mental unities, and express, at least partially, two personalities.

In the term 'personality' is included the concept of free will and the awareness of the continuity of a person in time and space, as the result of their own history. But what happens if parts of our own body, such as blood, heart or kidney, are surgically exchanged with an alien (from another human body) or artificial replacement? As we know, this has no influence on the personality. But what will happen if, in the future, a surgeon will be able to exchange an organ belonging to the brain, for example causing significant increase of the production of enkephalin, an endogenous neurotransmitter which mimics the effects of opiates (M.F. Perutz, 1996).

Personality also means individuality, and vice versa. What is the basis of such a strong differentiation between human individuals? One reason is each person's own nature—genetic differentiation and the resulting somatic peculiarities, especially in the brain. The other cause of differentiation is nurture—the influence of the social environment. The latter is memorized in the form of different bonus-malus labels for each memorized event. We are individuals because our bonus-malus system has been coloured in unique ways by our experiences.

7.3.2 *Intelligence: nature or nurture*

Just 20 years ago, the idea of genetic influences on complex human behaviour and intellectual abilities was anathema to many behavioural scientists. Now, however, the role of inheritance in behaviour has become widely accepted, though this does not mean that it should be exaggerated and used unfairly (Fig. 7-5).

Of more than 4000 single-gene effects in human beings investigated up to the present time, more than a hundred have shown a measurable effects in the direction of decrease of what we call cognitive abilities, gauged by means of intelligence quotient scores. The effect of genetics on the differentiation of the normal brain has been discussed in Section 4.6.

P.Medawar (1984): "Geneticism is the doctrine of the primacy of the genetic make-up in determining every aspect of the human mind, constitution and social behaviour: We must look to genetics for an explanation of tribalism and aggression, the rise and fall of nations, the stratification of societies into classes, the nature and degree of human intellectual process. Historicism is the doctrine that there exists, or can be propounded, a historical social science that makes possible ostensibly scientific predictions about the future condition of man. I cannot call to mind, and have never been informed of, any genuinely novel or illuminating insight that has come to us through socio-biology."

R.Plomin (1990): "Genetic influence on behaviour appears to involve multiple genes rather than one or two major genes, and nongenetic sources of variance are at least as important as genetic factors. ... This should not be interpreted to mean that genes do not affect human behaviour; it only demonstrates that genetic influence on behaviour is not due to major-gene effects."

Twins who were separated early in life and reared apart are a natural experiment in 'Nurture or nature'. Monozygotic twins (originating from the same fertilized egg) and dizygotic twins (originating from two different fertilized eggs) are the results of the same or very similar, genetic sets—the same 'nature'. Separately raised twins represent the influence of nurture. New results from the investigation of the psychological differences of twins brought up apart have been recently published.

T.J. Bouchard Jr (1990): "It is a plausible hypothesis that genetic differences affect psychological differences largely indirectly, by influencing the effective environment of the developing child. This evidence for the strong heritability of most psychological traits, sensibly construed, does not detract from the value or importance of parenting, education, and other propaedutic interventions.... The genes sing a prehistoric song that today should sometimes be resisted but which it would be foolish to ignore. If genetic variation was evolutionary debris at the end of the Pleistocene, it is now a salient and essential feature of the human condition... The correct formula is 'Nature via nurture'."

T.J. Bouchard Jr (1994): "Current thinking holds that each individual picks and chooses from a range of stimuli and events largely on the basis of his or her genotype and creates a unique set of experiences—that is, people help to create their own environment...Humans (are) dynamic creative organisms for whom the opportunity to learn and to experience new environments amplifies the effects of the genotype on the phenotype. Is also reminds us of our links to the biological world and our evolutionary history."

S. Rose (1995) who is well known as a strong critic of the overestimation of the connection between intelligence quotient and the genetic components wrote recently a paper entitled 'The rise of neurogenetic determinism'.

Social contacts are an almost self-evident example of the impact of the social environment on the human behaviour. For example neurons in crayfish whose response to the neurotransmitter serotonin differs dramatically depending on the animal's social status. In dominant animals, serotonin makes the neuron more likely to fire, while in subordinate animals serotonin suppresses firing (S.R.Yeh, 1996). Recently R.Adolphs et al (1994) studying the role of the amygdala on the ability to respond to the facial expression of communicating individuals claims "Facial expressions can convey both basic emotions whose expression and recognition may be partly innate, as well as subtler emotions whose meaning is partially determined by culture".

In consequence it seems to be possible to differentiate the various influences on human behaviour into a number of fairly well defined factors:

–Genetic factors, including genetic diseases, some sexual differences, and even some racial differences, which cannot be excluded a priori on ideological grounds,

– Somatic factors, such as the health of the mother during pregnancy and the kind of food eaten, and somatic diseases, especially in childhood,

– Social and cultural influences, beginning with the mother language and continuing through the parents' families, school, husband or wife, profession, political events, etc.

–Selfconsciousness and the feeling of free will, that is some 'chaotic' events.

J. Hogan (1996): "Optimists hope that in years to come, the sciences of the mind will coalesce around a new, more powerful paradigm, one that will transcend the schisms—nature versus nurture, drugs versus talk therapy. One proponent of such a shift is S. Hyman who declares: 'From the point of view of people who think about the brain and mental health, the traditional dichotomies are simply false'."

7.3.3 School and intelligence

A general accepted opinion is that for children the best way for improving their introduction into society and their 'intelligence' is the schooling system requiring a number of years. The phenomenon of 'schools' exists in number of species, beginning from fishes, birds and mammals.

One of the well known and not fully solved problems is how to improve the technique of learning, the didactic systems, the correlation between the individuals properties and abilities and the 'mass-production' of the school system.

7.3.4 Taboos, sexuality, placebos, hypnosis and psychosomatics

We have previously stated that intelligence, the highest stage of information processing, depends directly on the lower stages of instinct and reflex. Taboos, particularly those connected with food, such as the consumption of snails, by some individuals or even some nations, are a good example of the impact of intellect upon instinct. Some people will die (hunger as instinct) but don't eat snails because of quasi-taboo. It is trivial to say that the instinct of sexuality can only with great effort be controlled by intelligence and with strong social surveillance. A unique phenomenon which exists probably only amongst Homo sapiens sapiens, is pornography—the direct influence of sometimes sophisticated writings or two- dimensional, even very schematic black-and-white, drawings on sexual behaviour (Fig. 7-6).

On the other hand, control of the sexual instinct by intelligence can, in some cases, be positive. A good example is celibacy, promoted by religious motivation. Much more extreme is the phenomenon of martyrdom, when intelligence, normally in obedience to religious belief, suppresses the reflex of self-defence. Operating in the same direction is the modern phenomenon of the hunger strike, which acts against one of the strongest reflexes: the appeasement of appetite.

Much more complex is the so-called 'psychosomatic effect', exemplified by the phenomenon of the *placebo* (Latin: I shall please), an inert, inactive medicament or preparation, given for its psychological effects, but usually achieving real positive results. In this case, intelligence, through the conscious brain, influences the internal activity of some hormones or even transmitters, acting as a real medicament would do.

Everyday intelligence 281

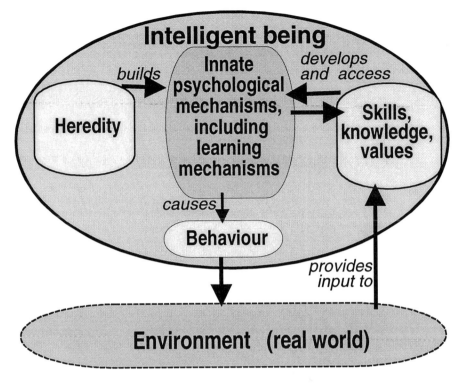

Figure 7-5 Nature via nurture
(according to S.Pinker, 1994)

Hypnosis refers to a condition in which the 'normal' individual becomes highly responsive to suggestions. Self-hypnosis may be self-induced, by trained relaxation, concentration on one's own breathing, or by a variety of monotonous practices and rituals, during which the focus of attention is withdrawn from the outside world and is concentrated on own feelings, simple arm moving etc.

7.4 THINKING: THE QUINTESSENCE OF INTELLIGENCE

7.4.1 *What do we think about thinking*

Descartes (1637-41) in 'Discourse on method and Meditations': "What then am I? A thinking being. What is a thinking being? It is a being which doubts, which understands, which conceives, which affirms, which denies, which wills, which rejects, which imagines also, and which perceives."

The term 'thinking' is notoriously vague, but we can all agree that it hinges on two kinds of properties. First, information must be represented internally;

and second, that information must be manipulated in order to make conclusions. Some students claim that thinking is no more than talking to oneself (S.M.Kosslyn, 1995). The controversy surrounding the nature of thinking, especially the use or non-use of words and of verbally formulated concepts, is as old as the study of selfconsciousness itself.

A.Einstein (1950): "The words or the language, as they are written or spoken, do not seem to play any role in my mechanism of thought. The physical entities which seem to serve as elements of thought are certain signs and more or less clear images which can be 'voluntarily' reproduced and combined."

R.Penrose (1989): "This is not to say that I do not sometimes think in words, it is just that I find words almost useless for mathematical thinking. Other kinds of thinking, perhaps such as philosophising, seem to be much better suited to verbal expression. Perhaps this is why so many philosophers seem to be of the opinion that language is essential for intelligent or conscious thought! No doubt different people think in very different ways."

G.M.Edelman (1989): "There is no language of thought."

According to S.Pinker (1995) J.A.Fodor claims that in the human brain 'propositions' are implemented in the brain as data structures organized like sentences, not in English, but in a language of thought 'mentalese'. Thinking is computation; mentalese symbols are rearranged by neural algorithms sensitive to the symbols' identities, including 'syntax'. Mental content is information in something like the mathematical sense. So when all goes well, the world causes us to have mentalese sentences whose contents are true. In some cases the same physical thing have different symbols in mentalese: e.g. 'morning star' and 'evening star'.

The deeper question is how knowledge of words is organized in the brain? (H.Damasio et al, 1996) may have provided a partial answer to this problem (Table 7-5). Their study was restricted to concrete concepts. Some questions are open: are abstract concepts (justice, evidence) also represented categorically? Are they, too, represented in the temporal lobe? And what about syntactic information? (A.Caramazza, 1996).

However, H.Damasio's results indicate that the normal retrieval of words which denote concrete entities depends not just on classical language areas (Wernicke's and Broca's areas), but also on regions in higher-order association cortices.

7.4.2 *Paradoxes and the principle of incompleteness*

Common sense contrary to its own concept exists side by side with paradoxes. A paradox arises when a contradiction follows from seemingly obvious premises (common sense) of logical and semantical systems.

Science progress goes through the solution of paradoxes. The interpretation of the nature of light swung back and forth between the concepts of particles and waves—until the quantum theory unified both. Physicists and engineers are familiar with 'liquid crystals', which are neither solid nor liquid. There are paradoxes which seem to be an inherent part of our system of thought, of our logic, and there is no hope that they can be resolved by a new theory.

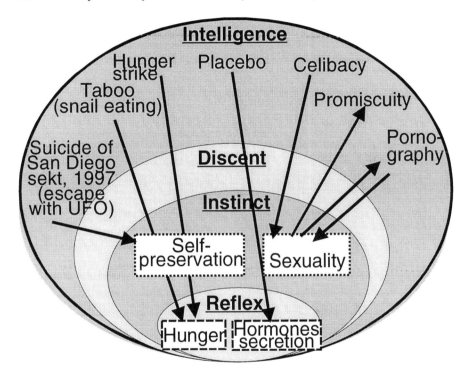

Figure 7-6 Taboos, placebo, celibacy, pornography, etc.

M.Chagall (1967): "Pictures consist of more doubts and silences than words could express."

The human way of thinking leads to a number of paradoxes. One of these is called Russell's paradox, which can be expressed in essentially commonplace terms. Imagine a library in which there are two catalogues, one of which lists just those books in the library which somewhere refer to themselves, and the other, precisely these books which make no mention of themselves. In which catalogue is the second catalogue itself to be listed? One could ask how paradoxical is the Russell's paradox?

Another paradox illustrates very well the situation of authors and readers of this book. The book appeals to human intelligence with the aim of investigating human intelligence itself.

Table 7-5 Brain's dictionary
(according to H.Damasio et al, 1996)

Area	Role in words organization	Content	Brain's area
One	Conceptual content of words	Meaning (semantic categories)	Distributed network involving structures in both left and right hemispheres
Two	Mediating between first and third	Lexical knowledge, modality independent	Left temporal lobe (cortex cerebri), outside 'language areas', divided into: inferior temporal lobe for non-manipulable *'animals'*, posterior regions for manipulable *'tools'* and for complex items as for *'people'*
Three	Phonological elements of words (speech production)	Sounds	Wernicke's and Broca's area on the left hemisphere

K.Gödel, in 1931, showed that any precise ('formal') mathematical system of axioms and rules of whatever procedure, provided that it is broad enough to contain descriptions of simple arithmetic propositions, and provided that it is free from contradiction, must contain some statements which are neither provable (able to be verified) nor disprovable (capable of being proved false) by means allowed within the system. The truth of such statements that cannot be proved is thus 'incompleteness'.

In 1963 P.Cohen showed that Gödel´s incompleteness proposition cannot be proved either. We can say, firstly that Gödel prepared an algorithm which described the incompleteness of all other systems of formal algorithms, and secondly, that Gödel´s proposition has also been formulated by means of some type of algorithm, and therefore cannot be proved or falsified itself. The human mind can agree that Gödel´s incompleteness principle is true, but it must be stressed that a computer, at present and in the future, at least in the near future, is incapable of dealing with such logic. Figure 7-7 illustrates a formal system in which one statement has been declared as incomplete. It is possible to find a larger system, of which this statement is only one part, and prove the decidability of the statement within the framework of the larger system. However, one could then discover that, in the larger system at least, one other statement must be declared as incomplete. Thus the principal problem of incompleteness remains.

It seems to be worth to cite this short anecdote according to one of the best mathematician of our century:

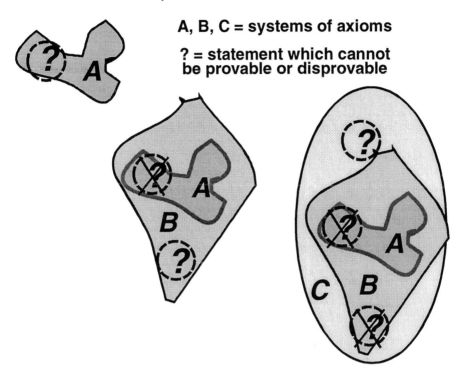

Figure 7-7 Gödel´s incompleteness

S.M.Ulam (1982, published 1987): "A prime number is an integer that is not divisible by any number except itself. The numbers 2,3,5,7,....41,43,47, et cetera, are all prime. The Greeks knew that there are infinitely many prime numbers. That is one of the oldest, greatest, and most beautiful discoveries in mathematics. Now certain pairs of prime numbers, such as 5 and 7, 11 and 13, 17 and 19, are called twins because they differ by only 2. The question is: How many twin primes are there, a fixed number or an infinity? Nobody knows the answer to this question, and it may be undecidable. I asked Professor Schmidt, a very famous number theorist, if he knew who first proposed this very old problem and whether he thought it may be undecidable. He did not know the answer to the former, and to the latter he answered. 'One might not be able to decide whether it is undecidable'."

Could the proposition of K.Popper concerning the criterion of 'falsifiability' (contrary to 'verification') be falsified itself?

A.Tarski (1931) phrased the traditional correspondence theory of truth in a general form: " 'It is snowing' is a true sentence if and only if it is snowing", then showed how it leads to the Liar's Paradox: " 'This is a false sentence' is a true sentence, if and only if, it is a false sentence."

Heisenberg's principle of uncertainty is equivalent to the principle of unpredictability, which in turn is equivalent to the principle of indeterminism. How certain is the principle of uncertainty?

Do we have enough clear and well founded information to describe what information (especially in our brain) is?

Is the human brain satisfying 'braining' to be able to investigate itself, that is our brain?

And the most important question: Do we have really enough 'free will' to research what the free will in reality is?

7.4.3 *Myths and rationality*

The human brain has been prepared by Nature for biological effective information processing. Rationality, something which is acceptable to reason, is not an inherent property of human thinking. If we refer to the opposite of rationality, not by the term 'irrationality', but as 'mythological thinking' then this term must be explained.

The original Greek term for *myth* (*mythos*) means 'word', in the sense of a decisive, final pronouncement. It differs from *logos*, the word whose validity or truth can be argued and demonstrated, because myths are simply unprovable or fiction stories, and thus the word myth has become a synonym for fable. Myth is a collective term used for one kind of symbolic communication.

Myths are accounts with an absolute authority that is implied rather than stated. Natural, social, cultural, and biological facts are explained by myth. Myths set the pattern for theoretical as well as practical instruction. The descriptive function of myths expresses a perennial human need and is a reminder that modern man does not have to regard a myth as something that comes to an end. The function of models in physics, biology, and other sciences resembles that of myths as paradigms (patterns) of the human world. Once a model has gained acceptance, it is difficult to replace, and in this respect it resembles a myth.

Models in science have their value only for specific scientists. In spite of the distance between scientific models and myths, there are reasons for speaking of mythological dimensions and components of science. M.Minsky, a prominent scientist in the field of artificial intelligence, very often claims that a number of terms used in the cognitive sciences, in neurobiology and especially in psychology are no more than myths.

In this book we have already mentioned Minsky's opinion that such terms as 'consciousness', 'creativity', and even 'intelligence' are all only myths and no more. A book by R.Weisberg (1986), with the title 'Creativity, genius and other myths', refers to the opinion of some scientists active in the field of the cognitive sciences.

It must be underlined that some terms used in this book have a more or less mythical character; selfconsciousness, free will, and mind are examples. It is not clear if our way of thinking, especially the function of understanding, is based on some inexplicable myths. However, there is no space here to discuss similar phenomena, such as religion, in spite of their significance in connection with human intelligence. The differences between the terms 'mythos' and 'logos' correspond to the two types of thinking: heuristic (also holistic) and algorithmic (also analytic). There is a clear link with the old theme: left (algorithmic) and right (heuristic) human brain hemispheres.

7.4.4 Deception, fraudulence, cheating

Deception, in the broad sense, begins at a rather low level of biotic evolution, at the stage of instinct. Viceroy butterflies, which are a tasty treat for birds mimic the markings of the foul-tasting Monarch butterfly to trick birds into not eating them. Cuckoos fool other birds by laying eggs in their nests. All these phenomena are controlled at the genetic level. Some forms of deception have been observed in the behaviour of chimpanzees which have been learned by experience. This corresponds to the stage of discent.

Cheating, fraudulence, lies are the most 'human' properties of communication with other individuals and with environment. Some scientists propose that one of the mind's most useful modules is dedicated to detecting 'cheating' by others. L.Cosmides (1995) in tests with volunteers showed that humans are much more adept at solving problems if the solution requires the detection of cheating rather than some purely logical, abstract chain of reasoning. Deception, in the narrow sense, seems to have fully evolved at the stage of intelligence. Men lie to women, and vice versa. The different advantages of deception do not stop with individuals. Some scientists argue, that the existence of human cultures has historically depended on what can be called 'noble lies'—myths or self-deceptions. From this point of view the theatre for adolescents and also the fairy-tales for children can be considered as an open, approved cultural deception.

Demosthenes (384-322 BC): "What a man desires, he also imagines to be true."

St.Thomas Aquinas (1225-1274): "The light of faith make us see what we believe."

J. Locke (1690) in 'An essay concerning human understanding' Dent. London 1691: " Referred to words as a 'perfect cheat'. "

Voltaire (1694-1778): "Humans use speech to hide own thoughts."

Royal Society (London) took for its motto: *"Nullius in verba"*; "In the words of no one"

He Zuoxiu (1996): "It's really strange that it is more difficult to tell the truth than to tell lies."

In Newsweek (6th January 1997) an article of S.Begley entitled 'Infidelity and the science of cheating' is a small but characteristic illustration of this side of human intelligence.

Throughout history men have imagined—and given even careful, meticulous and exact descriptions of—unicorns, dragons, giants, mermaids, gods, devils, angels, UFO's, people whose heads grew out of their chests and much else, to little or no purposes, so far as reality goes. In spite of all these old and common experience going through all cultures the people in American courts of law must swear to tell "the truth, the whole truth, and nothing but the truth", even if we all agree that this is impossible.

C.Sagan (1996): "It is simply beyond our powers. Our memories are fallible; even scientific truth is merely an approximation; and we are ignorant about nearly all of the Universe...The American system of jurisprudence recognizes a wide range of factors, predispositions, prejudices, and experiences that might cloud our judgement, or affect our objectivity—sometimes even without our knowing it."

The statement of Plautus (250 BC): *"Homo homini lupus est"* (Human being appear to another human being as a wolf') is well-known. Not so strong, but much more nearer to the truth, is the here proposed statement: *"Homo homini mendax est"* (An intelligent being appears to another intelligent being as a liar) (Latin *mendax* liar, adjective: mendacious). In biblical psalm CXV is to read: *"Omnis homo mendax"* (All humans are liars).

According to Newsweek (July, 1996) 48 per cent of Americans believe UFOs are real and 29 per cent thinks we've made contacts with aliens. Another 48 per cent thinks there's a government plot to cover the whole thing up.

Fairy-tales for children are a kind of teaching of deception in the human society, and therefore is of high importance. In the naive perception of 'Science' this is the only place under the Sun which is free of lies; it is the realm of truth. In the long history of science this picture was far from being realistic. The recent book of L.Grayson titled 'Scientific deception' (1995) consists of a precize review of 230 papers on misconduct in science, its perpetrators, their motives and methods.

There was some efforts to prepare in 1996 in the USA a scientific conference on scientific misconduct, especially the 'three deadly sins' of fabrication, plagiarism and falsification.

Moses Maimonides (1200): "It is forbidden to engage in astrology, to utilize charms, to whisper incantations... All of these practices are nothing more than lies and deceptions used by ancient pagan people to deceive the masses and lead them astray... Wise and intelligent people know better."

P. Picasso (1936): "Art is a lie that helps us see the truth."

The understanding of deception is usually taken as criteria in ascribing a 'theory of mind' to children and non-human primates. Children under four years of age are unable to recognize the very possibility of a false belief. By age four or five they understand that someone could hold a false belief and that such a belief would led the holder to fail to reach his goal. That understanding underlies a host of 'mental activities such as introspection, the understanding of surprise, deception, (possibility of tricking or misleading someone), the interpretation of ambiguity, as well as a variety of childhood games such as hide-and-seek (D.R. Olson, 1994).

And finally some questions: why our culture pays so high attention to the 'theory of truth' and so neglected the problem or even 'theory of lie'. One exception is the rather primitive 'lie detector'.

7.4.5 Humour is unique

One phenomenon which is unique to the stage of intelligence is humour. According to Webster's dictionary, humour is that quality in a happening, an action, or an expression of ideas which appeals to a sense of the ludicrous or absurdly incongruous elements; a comic or amusing quality. Humour, smiling and laughter are strongly linked. However, laughter can be observed in babies, and must therefore be at least partially controlled genetically. The strength of a smile increases due to social co-operation, firstly due to the smiling of the mother. Humans are humans, not only because of their ability to speak, but also due to their ability to smile and to deceive (J.S. Bystron, 1939).

To understand a joke one must know what a joke is, that is, one must have the appropriate concept. Young children tell faulty riddles, and laugh uproariously before they actually understand what a riddle is.

7.5 CULTURE: IDEAL ARTIFACT

7.5.1 *Culture: product and source of intelligence*

Intelligence exists exclusively inside an artificial man-made environment called culture.

Third New International Webster: "Culture...is the total pattern of human behaviour and its products embodied in thought, speech, action, and artifacts, and dependent on man's capacity for learning and transmitting knowledge to succeeding generations through use of tools, language and systems of abstract thought."

For centuries, the manufacture and use of tools have been taken as the hallmark of humanity. There is no evidence of tool use by species of Australopithecus, and the appearance of tools seems to have coincided with the emergence of Homo habilis, some 2 megayears ago. Only when Homo erectus appeared about 1.8

megayears ago, did standarization and sophistication in the form of the tools become the rule. This imposition of regularity of form following a mental representation of tool maker has been seen as the hallmark of culture (Ch.Boesch, 1996).

T.Dobzhansky, E.Boesiger (1983): "Culture is the totality of information and of behavioural patterns that are transmitted from individual to individual, and from generation to generation, by instruction and learning, and by example and imitation. Certainly, culture depends ultimately on the human genetic endowment; nonhuman animals at most have rudiments of cultural transmission. What human genes transmit is the potentiality for the acquisition of culture, not culture itself. The realization of this potentiality depends mostly on human symbolic languages."

D.R. Olson (1994): "Our literature, our science, our philosophy, our law, our religion, are, in an important way, literate artifacts. We see ourselves, our ideas and our world in term of these artifacts. As a result we live not in the world so much as in the world as it is represented to us in those artifacts."

7.5.2 *Psychoactive agents and culture*

Hundreds of plants, fungi, insects are bearer of substances which when consumed by higher animals, act on the brain, and heavily influence the 'normal' biotic information processing. The selection of these plants and fungi is going on since millions of years. Direct eating of these materials has rather weak 'psychedelic' activity. Therefore even apes know only a very limited number of such 'natural' drugs.

But the invention of fire using, coupled with its 'technology' of hot-water-extraction and inhalation of smoke of burned plants or fungi allowed hallucinogenous agents to be transported through the blood-brain barrier and to influence the action of the central nervous system. Maybe even Homo erectus and probably only Homo sapiens, particularly Homo sapiens sapiens discovered the possibility of fire and heat using for transformation of natural products into 'drugs' (Table 7-6).

Use of psychedelic substances (originated from more than 100 plants, fungi or even insects), has been 'discovered' probably by Homo sapiens sapiens about 40-50 thousands years ago. In the same time the artistic activity exploded. Probably the skin-painting, simple wood and stone ornaments and firstly astonishing cave paintings are intimately coupled with the refinement of use of some psychedelic drugs, and with the emergence of magic activities and ceremony. There is a number of hypotheses, or more exact, a number of 'suspicious' pseudo-hypotheses that the use of some psychedelic substances has had a rather significant impact on the development of the highest biotic information processing, human intelligence. It seems also probable that 'development' of intelligence facilitates the search, selection and use of some natural psychedelic drugs. There exists some kind of 'self-catalysis'. These activities could have influenced the sexual selection in some tribes of Homo sapiens, and even be a drive for territorial expansion.

It must be stressed, that the above formulated remarks are far from being well proved. But may be they illustrate the long, complex, and still hidden way of the development of intelligence.

The Incas were acclimatized to high altitudes in Andes, but the use of cocaine made it easier for them to accomplish strenuous work under such demanding conditions. One Georgia USA pharmacist designed in 1886 a drink including cocaine, called Coca-Cola. There is little doubt that these compounds endow the user with far greater mental endurance than he or she normally exhibits. Yet many brain researchers are interested in understanding why the psychedelic drugs (mescaline, psilocybin, LSD) affect the very core of the user's consciousness.

The two well-known addictive drugs, nicotine and cocaine, influence in a similar action the nucleus accumbens, but not the amygdala (E.M. Pich et al, 1997). Numerous (if not all) human cultures include the use of 'artificial' (man-made) means for more or less direct influence on the human brain. One of them are psychoactive drugs (S.H.Snyder 1986). Human beings appear to have a proclivity for chemically induced stimulation. Most people are apparently satisfied by chemicals such as the caffeine and theophylline found in coffee and tea, but others seek much more intense stimulation. The use of alcohol has been known for thousands of years. Although early users of more potent chemicals, such as alkaloids, also took them orally, current users compulsively self-administer them under conditions of maximal exposure to achieve the most intense sensations. The investigation of alcoholism shows some influence of genetic factors, though nongenetic, cultural factors are predominant. The position seems to be that nature provides facilities for nurturing such tendencies or overcoming them. In the framework of our scheme of intelligence, the bonus value for alcoholics increases when the amount of alcohol in the brain increases, while the malus value increases when the alcohol level drops. Almost all drugs of abuse, including cocaine, amphetamine, heroin, alcohol and nicotine influence the reward system (bonus-malus system) (A.H.Glassman, G.F.Koob, 1996).

S.Lem (1970), in his science fiction novel 'Futurological congress', describes society in the year 2039, when tens of giga of humans dwell on this planet and the material and spatial resources are almost exhausted, and when the world is ruled by a new social regime called by Lem 'crypto-chemo-cracy'. The entire world population is continuously fed by special drugs which make the real world, in deep poverty, with illness and a terrible climate, appear to be a good, brave, happy world. The drugs are distributed in the drinking water, or in the form of aerosols

Table 7-6 Psychoactive agents, drugs

Alcohol	produced by means of the fermentation of any sugar-containing concoction, such as one of grapes, fruits, or honey	probably discovered in the pre-agricultural gathering stage of human development, in numerous countries in the Near, Middle and Far East
Coffee	the seeds of a shrub	native to Ethiopia and Arabia
Cocaine	coca plant grows in the Andes: contains 1–2% cocaine:	the second most commonly used drug in USA intake of > 0.1 g is prohibitive
Tea	the leaves and buds of the tea plant	originating in China about 5,000 years ago, used on a global scale
Tobacco	the leaves of numerous species of Nicotiana plants, members of the potato family	cultivated in North America, used on a global scale, in various ways for smoking, sniffing, chewing and extraction of the nicotine
Opium, morphium, heroin	obtained as a milky juice from the unripe seed capsules of the poppy plant, synthetic codeine	native to Asia Minor, in China it was the mind-bending drug of choice, used on a global scale 6,000 years ago
Peyote (mescaline)	a desert cactus Lophophora williamsi	native to North America, Mexican highland psychotropic, during religious ceremony
Belladonna	a plant jimsonweeds	used by the Aborigines as a narcotic,
Cannabis (marijuana hashish)	marijuana plant, Cannabis sativa; wild in most countries. The most used illegal drug.	appears in the Egyptian papyrus of the sixtieth century BC. in Assyrian texts, and in Greek and Roman sources as minor drug
Ananda-mide (in chocolate)	cacao bean, include anandamide, an endogenous cannabinoid of the brain	brain cannabinoids in chocolate (E. Tomaso et al, 1996)
Psilocybin	hallucinogen from the mushroom Stropharia cubensis, growing on the excrement of grazing ruminants	on the grasslands of Africa, in Mexican highland perhaps also Thailand, since of thousands of years. Through hallucinations reinforced religious (shamans) practices
LSD	lysergic acid diethylamide, a semisynthetic preparation	derived from an extract from a fungus which grows as a parasite on wheat
Cathinon	leafs of Catha edulis	in Somalia chewed in large amount, up to 1 kg of leafs per day
Incense	grain of resins that burns with flagrant odour	Babylonians used it extensively while offering prayer or diving oracles
Myrrh	natural compound secreted by shrubs, common in Northeast tropical Africa	used by the Egyptians for embalming and by Jews as anointing oil. Known to have analgesic effects; interaction with brain opioid mechanisms
Ecstasy	synthesized in 1914, has become an essential component of modern drugs	amount of 0.1 gram of 'Saturday night', used on a global scale

dispersed by robots on the streets. The brain of the 'man-in-the-street' is consequently unable to recognize this deception, organized by a few experts.

Certain chemicals in a bar of chocolate contain anandamide, a brain lipid that binds to brain´s cannabinoid receptors with high affinity. Anandamide mimics the endogenous cannabinoid system of human brain. Cannabinoid drugs are known to heighten sensitivity and produce euphoria. Elevated anandamide levels could cooperate with other psychoactive components of chocolate: caffeine, theobromine, to produce a transient feeling of well-being (E. di Tomaso et al, 1996).

The effects of morphine for example, are not so much to suppress pain sensation, but to make it seem irrelevant. Hence consciousness of pain can clearly be modified as opposed to abolished, by brain´s own opiate system (the target for morphine). The question of why some people become addicted to these substances is often put in terms of the phrase 'Nature or Nurture'? Some hallucinogens are chemically related to neurotransmitters. The neurotransmitters dopamine and serotonin can be chemically converted to methylated analogues that are very known hallucinogens, such as amphetamine, mescaline, psilocin and LSD.

M.S.Gazzaniga (1997): "Until now, drug abuse research has been dominated by simple behavioural models that use outmoded theories of reinforcement to explain drug addiction, such as mistaken notion that users stay addicted because drug provide an intermittent schedule of reward. Neuroscience took up such ideas, and the field largely becomes locked into the view that if it could be determined exactly that neurotransmitter responds to what drug, the drug addiction problem would be solved. The importance of cognition is illustrated by the fact that the overall pattern of U.S. drug use has remained constant for years. Although many people experience drugs, only a small number become addicted. Specifically about 70 percent of Americans have tried illicit drugs, but less than 20 percent have used an illicit drug in the past year and only a few percent have done so in the past month. It is also relevant that drug use drops dramatically with age; past age 35, the causal use of illegal drugs virtually ceases. All these facts suggest that simple learning and reinforcement concepts do not explain the drug experience. Cognition is central to the pattern. Education, alternative choice, and competing temptations all play a role in the determining whether the user is seeking occasional reinforcement from drugs or heading for chronic use...most people eventually walk away from the hedonistic pleasure of illicit drugs."

In spite of the feeling of continuity which is the basis of personality, it is very well known how fast, simple and easy it is to change a personality by the use of some substances such as alcohol and drugs, or over a longer period by the administration of hormones. We are far from fully understanding how and where each psychoactive drug acts on the bonus-malus system.

Bonus-malus systems have developed over the course of evolution to reinforce useful behaviour, eliminate harmful conduct, and maintain and adaptively regulate a fine-tuned set of drives related to pleasure and pain, emotional and sexual satisfaction, hunger, thirst, and satiety. Addictive drugs act on the same systems by substituting for the natural neurotransmitters that act at different points in the circuitry, thus producing an artificial state of bonus (reward, euphoria, hedonia), with a powerful compulsion to sustain that state, but risking possible irreversible dysfunction of the bonus-malus system itself. Withdrawal symptoms result in dysphoria (anhedonia). It must be stressed once more that the bonus-malus system

Table 7-7 Similarity of some social action and sexual selection

Spectacle Partner	Sexual 'mental' selection Own sexual partner	Religious action Own God	Nationalistic action Own nation
Words as tools			
Bonus: partner	Verbal promises to be able in future to offer all 'goods' even own life for achieving benevolence		
Space	'Seventh heaven'	Paradise	'Fatherland'
Malus: enemy	Sexual competitor	Devil	Enemy
	Deception, mocking, frighten, curse, malediction, threaten		
State, space	Be refused	Hell	Slavery
Fire as 'tool'			
Bonus: partner	Candle, Firelight Firing love	Candle, Oil lamp	Torch, Fireworks
Malus: enemy	'Extinguished love'	Infernal fire	Shooting in air
Psychoactive agent			
Bonus: partner	Psychoactive agents alcohol, narcotics for self and for sexual partner:	Different 'psychedelics' for self, Myrrh, burnt offering	Psychoactive agent for self alcohol
Malus: enemy	'Hangover' loss of consciousness	For devil: Fumigating	Poisons, nervous gas?
Feelings			
Bonus, proliferation	Genetic proliferation	Religious 'proliferation'	Power proliferation
Malus, ending, death	Genetic 'death'	Lost of hope for 'life after death'	Slavery, lost of freedom

has a paramount impact on the biotic information processing, especially by humans. The broad use of such natural and, later, synthetic substances by almost the whole of mankind can be considered as a significant element of human culture. Often, some of these substances were used for cultic purposes and have played a significant role in the establishment of myths and religions, e.g. for Incas' aristocracy and priesthood (L.Pulvirenti, G.F.Koob, 1996).

Hallucinogenic natural substances have long been used in religious rituals by primitive societies to attain states of trance. By means of such ecstatic experiences, individuals such as shamans and medicine-men cured the diseased, accompanied the dead to the underworld, and acted as mediators between their community and its gods. The role of some hallucinogenic plants is well documented in the ecstatic rites of Siberian shamans, and there is strong indirect evidence for inebriating plants being used by the earliest Indo-European populations of India. In mediaeval Europe, hallucinogenic plants were used, especially among country-folk, in popular religious rites that were condemned by the church as witchcraft and heresy.

Newly developed compounds derived from the immune system may help combat cocaine abuse by destroying the drug soon after it enters the bloodstream (D.W. Landry, 1997).

Not only the biotic development of the brain and the emergence of lateralization, and later of speech-ability, but also the discovery of the effects of drugs, were impulses to the emergence of intelligence. We are far from suggesting that these psycho-active substances have played a role in the emergence of human culture, but on the other hand we ought not to underestimate their impact on our past, present and even future (Section 11.4.2).

R.J.Huxtable, (1995): "Most psychoactive agents—coffee, tea, heroin, cocaine—originate from underdeveloped countries. Coffee is the second (after oil) most important export from the developing world to the first, employing some 20 millions people. The combined trade in both legal and illegal psychoactive substances probably far outweighs the value of oil from developing countries. (some authors claim) that drugs made dangerous by myth, politics, illegality and social factors rather than by their psychoactive or pharmacological properties."

7.5.3 *Mental game*

For young animals (including human children) playing games are some of the significant techniques for learning a lot of things, such as effective patterns of behaviour; of physical skills, and for children, of mental skills.

Adult humans in almost all cultures, for thousands of years, have participated in precisely regulated forms of competition involving physical skills (from hunting to sports) as well playing games which are of a purely mental nature, such as chess, and cards, as well as crosswords, jigsaw puzzles, and riddles. Some of these depend more on chance, such as dice, dominoes, or cards.

The arbitrary, but precisely formulated and followed rules of a game, and the astonishing equality of both partners in some games, such as chess or dice, are very remote from the norms of everyday social contact. All players, rich and poor, are (in principle at least) equal on the playing field, and, for once in life, everyone knows the rules. The winning is almost always something that is not useful within the game itself. The equality of both partners in the face of the rules of a game and the unpredictability of the dice is unique in the realm of human social contact.

Both opponents playing chess or dominoes, in spite of the strict absolute and arbitrary rules of the game, have the feeling of free will, which allows them to make one move or another. Here, where personal freedom is so obviously regulated, the feeling of free will seems to be so strongly felt.

Playing cards is the 'school' for learning deception, as the 'best' strategy in social contacts. A propos, the role of riddles in mythology and religious tradition cannot be overestimated. Also remarkable is the ability to play some mental games alone, without a partner, having oneself as the opponent (crosswords), or to play 'against' a machine (computer games).

Brain activity peculiar to playing chess has been observed by PET. The result is that even though we would call playing chess a higher-order cognitive process, it is made up of lower-order functions. The involved right prefrontal cortex and left orbito-frontal cortex areas are a sort of information manager that organises working memory.

7.5.4 *Aesthetics, some remarks*

Intelligence is associated with the highest development of the bonus-malus system, which is present at all stages of biotic information processing, including reflex, instinct and discent. On the other hand, intelligence is inclined towards the highest of mental skills—the abstract concept. The overlapping of these abilities has enabled a rather abstract concept of beauty and the principles of aesthetics to be developed.

There are numerous examples of shared aesthetic feelings, such as the response of humans and female nightingales when hearing the song of the male nightingale, or the attraction of blossoming flowers (the sexual organs of many plants) for both humans and numerous insects. Humans and certain other species (apes, dolphins, birds and even bees) find symmetric patterns more attractive than asymmetrical ones. M.Giurfa et al (1996), investigated bees for learning ability of generalized symmetric and asymmetric preference. These preferences may appear in response to biological signals, or exploratory behaviour and human aesthetic response to pattern. It has been proposed that preferences for symmetry have evolved in animals because the degree of symmetry in signals indicates the signaller's quality. By contrast M.Enquist and A.Arak (1994) have shown that symmetry preferences may arise as a by-product of the need to recognize objects irrespectively of their position and orientation in the visual field. The existence of sensory biases for symmetry may have been exploited independently by natural selection acting on biological signals and by human artistic innovation. This may account for the observed convergence on symmetrical forms in nature (colours and symmetries in flowers, butterfly´s wings, coral reef fish) and decorative art.

Humans and certain other species, even on the stage of instinct (insects), and discent (birds) find symmetrical patterns more attractive than asymmetrical ones. This plays a significant role in sexual selection, because the degree of symmetry indicates the signaller's quality. However, the symmetrical body markings of wild animals are often lost or degraded in their domesticated descendants, for which the targeted, and not the sexual, selection is of significance. Nota bene: domesticated animals have very often brains about 1/5 smaller than the wild ascendants (M.Enquist, A.Arak, 1994).

What would a universal beautiful human be? May be the 'beauty' is averageness—the average value of the features of faces of the human population, because average physical properties have the best chance of survival. This may account for the observed convergence on symmetrical forms in nature and decorative art. The most astonishing is that the principle of symmetry (defined extremely broad and abstract) plays a decisive role in the most sophisticated ideas of modern physics.

T.Dobzhansky, E.Boesiger (1983): "We do not pretend that an aesthetic sense comparable to man's permits birds, butterflies, and many other organisms to develop those phenotypes which do bring us an aesthetic pleasure. Granted that man alone produces works of art, it is nevertheless unrealistic to claim that an absolute difference separates man and other organisms when art is a question of perceptions or even aesthetic sensations. Biological evolution leads to capacities of perception because they offered selective advantages."

The fusion of the bonus-malus system and the ability to think abstractly reaches the highest level in the activity of scientists. The connection between the highest intellectual activities and aesthetics is illustrated by the following quotations from some prominent scientists:

M.Copernicus (1540): "Among the many and varied literary and artistic studies upon which the natural talents of man are nourished, I think that those above all should be embraced and pursued with the most loving care which have to do with things that are very beautiful."

H.Poincare (1900): "The Scientist does not study Nature because it is useful to do so. He studies it because he takes pleasure in it; and he takes pleasure in it because it is beautiful."

W.A.Fowler (1983): "I have chosen the subtitle for this lecture: Ad astra per aspera et per ludum; which can be freely translated as: To the stars through hard work and fun. This is in keeping with my paraphrase of the biblical quotation from Matthew: Man shall not live by work alone."

S.Chandrasekhar (1983): "Study has convinced me of the basic truth of the ancient mottos: The simple is the seal of truth and beauty is the splendour of truth."

H.Weyl (1950): "When I must choose between truth and beauty, then I must choose beauty."

W.Heisenberg (1960): "Beauty is the conformity of the parts with the whole."

N.Chomsky (1983): "We like and understand Beethoven because we are humans, with a particular, genetically determined mental constitution. But that same human nature also means there are other conceivable forms of aesthetic expression that will be totally meaningless to us."

G.Paul (1988): "If Chomsky's hypothesis that there exists a universal grammar is acceptable, then a similar hypothesis concerning universal aesthetic rules is even more convincing."

P.A.M.Dirac (1970): "It is more important to have beauty in one's equations than to have them fit the experiments."

A.Einstein (1930): "That this insecure and contradictory foundation was sufficient to enable a man of Bohr's unique instinct and tact to discover the major laws of the spectral lines and of electron shells of the atoms, together with their significance to chemistry, appeared to me like a miracle—and appears to me as a miracle even today. This is the highest form of musicality in the sphere of thought."

The oldest wall painting (some hundreds) dating back 32,000 years (Ardeche, France) are particularly impressive for the aesthetical sophistication and complex technique used to present perspective and motion (Section 3.5.6). Taking into account that our brain processes information whose contents are often vague and far from the contents of the real world (visual illusions, acoustic hallucinations etc.) it could be assumed that the genial painters have painted so that the emerging illusions are nearer to our aesthetic taste than the naked realistic picture. The most abundant (in geographical, historical, social and religious scale) aesthetic phenomenon, coupled with almost all sensory channels is the to make good cheer. No other animal species is able to 'prepare' food in such complex, time consuming, and intensive way as humans.

7.6 BIBLIOGRAPHY

Adolphs R (1994) Impaired recognition of emotion in facial expressions. *Nature* **372** 669
Berk LE (1994) Why children talk to themselves. *Sci Am* **271** 11 60
Blumenschein RJ, Cavallo JA (1992) Scavenging and human evolution. *Sci. Am* **267** 4 70
Bouchard TJ (1994) Genes, environment and personality. *Science* **264** 1700
Bystron JS (1939) *Komizm.* Ksiaznica-Atlas, Warszawa
Calvin WH (1994) The emergence of intelligence. *Sci Am* **271** 4 78
Caramazza A (1996) The brain's dictionary. *Nature* **380** 485
Cavalli-Sforza LL, Cavalli-Sforza F (1995) *The great human diasporas.* Addison-Wesley, Reading, Mass.
Ciba Found.Symposium (1993) *The origins and development of high ability.* GR.Bock, K.Ackrill (eds) Wiley, Chichester
Cromer A (1993) *Uncommon sense. The heretical nature of science.* Oxford Un Pr N.York
Cohen G (1983) *The psychology of cognition* (sec.ed.) Academic Press, London
Damasio AR, Damasio H (1992) Brain and language. *Sci Am* **267** 3 63
Damasio H et al (1996) A neural basis for lexica retrieval. *Nature* **380** 499
Descartes R (1637) *Discourse on method and Meditation* Paris
Enquist M, Arak A (1994) Symmetry, beauty and evolution. *Nature* **372** 169
Feyerabend P (1995) *Killing time.* Univ. Chicago Press, Chicago
Frith U (1997) Autism. *Sci. Am. Spec.Iss. Mysteries of Mind*, Vol **7**,92
Gazzaniga MS (1997) Brain, drugs and society. *Science* **275** 459
Gershon ES, Rieder RO (1992) Major disorders of mind and brain. *Sci Am* **267** 3 89
Gibbons A (1991) Deja vu all over again: chimp-language wars. *Science* **252** 1561
Giurfa M et al (1996) Symmetry perception in an insect. *Nature* **382** 458
Goldman D (1996) High anxiety. *Science* **274** 1483.
Gottesman II (1991) *Schizophrenia genesis. The origins of madness.* Freeman. N.York
Grayson L (1995) *Scientific deception.* British Library 24, 5
Herrnstein RJ, Murray Ch (1994) *The Bell curve: Intelligence and the class structure in American life.* Free Press. N.York
Horgan J (1996) Why Freud isn't dead. *Sci Am* **275** 6 74
Huxtable RJ (1995) Usage and abuse. *Nature* **377** 24
Jamison KR (1997) Manic-depressive illness and creativity. *Sci. Am. Spec.Iss. Mysteries of Mind*, Vol **7**,44
Kamin LJ (1995) Behind the curve. *Sci Am* **272** 2 82
Kleinnman A, Cohen A (1997) Psychiatry's global challenge. *Sci Am* **276** 3 74.
Kolb B, Whishaw IQ (1985) *Human neuropsychology.* Freeman, N.York
Landry DW (1997) Immunotherapy for cocaine addiction. *Sci Am* **276** 2 28
Leenders KL et al (1986) Brain dopamine metabolism in patients with Parkinson's dis ease measured with positron emission tomography. *J Neur Neuro Psych* **49** 853
Leenders KL et al (1990) The nigrostriatal dopaminergic system assessed in vivo by posi tron emission tomography in healthy volunteer subjects and patients with Parkinson's disease. *Arch of Neurology* **47** 1290
Lenat DB (1984) Theory formation by heuristic search. *Artif Intell* **21** 31
Liberman AM (1989) A specialization for speech perception. *Science* **243** 489
Lindley D (1993) *The end of physics-The myth of a unified theory.* Basic Books, N.York

Morris JS et al (1996) A differential neural response in the human amygdala to fearful and happy facial expressions. *Nature* **383** 812.

Miller GA (1991) *The science of words.* Sci Am Libr, N.York

Penrose R (1997) *The large, the small and the human mind.* Cambridge Univ Press Cambridge, UK

Pich EM et al (1997) Common neural substrates for the addictive properties of nicotine and cocaine. *Science* **275** 83

Pinker S (1995) Beyond folk psychology. *Nature* **373** 205

Rose S (1995) The rise of neurogenetic determinism. *Nature* **373** 380

Sagan C (1996) *The demon-haunted world.* Random House, N.York

Schlaug G et al (1995) In vivo evidence of structural brain asymmetry in musicians. *Science* **267** 699

Sergent J (1992) Distributed neural network musical sight-reading. *Science* **257** 106

Silbersweig DA et al (1995) A functional neuroanatomy of hallucinations in schizophrenia. *Nature* **378** 176

Snyder SH (1986) *Drugs and the brain.* Sci Am Library, N.York

Sternberg RJ (edit) (1990) *Wisdom. its nature, origins, and development.* Cambridge Univ. Press, N.York

Ulam SM (1987) Reflections on the brain's attempts to understand itself. *Los Alamos Sci Spec Iss* Oct, 283

Tomaso E et al (1996) Brain cannabinoids in chocolate. *Nature* **382** 677

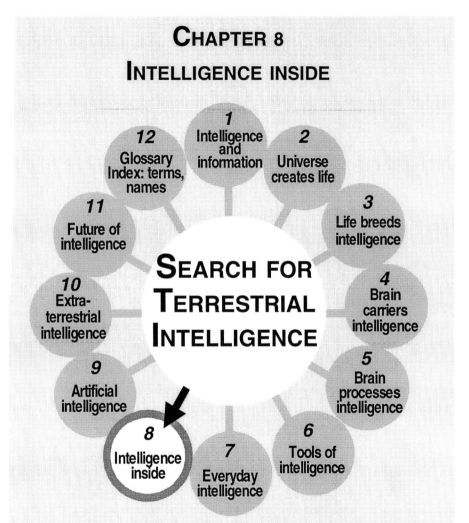

CHAPTER 8
INTELLIGENCE INSIDE

Self as supreme concept. Selfconsciousness: definition, origin and evolution? Are animals conscious? Thinking and intuition.

Unconsciousness and natural sciences. Dreams, significant or not. Para-psychology? Vagueness of information: illusion, hallucination. Necessity, chance, free will.

Free will and psychic force? Free will and natural sciences. Intelligence and morality. Genocide, homicide.

Consciousness about own death and altruism.

8 INTELLIGENCE INSIDE 302
8.1 SELFCONSCIOUSNESS, OBJECTIVE OR MYSTERIOUS? 302
8.1.1 Self as the supreme concept *302*
8.1.2 Selfconsciousness must be defined *303*
8.1.3 In asymmetric brain the consciousness could also be asymmetric *307*
8.1.4 Thinking and intuition *308*
8.1.5 Selfconsciousness: origin and evolution *311*
8.1.6 Are animals conscious? *312*
8.2 CONSCIOUSNESS FROM DIFFERENT POINTS OF VIEW 314
8.2.1 Unconsciousness and the natural sciences *314*
8.2.2 Para-psychology etc. *315*
8.2.3 Vagueness of information; illusions *317*
8.3 FREE WILL, THE REAL MYSTERY OF HUMANS 319
8.3.1 Necessity, chance and free will *319*
8.3.2 Hypothesis of free will mechanisms *320*
8.3.3 Free will and the natural sciences *323*
8.3.4 Free will and psychic force? *325*
8.4 INTELLIGENCE AND MORALITY 326
8.4.1 Intelligence and morality *326*
8.4.2 Genocide, homicide and crimes *329*
8.4.3 Consciousness about own death and altruism *331*
8.5 BIBLIOGRAPHY 333

8 INTELLIGENCE INSIDE

8.1 SELFCONSCIOUSNESS, OBJECTIVE OR MYSTERIOUS?

8.1.1 *Self as the supreme concept*

The only example of intelligence which we have is terrestrial human intelligence. In this case, the phenomenon of intelligence is inseparably linked to some very abstract concepts; in the first place, the concept of Self. Without going into detail it is clear that the contrary is Non-self. Non-self consists of all things and events which are not included within Self. G.M.Edelman (1989) has defined the bifurcation of the cognitive processes as 'Self' and 'Non-self'.

Let us begin with a very simple monologue: 'I', 'I am', 'I know something', 'I know that I know something', 'I know that there are also others', 'I know that others also know that they know', 'I know that....'. This condition of internal information concerning the state of internal information processing can be called selfconsciousness. How do we know that other people have conscious minds (as opposed to being 'zombies' or 'androids' that resemble humans in every respect—except that they do not have a conscious experience of the world) or robots, and that those minds are anything like our own? (The so-called 'Other Mind Problem').

The most frequently spoken word is the first person singular pronoun; on average every sixteenth word is 'I'. There is no doubt that most of our sentences begin with the term 'I'. 'Self', which is the foremost and the most complex concept in the human mind, occupies both the sensor and the motor halves of the brain, both the left and right hemispheres. This therefore requires the existence of fully developed contralateral connections (primarily the corpus callosum), especially when we take into account the asymmetry of both hemispheres.

D.R. Hofstadter (1979): "In fact, the symbol for the self is probably the most complex of all the symbols in the brain."

M.Minsky (1986): "A paradox: perhaps it's because there are no persons in our heads to make us do the things we want—nor even ones to make us want to want—that we construct the myth that we're inside ourselves. .. The concept of 'Self' is without doubt the crystallization grain of the whole process which is called 'consciousness'."

Do we know when a child begins to use the word 'I'? How this depends from culture? In all languages, even in the most primitive, the concept of 'Self' is overwhelming. From the individual mental development including extreme mental retardation? Do some brain injuries, or commissurotomy cause the lost of the feeling of 'I'?

It could be claimed that the experience of 'Self' (personality) includes the experience of free will. But what is the experience of free will. In some opinion free will is a product of sclf-deception, therefore 'Self' is maybe also a product of self-deception. Could we be conscious of self-deception?

8.1.2 *Selfconsciousness must be defined*

Speech enabled the emergence of an important invention (or discovery?) in human evolution—the invention of consciousness and selfconsciousness. In our model of intelligence, selfconsciousness plays a unique role (A. Scott, 1995). The relationship intelligence—consciousness is not only terrestrial, but we assume that it has cosmic relevance.

Our proposition for description of the term 'consciousness': Consciousness is the ability to process (and to speak about) the internal information, concerning the processing of internal information, including consciousness itself.

M.S.Gazzaniga (1995): "It is curious that the problem of consciousness is studied even though there is no general agreement about what is meant by the term. If you asked 20 students of the problem to finish the sentence, 'Consciousness is...', 20 different definitions would result. Yet, most of us know what is really meant by the term. It refers to that subjective state we all possess when awake. In short, it refers to our feelings about our mental capacities and functions."

M.Minsky (1986): "The word 'consciousness' is used mainly for the myth that human minds are 'self-aware' in the sense of perceiving what happens inside themselves... Many people seem absolutely certain that no computer will ever be sentient, conscious, self-willed, or in any other way 'aware' of itself. But what makes everyone so sure that they themselves possess those admirable qualities?"

We do not agree with Minsky and will now explore the phenomenon of human consciousness. However, we cannot ignore the remark of P.S. Churchland (1983) that consciousness is theoretically so badly defined that it is similar to the problem of counting angels on the tip of a needle. It has been said that when one has explained in a comprehensible way what consciousness is, one has falsely explained it (Fig. 8-1).

I.B. Farber, P.S. Churchland (1995): "..(We) suggest that 'consciousness', while admittedly a fragmented and vague concept, should not presently be subjected to more precise definition....After all, it is clear that, even just within science, many people mean different things when they talk about consciousness, and using the same word for so many different things is bound to produce a certain amount of confusion and pointless argumentation."

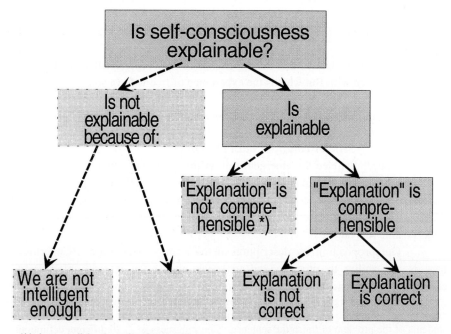

*)It is possible that the future Homo 'super' sapiens will be intelligent 'enough'

Figure 8-1 Is selfconsciousness explainable?

P.M. Churchland (1995) claims that: "If consciousness is our explanatory target, let us try to identify some of its more salient features.... Consider, then, the following salient dimensions of human consciousness (Remark: strongly shortened). Consciousness:

- involves short-term memory
- is independent of sensory inputs
- displays steerable attention
- has the capacity for alternative interpretations of complex or ambiguous data
- disappears in deep sleep
- reappears in dreaming, at least in muted or disjointed form
- harbours the contents of several basic sensory modalities within a single unified experience".

Our remark: P.M. Churchland does not couple the phenomenon of consciousness with the experience of 'Myself', 'I' and 'selfconsciousness'.

D.J.Chalmers (1996): "Consciousness possesses the most baffling problems in the sciences of the mind. There is nothing that we know more intimately than the consciousness experience, but there is nothing that is harder to explain."

The question arises of whether consciousness, and in particular selfconsciousness, can be the object of scientific study and investigation. N.Bohr, discussing with W.Heisenberg, affirmed that each branch of the natural sciences concerned with living beings must consider the phenomenon of consciousness, because it is a part of reality. And R.W.Sperry (1981) claimed that the problem of how consciousness is generated by the brain is currently the number one quest of the natural sciences and one of the most mysterious phenomena in the whole of science.

I.Prigogine (1983):"We see a convergence between the world outside and the world inside us. Consciousness plays an essential role because we construct reality through mathematical concepts. If our consciousness had a different structure we probably could not use the same type of constructs that we do. That's not to say physics is subjective; there must be a relation between our physics and reality. To me this coexistence of unconscious and conscious activity—of opacity and transparency—will ultimately lead to a new cultural unity."

G.Wald (1988):"I think that probably all mammals are conscious. I think that probably birds are conscious—why else would they sing?"

J.Searle (1984):"Because it seems to me that I am conscious, I am conscious".

S.Freud (1856-1939) wrote in his posthumous 'Outlines of Psycho-Analysis': "The starting point of this investigation is provided by a fact without parallel, which defies all explanation or description—the fact of consciousness."

It is probably obvious that in this book both viewpoints must be stressed; on the one hand the significance of selfconsciousness is important for the description and study of intelligence, and on the other hand we need to develop a scientific (physical and biological) theory of consciousness.

R.W.Sperry (1983):"Instead of renouncing or ignoring consciousness, the new interpretation gives full recognition to the primacy of inner conscious awareness as a causal reality. Consciousness is put to work and given a use and a reason for having been involved in a physical system."

J.C.Eccles (1994) suggests that in the neuron bundles, called 'dendrons' with hundreds of fibres, a corresponding number of 'psychons', the units of mental activity, interact with the dendrons, and that these are the site of the mind, controling the brain.

This point of view has strong support from modern scientists, as can be seen in the book by G.M.Edelman (1989), which has the subtitle, 'A biological theory of consciousness'

G.M.Edelman (1989):"Consciousness is something, the meaning of which 'we know so long as no one asks us to define it'. Like most of my colleagues, I considered consciousness a danger subject. Consciousness is bound up to some degree with volition and decision. I mean that a consciousness model based on ... a scientific theory of the mind must either be testable by experiment or, if not, at least be consistent with other brain models that are testable by experiment... I consider that consciousness (in humans) .. depends upon language. Both primary consciousness and higher-order consciousness provide the means of freeing animal behaviour from the tyranny of ongoing events. One of the most striking features of consciousness is its continuity. No special addition to physics is required for the emergence of consciousness."

F.Crick (1994): "You, your joys, your memories and ambitions, your sense of personal identity and free will, are, in fact, no more than the behaviour of a vast assembly of nerve-cells. As Lewis Carroll's Alice might have phrased it: 'You're nothing but a pack of neurons'."

According to F.Crick the general principles of the neural correlate of consciousness has now been abbreviated to NCC.

Some scientists claim that consciousness has so complex a nature that explanation must be based on a 'new physics', for example on the future 'superunified theory' which will include both quantum mechanics (theory of elementary particles and elementary forces) and general relativity (theory of gravitation).

R.Penrose (1994): "I think of consciousness, it is not just the sum of parts, but some sort of global capacity which allows us to take into account the whole of the situation at once. And this is one of the reasons why I think that there must be some connection with quantum mechanics, because one can have things in quantum mechanics which are like this, global states which have to be thought of as things on their own, and not just the result of all the little bits.......My own understanding of quantum mechanics,..., leads me to believe that quantum mechanics is not a complete theory, because it does not explain the large-scale world.... So there is something profoundly missing there. So my view is not quite that we look to quantum mechanics to explain consciousness; it is that we need to look beyond quantum mechanics. And the basic difference is that I introduce gravity effects."

It has been proposed that the phenomenon of consciousness is not only coupled with 'new physics' but even possess its own 'habitat', the microtubules (Section 4.2.6).

S.R.Hameroff (1994): "The emergence of consciousness in our brain (during each conscious moment, during evolution and during the development of each human being) may be linked to new properties of materials which develop from microscopic or quantum-level events.... in and around a specific class of neurobiological microstructure: cytoskeletal microtubules within neuron throughout the brain."

Consciousness has been superimposed on a brain what was never designed for it. The mind did not grow the brain, but the brain developed a mind of its own. We are conscious of one thing or experience after another. Consciousness is 'gappy'. Sometimes we do not know what we are thinking of.

Conscious states correspond not to some loosely defined global state of the entire brain, but much more precisely to the activities of specific populations of neurons.

F.Crick (1994): "To repeat: Consciousness depends crucially on thalamic connections with the cortex. It exists only if certain cortical areas have reverberatory circuits, that project strongly enough to produce significant reverberations."

F.Crick (1996): "I believe.... we first need to discover the neural correlates of consciousness. Consciousness is now largely a scientific problem. It is not impossible that, with a little luck, we may glimpse the outline of the solution before the end of the century."

The relationship between mind and brain has been extensively discussed in contemporary philosophy and psychology, without any decisive resolution. One heuristic solution, therefore, is to adopt the position that the mind is the expression of the activity of the brain and that these two are separable fur purposes of analysis and discussion but inseparable in actuality. That is, mental phenomena arise from brain, but mental experience also affects the brain, as is demonstrated by the many examples of environmental influences on brain plasticity. The aberrations of mental illnesses reflect abnormalities in the brain/mind's interaction with surrounding world; they are diseases of psyche (or mind) that resides in that region of the soma (or body) that is the brain (N.C. Andreasen, 1997).

8.1.3 *In asymmetric brain the consciousness could also be asymmetric.*

M.S.Gazzaniga (1995): "Of course, viewing consciousness as a feeling about specialized abilities would lead to the prediction that the quality of consciousness emanating from each hemisphere should differ radically. While left-hemisphere consciousness would reflect what we mean by normal conscious experience, right-hemisphere consciousness would vary as a function of the kinds of specialized circuits that a given half brain may possess. Mind left, with complex cognitive machinery, can distinguish between the states, say, sorrow and pity, and appreciates the feelings associated with each state. The right hemisphere does not have the cognitive apparatus for such distinctions, and as a consequence has reduced state of awareness..... The left hemisphere is busy differentiating the world whereas the right is simply monitoring the world......(Consciousness) is certainly not learned. It is an efficient way the brain puts heat to our cold specialized capacities. It is nothing more and nothing less. It is what makes it all worthwhile."

A short summary of recent discussion (Second multidisciplinary meeting "Towards a science of consciousness" April, 1996, Tucson, Arizona) about the problems of the problem of consciousness is given below.

Anonymous (1996): "There are deep disagreements about how science should deal with the problem of consciousness. But neuroscientists should soon be in a position to set the agenda. The result is likely to be rapid progress. Perhaps the most basic question is whether existing physical principles are sufficient to explain the brain. ... Brain performs feats of understanding that cannot be computed and so cannot be implemented by any system based on known physical laws.... Explaining the brain will require a new theory linking events at the quantum level with those of the macroscopic world .... Consciousness may simply be an epiphenomenon of certain arrangements of matter, or indeed of all matter,.... perhaps even an atom feels a spasm or relief as an electron falls to a lower orbital. ...... Complex brain would presumably generate more complex forms of consciousness as by-product of their more sophisticated computations."

8.1.4 Thinking and intuition

What is thinking? What is conscious thinking, and what is unconscious thought? R.Penrose proposed the following mechanism. Conscious thought is effective only in some instances, such as thinking about simple things and only dealing with one problem at a time. Also, 'common sense' judgements are realized consciously. Judgements about the truth and value of art are the next fields of conscious thought. Other aspects, such as thinking about relatively complex problems, or thinking about more things at once, are the domain of the unconsciousness. Here belong automatic and algorithmic processes, too (Fig. 8-2).

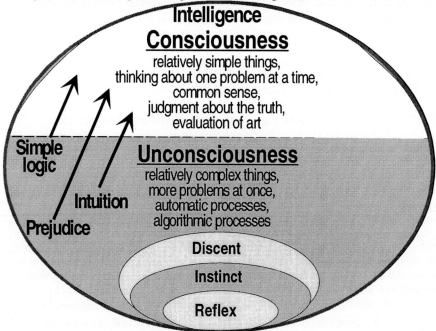

Figure 8-2 Thinking is conscious or unconscious?

This type of unconscious information processing is called 'intuition' or 'prejudice'. Intuition is the knowledge of a concept, truth, or solution to a problem, which is arrived at apparently spontaneously, without conscious steps of reasoning or inquiry.

H.A.Simon (1969):"What about the sudden flashes of 'intuition' that sometimes allow the expert to arrive immediately at the answer that the novice can find (if at all) only after a protracted search? (I put 'intuition' in quotes to emphasize that it is a label for a process, not an explanation of it.) Intuition is a genuine enough phenomenon which can be explained rather simply: most intuitive leaps are acts of recognition. Let me illustrate this. A feature-testing system capable of discriminating among 50,000 different items might make its discrimination quite rapidly. If each test were dichotomous, only about 16 tests would have to be performed to achieve each recognition (2 to the power 16 = 65,536). If each test required 10 milliseconds, the whole process could be performed in less than 200 milliseconds—well within the time limits of human recognition capabilities."

M.Gell-Mann (1985):"A modern physicist must develop a quantum-mechanical intuition—it's of critical importance in creative work."

This does not mean that intuition comes from outside the human brain, that it is a form of knowledge that is akin to divine empathy. In this book, intuition is more precisely defined as a method of processing concepts by working memory outside the control of consciousness.

G.Vogel (1997): "Intuition may deserve more respect than it gets these days. Although it's often dismissed along with emotion as obscuring clear, rational thought, a new study suggests that it plays a crucial role in humans` ability to make smart decisions. A.Bechara et al (1997) set out to shed light on the role of intuition and emotion in normal decision-making by studying a group of brain-damaged individuals who seem unable to make good decision."

When speaking about internal experience or introspection, even people who believe in the unlimited possibilities of the evolution of man-made artificial intelligence use some very human terms, such as 'intuition'.

In the early period of man's development intelligence created numerous representations (concepts) of its natural and social environment. The complexity and the number of concepts seems to be unlimited.

Later intelligence generated concepts concerning the phenomenon of life, human body, brain, and all the material carriers of intelligence. The complexity and the number of concepts seems to be unlimited.

A.Bechara (1997): "Deciding advantageously in a complex situation is thought to require overt reasoning on declarative knowledge, namely, the facts pertaining to premises, options for action, and outcomes of actions that embody the pertinent previous experience".

One of the authors of the cited paper, believes that even healthy brain acts against own best interests: „We've all had that feeling that we know the right thing to do, but are incapable of doing it." All these experiments demonstrate the impact of different elements of real-life decision-making such as factors of uncertainty, rewards and penalties (Bonus-malus) (Section 1.3.2) and memory of previous emotions (Section 6.3.1) being part of 'intuition'.

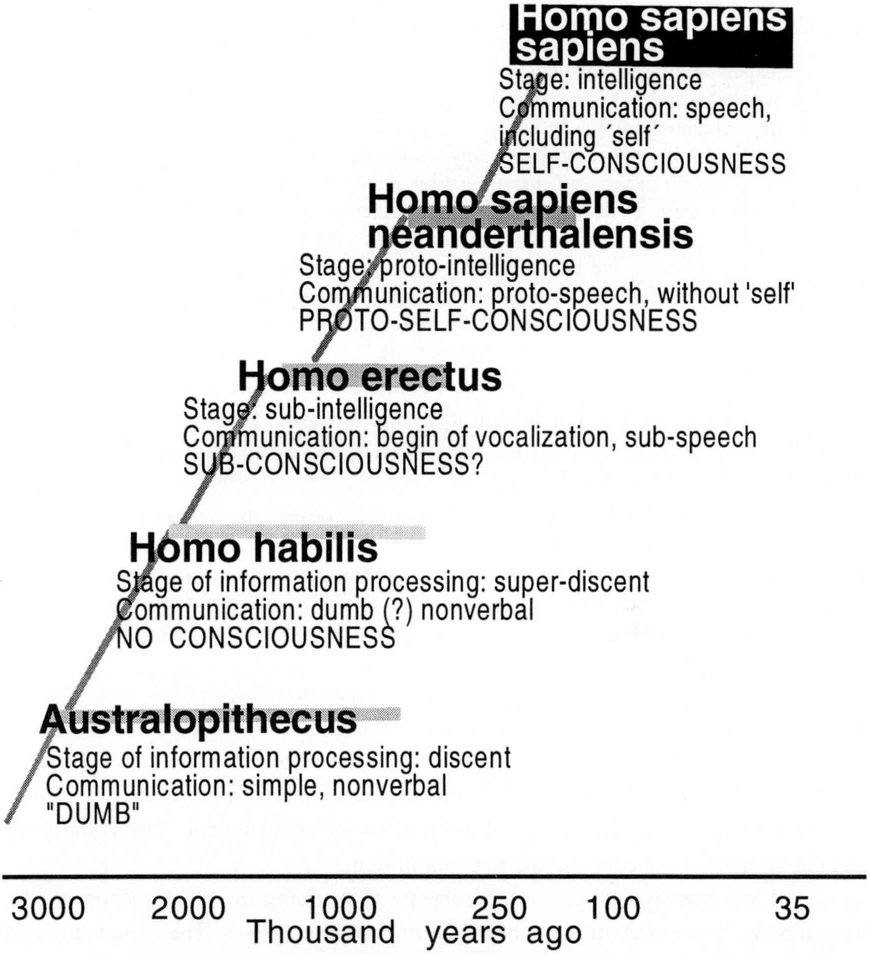

Homo sapiens sapiens
Stage: intelligence
Communication: speech, including 'self'
SELF-CONSCIOUSNESS

Homo sapiens neanderthalensis
Stage: proto-intelligence
Communication: proto-speech, without 'self'
PROTO-SELF-CONSCIOUSNESS

Homo erectus
Stage: sub-intelligence
Communication: begin of vocalization, sub-speech
SUB-CONSCIOUSNESS?

Homo habilis
Stage of information processing: super-discent
Communication: dumb (?) nonverbal
NO CONSCIOUSNESS

Australopithecus
Stage of information processing: discent
Communication: simple, nonverbal
"DUMB"

| 3000 | 2000 | 1000 | 250 | 100 | 35 |

Thousand years ago

Figure 8-3 Evolution of consciousness

Recently, intelligence began to investigate intelligence itself and encounters enormous problems. The number and the kind of concepts indispensable to represent the complexity of these phenomena seems to be not only unlimited but also often not clear enough described and defined. This is an inherent property of the nature of intelligence, as self-describing process of processing of representations. Intelligence is an information processing of information processing. Depth and broadness of concepts describing components of intelligence is not only very large but, and this is the crucial problem, always in continuous change and development.

W.Schultz (1997): "The capacity to predict future events permits a creature to detect, model and manipulate the causal structure of its interactions with its environment. Behavioural experiments suggest that learning is driven by changes in the expectation about future salient events such as reward and punishment (Remark in this book: 'bonus-malus'). Physiological work has recently complemented these studies by identifying dopaminergic neurons in the primate whose fluctuating output apparently signals changes or errors in the prediction of future salient and rewarding events. .... The experimental data suggest that the dopamine system provides information about appetitive stimuli (Remark: 'bonus') not aversive stimuli (Remark: 'malus'). It is possible however that the absence of an expected reward is interpreted as a kind of 'punishment' to some other system to which the dopamine neurons send their output. .. It was long ago proposed that rewards and punishments represents opponent processes and the dynamics of opponency might be responsible for many puzzling effects in conditioning."

The concept of intelligence is created by intelligence. All biotic information processing is controlled by the bonus-malus system. A positive description of the concept of intelligence is rewarded by the bonus-malus system. And in contrary a negative description, showing the dark side of intelligence, is pushed away, is discriminated, tends to be forgotten, to be resting in shadow.

The self-defence of consciousness relies on a rather sophisticated tactics and strategy to omit and to glance over the real shadow-sides of activity and even thoughts of myself.

S.Weinberg (1995) according Horgan (1996); "If neuroscientists ever explain consciousness, for example, they will explain it in terms of the brain... and the brain is what it is because of historical accidents and because of universal principles of chemistry and physics."

D.J.Chalmers (1997): "Conscious experience is at once the most familiar thing in the world and the most mysterious. There is nothing we know about more directly than consciousness, but is extraordinarily hard to reconcile it with everything else we know. Might consciousness be an irreducible feature in nature, as basic as mass or electrical charge? Making that radical assumption, might be the only way for science to make sense of the subjective experience of self."

8.1.5 *Selfconsciousness: origin and evolution*

If selfconsciousness did not fall from heaven, it must have had a more or less definable origin and some, perhaps vague, evolutionary steps. Without reference to any other published work, we will try to formulate some naive and rudimentary ideas concerning the emergence and evolution of selfconsciousness.

It must be stressed that, in this book, selfconsciousness is directly and indivisibly associated with the origin and evolution of human speech (Fig. 8-3).

F.Crick and Ch.Koch (1992) in their article 'The problem of consciousness': "It is not easy to grasp exactly what we need to explain... We did not attempt to define consciousness itself because of dangers of premature definition. In this article, we will attempt to.. attack this profound and intriguing problem."

Let us define the previous stage on the evolutionary path as 'proto-consciousness', which corresponds to the 'primary consciousness' postulated by G.M.Edelman (1989). Proto-consciousness emerges from the interaction of value-oriented (hedonically labelled; in this book the 'bonus-malus system') memory and ongoing (current or present) perceptual categorization. Possessing proto-consciousness alone, animals are bound to the small time intervals ordained by short-term memory; they have no concept of the past.

G.Baumgartner (1990):"If I have such a complicated structure as our brain, which is in my opinion the end point of the universal, then nature has obviously invented something like consciousness. Why did she invent consciousness I am not quite sure, but maybe it's a probe which makes it easier to sweep around, to focalize, to get out information from many different points. I think we will never know. We certainly will never understand it in terms of real understanding. It is, in my opinion, a question which is much more than our brain can access."

D.L.Schacter (1995): "Mention the term 'consciousness' to a cognitive neuroscientist, and you will probably elicit one of two very different reactions. Some will likely shrug or groan, mumbling uncomfortably that consciousness is a construct that we do not know how to approach sensibly or to investigate productively.... Others, however, will probably respond with enthusiasm, even excitement, citing the latest experimental dissociation between conscious and unconscious processes... as evidence that cognitive neuroscience is on the verge of finally cracking the riddle of consciousness. Weiskrantz (1991) has noted that this enthusiasm is not without historical precedent."

The discussion concerning the emergence of human being during the process of fertilization achieves at the end of the 20th century non only a 'theoretical' or a religious significance but also a very practical meaning in the global control of population increase. It is clear that both the egg and the sperm are alive, and that their life is human, and that life is continuous from one generation to the next. Yet life, at this stage, is not yet that of a person.. There is no biological discontinuity that would be a candidate for a moment of unique philosophical significance, whether the advent of the individual person or the infusion of the soul (J. Godfrey, 1995). Many fertilized eggs fail of implant quite natural. Is each of these to be considered the death of a person?

8.1.6 *Are animals conscious?*

The argument that dolphins and whales have cerebra which are as large as or larger than our own and therefore have a feeling of awareness, and even that this awareness may be an expression of the continuity of consciousness, is, we believe, unacceptable (Section 6.3.2). Although the brain of a dolphin is larger than that of a human, it is evidently simpler and less developed, much less even than that of an ape.

S. Savage-Rumbaugh (1991) claims that pygmy chimpanzees, bonobos, (9-year-old called Kanzi) seem to be capable of cognitive skills. Like young children (2 years old) they can learn human language by observing its use and they seem to have an understanding of both reference and simple syntax. Yet they lack the ability to co-ordinate their actions among one another with regard to past or future activities. Such limitations would conceivably make it difficult to contemplate previous actions as a source of information while contemplating future actions.

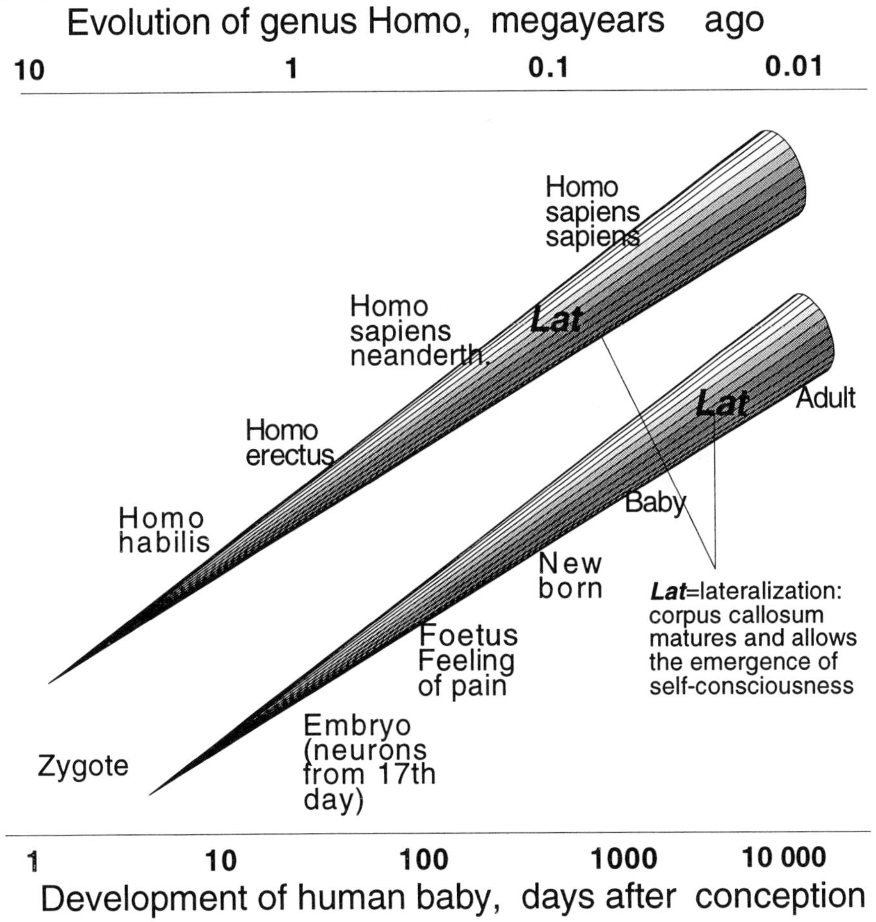

Evolution of genus Homo, megayears ago

10　　　　　　　1　　　　　　0.1　　　　　　0.01

Homo sapiens sapiens

Homo sapiens neanderth.

Lat

Lat Adult

Homo erectus

Baby

Homo habilis

New born

Lat=lateralization: corpus callosum matures and allows the emergence of self-consciousness

Foetus
Feeling of pain

Embryo (neurons from 17th day)

Zygote

1　　　　10　　　　100　　　　1000　　　10 000
Development of human baby, days after conception

Figure 8-4 Emergence of soul?

Perhaps, as R.Penrose suggests, by denying apes entry to the verbalizer's club, some would hope to exclude them from the club of conscious beings. But exclusion cannot be based upon the animals' brain anatomy, which does not preclude the possibility of acquiring speech, in a rudimentary form. For achieving the level of consciousness, the brain and its functions must be on a very high level, at present found only in the case of humans. However, this opinion should not be regarded as 'human chauvinism', or a kind of racism 'speciesm', directed against animals, or especially apes.

8.2 CONSCIOUSNESS FROM DIFFERENT POINTS OF VIEW

8.2.1 *Unconsciousness and the natural sciences*

The existence of consciousness implies the possibility of the existence of unconsciousness. The unconscious state cannot be entirely divorced from the conscious, because then rational discussion would be impossible. It seems to be reasonable to differentiate between the following four stages of consciousness:

1) Active (full)(self-knowing) consciousness, which involves awareness of self and personal experience extended in time,

2) Latent (non-knowing) consciousness, which entails simple awareness of external stimuli which can, in a given situation, be accessible to consciousness,

3) Unconsciousness (or subconsciousness), which can be expressed only in dreams (according to S.Freud) or in a psychic disorder Unconsciousness may partially correspond to lower stage of human information processing: instinct,

4) Non-consciousness, which is a rather hypothetical state and which, by definition cannot be proved by conscious activity. Unconscious states can contribute to conscious states, but their mechanisms can never be brought to awareness. Unconscious mental processes resemble conscious ones in many ways, but cannot be directly accessed by consciousness.

Table 8-1 Consciousness versus unconsciousness
(According to R.Penrose, 1989)

Consciousness needed:	Consciousness not needed
Common sense	Automatic
Judgement of truth	Following rules mindlessly
Understanding	Programming
Artistic appraisal	—
In this book corresponds with **Intelligence**	In this book corresponds with **Reflex, instinct, discent**

So-called unconsciousness can, according to S.Freud, be easily and frequently transformed into conscious states. It must here be said that thousands of books, and probably hundreds of thousands of papers, have been written about unconsciousness; many more than about consciousness. There are many people who are ready to pay substantial amounts of money in the quest for their hidden unconsciousness, instead of trying to do the best they can with their accessible consciousness. Non-consciousness may partially correspond to the lowest stage of human information processing: reflex.

8.2.2 *Para-psychology etc.*

In the considerations about intelligence and some psychological aspects of this phenomenon some concepts which play a role in social life can't be excluded, even if they are not an object of scientific investigation.

J.Beloff (1994): "The interactionist argument supports the relevance of parapsychology to the problem of consciousness.."

Para-psychology is primarily thought of as 'research' of the ability of the 'mind' to perform 'psychic acts'. These psychic activities fall into two categories:

-Extra Sensory Perception (ESP). It is divided into two sub-categories: telepathy (remote exchange information) and clairvoyance,

-Psychokinesis: the ability to move or alter animate and inanimate matter by thought alone.

There is no evidence for the existence of such phenomena. This opinion seems to be valid, in spite of the known principle, that absence of evidence is not evidence of absence. It is clear that existence of such phenomena, as telepathy, psychokinesis or clairvoyance would break the fundamental physical laws. And there is no evidence on the basis of para-psychological 'experimental data'. The reason why people believe in para-psychological phenomena is that they want to believe in them, because they want to think there is something more to human beings than just the matter comprising them (T. Crane, 1996). Para-psychology is one more example of the social effect of self-deception.

As we have mentioned according to P.M. Churchland (1995) consciousness disappears in deep sleep and reappears in dreaming, at least in muted or disjointed form (Section 8.2.1).

Dream states represent a state of consciousness in which there is a sharp decrease in world-outside input. Dreams are more perceptual than conceptual: things are seen and heard rather than being subjected to thought. Only smaller parts of the sleep period are connected with dreams (Section 5.3.4). Dream periods usually last from 5 to 20 minutes.

Modern research does not suggest that dreams have no meaning. Dreams are meaningful mental products, just as thoughts. They express important wishes, fears, concerns, and worries of the dreamer. Undoubtedly the study and analysis of dreams can sometimes be a useful procedure, revealing different aspects of a person's mental functioning.

There are many different hypotheses which try to elucidate the still 'mysterious' phenomenon we call dreaming:

–Freud's theory of dreams. The role of subconsciousness is the essential binding principle of all of Freud's psychological theories. Unlike the fundamental unavailability of nonconscious to conscious states, however, Freud believed that subconscious states, although inaccessible at any particular time, could be made accessible. Freud (1900): 'Dreams are the 'royal road' to the unconscious' It is important to notice that subconscious states derive originally from the interactions of basic value-dependent bonus-malus system. S.Freud (1914) also prophesied that one day we will discover the 'organic (chemical) substances' which are responsible for human sexuality and behaviour, and also for the unconscious.

–Crick's hypothesis of 'reverse learning' (F.Crick, 1983), in which sleep results in erasure, or unlearning, of false information; 'We dream to forget'. This erasure of parasitic thoughts accounts only for bizarre dream content. Nothing can be said about the narrative dream. This hypothesis corresponds to our general principle of overabundant generation (of information) and selective elimination (erasure, unlearning).

Table 8-2 Sleep and dreams

	Wake	NREM sleep*)	REM sleep
Behaviour	Posture shifts	Immobility	Immobility inhibition
Electromyogram Electroencephalogram	Smooth	Extremely variable	Smooth
Sensation	Vivid externally generated	Dull or absent	Vivid, internally generated
Thought	Logical progressive	Logical perseverative	Illogical, bizarre
Movement	Continuous voluntary	Episodic involuntary	Commended but inhibited
Verbal report length after awaking		Median length 20 words	Median length 150 words
REM: Rapid Eye Movement NREM: Non Rapid Eye Movement			

–The hypothesis of 'activation-synthesis', which claims that although dreams might have a psychological content they are inherently meaningless.

–Winson's theory that dreams may reflect a fundamental aspect of mammalian memory processing. 'We dream to consolidate the memory.' Crucial information acquired during the waking state may be reprocessed during sleep. According to J. Winson (1990), dreams may reflect a memory processing mechanism inherited from lower species. This reprocessed information may constitute the core of our unconsciousness, in a non-verbal form. Because animals do not possess language, the information they process during active periods of sleep (REM periods) is necessarily sensory.

J.Winson (1997): "Strangely meaningful images and bizarre flights of fancy may all be part of the dreaming brain's efforts to review memories, evaluate recent experiences and plot new strategies for surviving challenges in the waking world."

The neurophysiology of dreams makes progress. J.A.Hobson and R.Stickgold (1995) suggest that the bizarre cognition and amnesia of REM sleep result from the withdrawal of aminergic modulation of the forebrain.

8.2.3 Vagueness of information; illusions

To be conscious of own experience includes knowledge about the 'reality' of incoming signals, or more precisely about correctness of primary information. The physical nature of signals and physical and biotic character of nervous signal and information results in the vagueness of all these phenomena. The information processing apparatus is structured and prepared for processing of vague inputs and to produce vague output. Only very 'artificial' information processing, called 'science' tries to purify the input information in a non-vague, more or less clear and univocal system of logic manipulation and clearly defined objects.

F.Crick (1994): "The eye and the brain must attempt to analyze the incoming light so that it provides all this important information. How does it do this?..(L)et me make three general remarks: 1. You are easily deceived by your visual system, 2. The visual information provided by our eyes can be ambiguous, 3. Seeing is a constructive process. ... What you see is not what is really there, it is what your brain believes there."

Crick speaks about 'ambiguous visual information". In this book the term 'vague' is used. Crick speaks about 'deceit' in the visual system. In this book this phenomenon is applied to all sensory activity of animals and humans (Fig. 8-5).

R.Melzack (1997): "Under normal circumstances the myriad qualities of sensation people experience emerge from variations in sensory input. Yet even in the absence of external stimuli, much the same range of experiences can be generated by other signals passing through the neuromatrix—such as those produced by the spontaneous firing of neurons in the neuromatrix itself or the spinal cord. People who have lost an arm or leg sometimes still 'feel' the missing part Neurobiologists are beginning to understand more fully what creates this disturbing illusion. Phantom seeing often coexists with a limited amount of normal

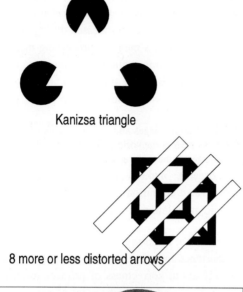
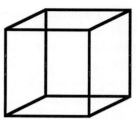

Figure 8-5 Illusions

vision. People usually describe the visual phantoms as seeming real despite the obvious impossibility of their existence. Phantom seeing occurs most among the elderly, presumably because vision tends to deteriorate with age. Phantom sounds are also extremely common, although few people recognize them for what they are "

Even relative clearness of information is additionally destroyed by the very weighty bonus-malus system, present in all animals and arriving at the summit in humans. In humans the clearness of information is intentionally fogged by the 'usefulness' and easiness of lie, cheat and deception in the social world. Certainly, we could assume that there is no such thing as certainty (Fig. 8-5).To repeat: intelligence, as the highest stage of information processing, seems to be indivisible from free will, or at least from the subjective feeling of the existence of free will. Before going further, we must define the related terms of necessity, chance, and free will.

8.3　FREE WILL, THE REAL MYSTERY OF HUMANS

8.3.1　*Necessity, chance and free will*

Necessity is the expression of the absolute power of the processes going on in Nature whose representations in human brain are called 'Laws of Nature'. There exists no phenomenon in the real world which is not governed by these laws. This is the way in which human brain produces the concept of 'Laws of Nature', knowing that it itself is a product of these Laws, and hoping that this concept enables human existence and human knowledge of the Nature.

Chance; the very essence of processes on the atomic and sub-atomic level is their unpredictability—the principal (at least at present state of quantum mechanics, the theory of elementary particles and elementary forces) impossibility of predicting where and when a particular elementary event could occur. For example, in a sample of radioactive atoms it is impossible—even forbidden—to predict when one selected atom will decay, either in the next second or in some giga (billion) years, in spite of knowing the exact value of the half-life (the period after which the number of decayed atoms equals half the initial amount). Another and not less significant source of 'chance, contingency, accident' is the lack of knowlegde of own and other persons or animal reactions and thoughts.

Freedom is an abstract and ill-defined term. It can be used only for some events which take place in the human mind and in its products—thoughts, ideas, concepts, and the appropriate means for their transmission, such as writing and drawing. The subjects of freedom and unexpected events are linked to the phenomenon called 'chaos'. It is true that the term 'chaos' has become very popular in recent times. But it is also true that some properties of a deterministic system can be better explained by taking into account the idea of 'unexpected' jumps, surprises, or even catastrophes.

In any deterministic system—that is, a system which can principally be described by means of a set of clearly formulated mechanisms, but which are very complex and have non-linear relationships—unexpected changes emerge when the boundary conditions are unknown. This results in 'chaos'.

The question also arises as to how a butterfly's wings (macroscopic), which causes the hurricane, depend upon atomic or even sub-atomic (microscopic) events (Section 1.2.2). At the lowest level, such as the movement of electrons in atomic orbits, fundamental uncertainty exists (Heisenberg's Uncertainty Principle, see Section 2.2). Does this uncertainty influence the unpredictable flapping of a butterfly's wings, with all its far-reaching consequences, due to changes in the boundary conditions of the butterfly's neuronal network?

Recently some experiments in mesoscopic scale (of about 0.1 micrometer) may provide significant insight into the vague boundary between classical macroscopic events and microscopic quantum worlds (C.Monroe et al, 1996, paper in 'Science' entitled 'A 'Schrödinger Cat' superposition state of an atom').

The boundary conditions of every nervous network, especially very complex ones, such as those in the human brain, are not only unknown but probably continuously change in an unpredictable manner. The deaths of neurons and synapses, especially, seem to be partially unpredicted, or almost chaotic, events. If the neuronal network 'seeks' a final decision, resulting from very numerous non—linear links and influenced by changeable 'boundary conditions', in the form of stored information which is itself influenced by a changeable 'bonus-malus system', the final result—the decision of the whole mental process—can be surprising, and subjectively felt by the individual as the expression of 'free will'. Chance, an unexpected event, is an inherent property of all physical events on the sub-atomic level. This is not due to human ignorance, but is caused by the real nature of elementary phenomena. The radioactive decay of natural radio-isotopes, such as tritium, carbon-14 and potassium-40, is an example of the real influence of chance on the functioning of every material structure, especially on intellectual processes in the human brain, assuming that some of the energy of radioactive decay could significantly influence nervous activity (which is a weak hypothesis only).

8.3.2   *Hypothesis of free will mechanisms*

Our hypothesis of the nature of free will can be summarised as follows:

–All internal information processing mechanisms rely more or less on known electro-chemical processes taking place in neuronal networks. However, this does not mean that, in the future, we will be able to explain free will by reducing

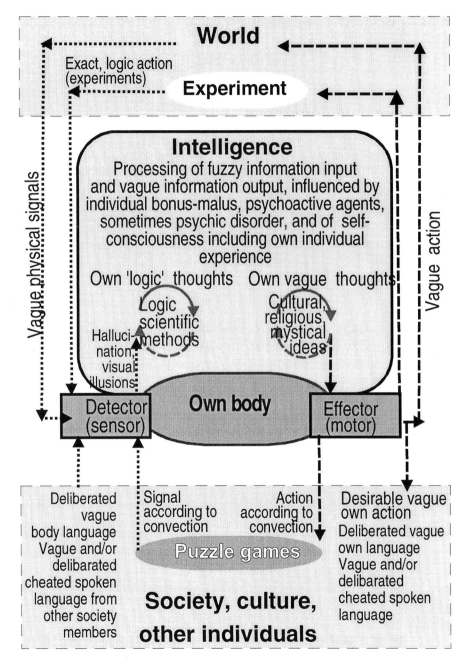

Figure 8-6 Mind and vagueness

this extremely complex phenomenon to a suitably large number of simple processes,

–A bonus-malus system labelling each processed information with subjective quality value is of significance in biotic information processing, and achieves its highest level in humans,

–The bonus-malus system possibly uses a 'scale', which labels a plus-value for maximum bonus and a minus-value for maximum malus. The resulting bonus-malus value is the 'algebraic sum' of both qualities—the bonus and the malus,

–After a longer time period all pictures fade and the absolute bonus- malus value decreases,

–The most important concept in the mind is Self (Section 8.1.1). The Self is also marked with a bonus-malus label. In the case of a positive value, with a label greater than zero, I can say: 'I am good', 'I feel good'. With a negative label, the opposite is true: 'I am bad', 'I feel bad',

–At each step of information processing, the conscious and 'unconscious' concepts are processed in the 'working memory' (Section 5.2.1), co-processed with, and influenced by, other concepts, each with its own bonus-malus label. During processing, the value labels influence one another, which results in a larger or smaller increase or decrease of overall bonus or malus. The most important concept is the Self, having at any instant its own bonus-malus value, which therefore influences other concepts processed at that moment,

–After processing, the concept is storage in the networks memory along with the latest bonus-malus value,

–A large number of events inside the human brain significantly influence the current state of the neural networks, and thus influence the formation of free will. These events include: the uncontrolled death of some neurons (in the adult human brain some hundreds die per second) (Section 4.3.2), the uncontrolled death of synapses (in an adult human brain some hundreds of thousands die per second), (Section 4.3.3), may be also radioactive decay of natural radioactive substances and impact of external radiation, and the resulting changes in the networks, as well as some so-called chaotic, (that is, unpredictable, because strongly depending from beginning conditions) 'spontaneous' neuronal signals,

–At the moment when free will is being expressed, concepts, including the Self, are used in a more or less conscious form in the 'working memory' and the resulting decision is compared with the expected effect of proposed activity. If the result is positive, the bonus value is greater than the malus, the decision on the neuronal network level is made, and the activity is carried out,

–Free will depends not only on conscious thinking, mostly at the intelligence stage of information processing, but also at the unconscious processing at the stages of discent, instinct, and reflex,

–The feeling of free will is, in normal people, a positive feeling with a relative positive bonus-malus label. But what advantage does consciousness give to an organism, so that we can see why it has successfully evolved?

–The feeling of free will is the most significant basis for the feeling of selfconsciousness. Without this feeling, selfconsciousness is not possible at all. Thus, to be selfconscious means having the feeling of free will. Not to forget that free will is in large a product of self-deception.

Mankind is the product of the subtle feeling of free will.

8.3.3 *Free will and the natural sciences*

Some prominent scientists, are of the opinion that the phenomenon of free will has nothing to do with the natural sciences.

M.Minsky (1986): "Freedom of will: the myth that human volition is based upon some third alternative to either causality or chance."

A growing number of scientists are of the opposite opinion; that is, they not only accept the objective existence of free will, but are ready to investigate it using all available techniques of the natural sciences.

J.H.Fremlin (1986): "Like most people, I feel I have some freedom of will. Our consciousness is, in biological terms, very expensive. When we have gained a full understanding of the way in which some ten thousand million cells, with their complex multiple interconnections, can support a single apparently unitary conscious awareness, and when we have learned how the brain so promptly finds and presents the relevant memories to the consciousness to make optimum choices possible, we can begin again seriously to discuss free will."

It is our viewpoint that the investigation of free will by all available scientific methods is the greatest and the most significant challenge of the present day, because it has an enormous influence on the moral development of mankind. Without such development, the future of intelligence on this planet could be threatened. However, it cannot be excluded that the concept of 'free will' is a product of longer cultural self-deception. The largest self-deception in human ability to prepare false information.

N.Chomsky (1983): "Free will is simply an obvious aspect of human experience. I don't think that there's any scientific grasp, any hint of an idea, as to how to explain Free will. ... We have a body of scientific knowledge that simply doesn't appear to connect with the problem of Free will in any way. People have been trying to solve the problem of Free will for thousands of years, and they've made zero progress. They don't even have bad ideas about how to answer the question. My hunch—and it's no more than a guess—is that the answer to the riddle of Free will lies in the domain of potential science that the human mind can never master, because of the limitations of genetic structure. I'm not sure that I want Free will to be understood."

R.W.Sperry (1983): "The seemingly irreconcilable dichotomies and paradoxes that formerly prevailed with respect to mind versus matter, determinism versus free will, and objective fact versus subjective value become reconciled today in a single comprehensive and unifying view of mind, brain, and man in nature..... The proposed brain model provides in large measure the mental forces and abilities to determine one's own actions. It provides a high degree of freedom from outside forces as well as mastery over inner molecular and atomic forces of the body. In other words it provides plenty of free will as long as we think of free will as self-determination. A person does indeed determine with his own mind what he is going to do, and often from among a large series of alternative possibilities. This does not mean, however, that there are cerebral operations that occur without antecedent cause. Man is not free from the higher forces in his own decision-making machinery. In particular, our model does not free a person from the combined effects of his own thought, his own impulses, his own reasoning, feeling, beliefs, ideals, and hopes, nor does it free him from his inherited makeup or his lifetime memories. All these and more, including unconscious desires, exert their due causal influence upon any mental decision, and the combined resultant determines any inevitable but nevertheless self-determined, highly special, and highly personal outcome. Thus the question: Do we really want free will, in the indeterministic sense, if it means gaining freedom from our own mind, from our own self and inner being?"

J.Searle (1984): "For reasons that I don't really understand, evolution has given us a form of experience of voluntary action where the experience of freedom is built into the very structure of conscious, voluntary, intentional human behaviour. That is, it looks as if we would have to contain some entity that was capable of making molecules swerve from their paths. I don't know if such a view is even intelligible, but it's certainly not consistent with what we know about how the world works from physics. And there is not the slightest evidence to suppose that we should abandon physical theory in favour of such a view."

N.Bohr (1958): "Quite apart from the extent to which the use of words like 'instinct' and 'reason' in the description of animal behaviour is necessary and justifiable, the word 'consciousness', applied to oneself as well as to others, is indispensable when describing the human situation... The use of words like 'thought' and 'feeling' does not refer to a firmly connected causal chain, but to experiences which exclude each other because of different distinctions between the conscious content and the background which we loosely term ourselves... We must recognize that psychical experience cannot be subjected to physical measurements and that the very concept of violation does not refer to a generalization of a deterministic description...In an objective description of our situation, the use of the word 'violation' corresponds closely to that of the words 'hope' and 'responsibility', which are equally indispensable to human communication."

G.Wald (1965): "I believe that we have free will and that it comes from our uniqueness as individuals, possibly pre-dominated, but to a certain degree unforeseeable. Free will is frequently inefficient, unconvincing and always unreliable. That is the characteristic of freedom."

O.Creutzfeldt (1990): "In my opinion (the free will), it will remain a mystery or—in scientific terms—an unexplainable property of the brain."

F.Crick (1994): "The actual cause of the decision may be clear cut, or it may be deterministic but chaotic-that is, a very small perturbation may make a big difference on the end result. This would give the appearance of the Will being 'free' since it would make the outcome essentially unpredictable. Of course, conscious activities may also influence the decision mechanism... Free Will is located in or near the anterior cingulate sulcus. Other areas in the front of the brain may also be involved."

R.Penrose (1994): "One appears to conclude(...) that:(i)the conscious act of 'free will' is a pure illusion, having been, in some sense, already pre-programmed in the preceding unconscious activity of the brain; or (ii)there is a possible 'last-minute' role for the will, so that it can sometimes (but not usually) reverse the decision that had been unconsciously building up for a second or so before; or (iii)the subject actually consciously wills the finger-flexing at the earlier time of a second or before the flexing takes place, but mistakenly perceives, in consistent way, that the conscious act occurs at the much later time, just before the finger is indeed flexed."

E.T.Rolls (1995): "Free will would ... involve the use of language to check many moves ahead on a number of possible series of actions and their outcomes, and with this information to make a choice from the likely outcomes."

8.3.4 *Free will and psychic force?*

What is free will? In the simplest terms, and in everyday language, it is something which everybody can use to move his own finger, in a way which is only restricted by its anatomical construction. This 'something' seems to be a 'thought' and therefore of non-material, ideal, psychic, or spiritual nature. The opposition 'free will'—'matter' seems to be obvious and without question.

A more detailed scenario for the interaction of free will and the finger could be as follows: I think, therefore I will move my finger. My decision is of a non-material nature, it is only a thought. At least, this is a common opinion about what thought is. In an unclear way, this thought is transformed into neuronal signals sent from my brain to my finger. The nervous signals cause the contraction of the appropriate muscles and my finger moves. Evidently, the most critical point is the transformation of my thought into the first nervous signal in my brain, the transformation of will into material movement.

The simplest way to find a solution to this enigma is to postulate a kind of special force, let us say a 'psychic force', responsible for this mediation. However, 'Ockham's razor' prohibits making assumptions which result in the emergence of new entities, especially postulated for one situation alone. There exists no independent evidence for the existence of a force, in this case a 'psychic force', which cannot be investigated by means of physical methods, but which acts inside the brain as mediator between the realms of thought and matter. This book is based on the principle of 'physicality' and we cannot resign even at this point. However without such a special mediating force we cannot explain how an 'immaterial thing', such as thought, is able to influence the movement of electrons in neuronal atoms, thereby initiating neuron firing.

How can free will, having apparently all the properties of immateriality, influence the movement of electrons in atoms? The only possible solution is very radical. Free will must be borne by well-defined material carriers, and only at the last instant appear as a subjective 'immaterial' phenomenon. Some years ago, B.Libet published the results of his experiments with humans concerning free will.

B.Libet (1986): "The conscious mind doesn't initiate voluntary actions. I propose that the performance of every conscious voluntary act is preceded by special unconscious cerebral processes that begin about one-half second or so before the act. Yet free will is not just a noble illusion. There is about a quarter of a second between conscious awareness of an impending action and its actual occurrence, time enough to permit or cancel the intention. Free will is traditionally viewed as following slower deliberations, but, no matter how much silent choice—making you engage in, the same unconscious processes come into play just before you act. The two-tenths of a second delay between awareness of an intention and actual muscle action provides an opportunity for conscious control. Neurally that's a long time for the selection of events to occur. This suggests that the conscious veto interfered with the final development of 'readiness potential' processes leading to action and agrees with our common intuitive experiences of self-control over urges to act. If you want a philosophy that allows conscious control, and even free will, the experimental demonstration of vetoing an intended act provides an opportunity for it. I ... believe free will exists. If it does exist, our experimental data are pointing toward the brain processes that would be involved."

Libet's theory (1990) states that the transition from an unconscious to a conscious mental function is determined, at least in part, by an increase in duration of the appropriate neural activities.

B.Bridgeman made an observation which exactly corresponds to our own point of view:

B.Bridgeman (1986): "Libet's findings are important, but he has yet to forge a 'physiology of free will'."

The question of free will must be solved on the basis of physicality and biogeneity (Section 1.1.2); that is, on the basis of modern natural sciences.

8.4   INTELLIGENCE AND MORALITY

8.4.1   *Intelligence and morality*

Is the link between information processing and such abstract terms as morality (behaviour rules of individuals in society) and ethics (general validity, rational, natural theory of good and wrong) the relation between right and wrong, good and bad, so strong that it must be discussed in this book? The answer is, 'Yes'. One of the principal concepts in this book is the quality label of information, the bonus-malus system, which has evolved through the whole of biotic evolution (Fig. 3-10). The highest level of evolution is reached at the stage of intelligence, and it is here that the bonus-malus system also reaches its highest level. The concepts of right and wrong, at the abstract level, belong to the realms of 'morality' and 'ethics'. Intelligence cannot be separated from these concepts.

T.Dobzhansky, E.Boesiger (1983): "The concept of ethics is meaningless, unless the following conditions exist: a) there are alternative modes of action; b) Man is capable of judging the alternatives in ethical terms; and c) he is free to choose what he judges to be ethically good. The animal ancestors of the human species had no ethics. The question inevitably arises, then, 'What is the biological foundation of the capacity to have ethics?' and ' How has this foundation been shaped during evolution?'...Ethical codes are parts of cultural inheritance, not of biological heredity. It is not ethics themselves but the capacity to have ethics that has a biological foundation. Is man by nature good or evil? Great ethologists such as Lorenz and Tinbergen and their students believe that man has inherited from his animal ancestors aggressive drives, tendencies toward violence, and other unpleasantness. Our own position is this: Man is born neither good nor evil but with the capacity to become either. The manifestation of behavioural tendencies is influenced by the social environment.. ..by training and education...and, in third place, an individual may be able with some effort to control certain unwanted drives."

So what are the basic conditions necessary for the evolution of morals? De Waal (1996) postulates two: an organism must live in groups on which it depends for subsistence and defence, and these group members must co-operate even though they have disparate individual interests. It is from the resolution of conflicts that morality emerges. Among the chimpanzees (humankind's nearest living relations) adult males spent an average 30 percent of their waking hours alone, and mothers and their offspring spent 65 percent on their own (W.C. McGrew, 1996).

Our morality cannot be directly influenced by our ancestors, because they never faced such crowded conditions as we have at present.

P.M. Churchland (1995): " Moral knowledge is real knowledge precisely because it results from the continual readjustment of our convictions and practices in the light of our unfolding experiences of the real world, readjustment that lead to greater collective harmony and individual flourishing."

Science, as one of highest and most abstract products of intelligence, is of course intimately links to the problems of morals and ethics.

R.W.Sperry (1983): "The role of science is envisioned in which science is upheld, not because it begets improved technology, but because of its unmatched potential to reveal the kind of truth on which faith, belief, and ethical principles are best founded. Recent developments in the mind-brain sciences eliminate the traditional dichotomy between science and values and support a revised philosophy in which modern (non-reductive) science becomes the most effective and reliable means available for determining valid criteria for moral value and meaning... Human value priorities stand out as the most strategically powerful causal agents now shaping events on the surface of the globe. More than any other causal system with which science now concerns itself, the human value factor is going to determine the future."

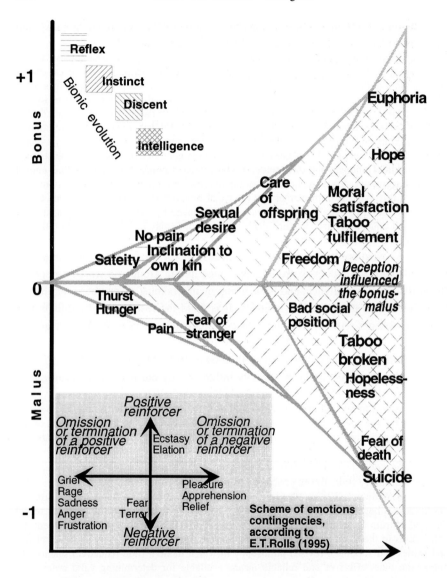

Figure 8-7 Emotions and morals

Charles Darwin himself speculated that natural selection might favour groups whose members engaged in altruistic behaviour, even if the individual harmed their own fitness or even life. A contrary concept presents Richard Dawkins with his model of 'selfish gene'. Some authors defend the idea of positive influence of altruism, arguing that separate organisms can be viewed as collections of mutually dependent genes (hives, herds, clans, may be also mankind?). Within the same hive, bees certainly compete with one another, but they may also band together to compete with other hives for resources, thus acting like one organism (J.Horgan, 1996).

R.J.Blumenschein, J.A.Cavallo (1992): "Altruistic behaviour and social and sexual co-operation (males searching the scavengeable meat; females gathered fruits and tubes nearer home and families) shared the fake. This began to select for intelligence, language and culture."

There exist some questions which seem to rest unanswered: would Homo erectus have the same rights as Homo sapiens in our modern society? Could in future an 'intelligent' computer find the solution of the ethics problems?

In the modern society the term 'conscience', a personal sense of the moral goodness or blameworthiness of one's own conduct, intentions, or character with regard to a feeling of obligation to do right or be good plays an insufficient role. Conscience is thus generally understood to give intuitively authoritative judgements regarding the moral quality of single actions. Why are there so few considerations concerning the coupling between intelligence and conscience? Even in the time when the link between intelligence and consciousness is written with large characters.

8.4.2 *Genocide, homicide and crimes*

The human being has the 'art of killing' degenerated to 'art of killing of own species'. It is a self-evident duty not to neglect this terrible property of intelligence.

Where are the sources of this 'human' behaviour? Some anthropologists think that the first warriors battled over women instead of food, water holes or land. Similar theories about the influence of sexual competition on the evolution of male behaviour are not rare. The human-human wars began according to these opinion in the Pleistocene, hundreds of thousands years ago. The victors would produce more offspring. That would lead to the selection of certain complex cognitive mechanisms in later generations, such as the ever-improving ability to form groups (hierarchically organized?) to prepare special weapon and sophisticated tactics and go to battle. The battle against 'intelligent' but defeated enemy and to rape his women is the best selection for the 'more intelligent' victor. Is this opinion not an accusation against the role of intelligence in the biotic evolution?

J-P. Changeux (1985): "Were the first .. stones of Australopithecus really weapons of war, as much as hunting tools or farming instruments? A frightening thought is that the genetic evolution of the human brain may be a consequence of the ability to murder one's fellows."

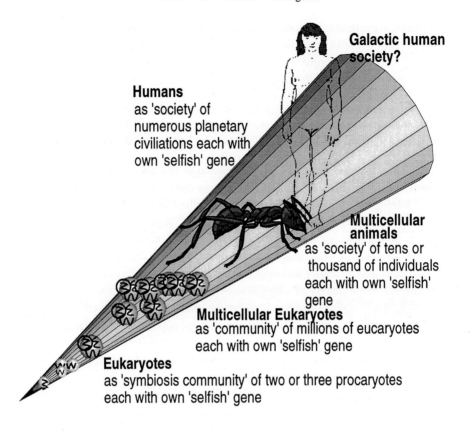

Figure 8-8 Evolution of altruism

The oldest images of humans spearing each other have been found recently in northern Australia among hundreds of old rock paintings dating back as far as 6000 years, corresponding to the end of the last ice age.

Another black side of human intelligence is the killing of human children, especially of parent-child homicide, especially by stepparents.

There exists since years a discussion, based on rather weak and insignificant data, that some inclination to violence is connected with some peculiarities in the brain (brain scans of murderers showing frontal lobe dysfunction, or serotonin level change) and has a genetic basis (H. Hemminger, 1994).

W.W.Gibbs (1997): "As medical researches have teased out a few tantalizing links between brain chemistry, heredity, hormones, physiology and assaultive behaviour, some have become emboldened. 'Research in the past 10 years conclusively demonstrates that biological factors play some role in the etiology of violence.' The importance of that role is still much in doubt. Trace element deficiencies are just one of many frequently cited but poorly demonstrated claims that nutritional problems can cause criminal and violent behaviour."

Is genocide a unique property of humans? In cave paintings, and in aborigines rock paintings, a man-to-man struggle seems to be absent. But, absence of evidence is not evidence of absence. Only after emergence of agriculture, field irrigation, seasonal storage of seeds began the period of more-or-less organized man-to-man struggles, wars. The question is whether this behaviour influenced the improvement of intelligence.

Ch. Darwin (1871) in 'Descent of Man': "Now, if some one man in a tribe, more sagacious than the others, invented a new snare or weapon, or other means of attack or defence, the plainest self-interest, without the assistance of much reasoning power, would prompt the other members to imitate him; and all would thus profit. The habitual practice of each new art must likewise in some slight degree strengthen the intellect.....If such men left children to inherit their mental superiority, the chance of the birth of still ingenious members would be somewhat better, and in a very small tribe decidedly better."

A.K.Turner (1992): "Violence is just one form of aggressive behaviour in primates. Sometimes the aggression escalates to ... such as hitting and biting and the combatants can suffer serious injury. In many species, adult males have higher mortality rates than females, in part because of fatal injuries in such fights. (In some species) males...may kill young infants already in the group. Chimpanzees sometimes chase and kill members of other community. Contrary to Lorenz's ideas and to popular belief animals, including primates, do not have an aggressive drive (instinct in this book) that builds up inside them."

R.S.Meindl (1992): "There is some evidence of violent death at the archaeological sites including a few bone-imbedded flint projectile points... but the number of individuals who met premature deaths at the hands of other will never be known."

Encyclopaedia Britannica (1978): "Knowledge of primitive, preliterate warfare is limited... Primitive warfare seemed to have functioned mainly to preserve the social group by increasing solidarity. It was conducted with very primitive weapons, and its conduct was frequently ritualized, which decreased the damage done. The establishment of the first cities, was followed by the establishment of first professional armies."

For many centuries, infanticide was an accepted practice for disposing of unwanted babies. An archaeological evidence have been obtained in Roman Ashkelon (Israel) where skeletal remains of around 100 babies, mostly male, were discovered in a sewer, which may be have been unwanted offspring of courtesans working in this bathhouse (M. Faerman et al, 1997).

8.4.3 *Consciousness about own death and altruism*

Some people think that the most human property is the knowledge about own death, being unique to our species, among all animals. The fear of own death cannot go traceless through our thoughts. Each individual has his own 'strategy' to think about his own end. The impact of this strategy influenced strongly his bonus-malus system. Often this strategy used sophisticated self-deception.

Altruism, the ability to sacrifice own life for a 'higher' aim is expressed mostly among the social insects (Fig. 8-7). Examples of real altruism among the humans exist not only in the Bible but also in the past. In the present time they are relatively numerous, but cannot match the terribly long list of homicide acts.

8.5 BIBLIOGRAPHY

Andreasen NC (1997) Linking mind and brain in the study of mental illness: a project for a scientific psychopathology. *Science* **275** 1586
Barron F (1981) Creativity, intelligence, personality. *Ann Rev Psych* **32** 439
Bechara A et al (1997) Deciding advantageously before knowing the advantageous strat egy. *Science* **275** 1293
Beloff J (1994) Minds and machines: a radical dualist perspective. *J Consci Stud* **1** 1,32
Boesch Ch (1996) The question of culture. *Nature* **379** 207
Chalmers DJ (1995) Facing up to the problem of consciousness. *J Consci Stud* **2** 3 200
Chalmers DJ (1996) Das Rätsel des bewussten Erlebens. *Spekt Wiss* **2** 40
Chalmers DJ (1997) The puzzle of conscious experience. *Sci. Am. Spec.Iss. Mysteries of Mind*, Vol **7**,30
Churchland PM (1995) *The engine of reason, the seat of soul.* MIT Pr, Cambridge, Mass
Chomsky N (1980) *Rules and representation.* Columbia Univ Press, N.York
Crane T (1996) Beyond reasonable doubt. *Nature* **379** 685
Crick F (1994) *The astonishing hypothesis.* Scribners, N.York
Crick F (1996) Visual perception: rivalry and consciousness. *Nature* **379** 485
Crick F, Koch C (1992) The problem of consciousness. *Sci Am* **267** 3 111
Crick F, Mitchison G (1983) The function of dream sleep. *Nature* **304** 111
Cromer A (1993) *Uncommon sense. The heretical nature of science.* Oxford Un Pr, N.Y.
Dobzhansky T, Boesiger E (1983) *Human culture: A moment in evolution.* Columbia Univ Press, N.York
Dolara P et al (1996) Analgesic effects of myrrh. *Nature* **379** 29
Eccles JC (1994) *How the Self controls the brain.* Springer, Berlin
Eigen M (1985) Mozart-oder unser Unvermögen, das Genie zu begreifen. *NatRund* **9** 355
Enquist M, Arak A (1994) Symmetry, beauty and evolution. *Nature* **372** 169
Feyrabend P (1976) *Wider den Methodenzwang. Skizze einer anarchistischen Erkenntnis theorie.* Suhrkamp, Frankfurt
Feyrabend P (1984) *Wissenschaft als Kunst.* Suhrkamp, Frankfurt
Feyrabend P (1995) *Killing time.* Univ Chicago Press, Chicago, Ill.
Gardner H (1984) *Frames of mind: theory of multiple intelligences.* Heinemann, London
Gawin FH (1991) Cocaine addiction: psychology and neurophysiology. *Science* **251** 1580
Gazzaniga MS (1995) Consciousness and the cerebral hemispheres, in Gazzaniga MS (edit) *'The cognitive neurosciences'*. MIT Press, Cambridge, Mass
Gibbs WW (1997) Seeking the criminal element. *Sci. Am. Spec.Iss. Mysteries of Mind*, Vol **7**,102
Glassman AH, Koob GF (1996) Psychoactive smoke. *Nature* **379** 677
Gould SJ (1983) *Der falsch vermessene Mensch.* Birkhäuser, Basel
Hameroff SR (1994) Quantum coherence in microtubules: a neural basis for emergent consciousness. *J Conscious. Stud* **1** 1 91
Kostek T (1983) *Das neue Gesicht.* Causa, München

Lem S (1984) *Phantastik und Futurologie*. Suhrkamp, Frankfurt
Lem S (1981) *Summa technologiae*. Suhrkamp, Frankfurt
Libet B (1994) A testable field theory of mind-brain interaction. *J Consci Stud* **1** 1 119
Lumsden CH, Wilson EO (1982) Precis of 'Genes, Mind, and Culture'. *Beh Brain Sci* **5** 1
McGrew WC (1996) Moral kin? *Sci Am* **275** 3 137
Melzack R (1997) Phantom of limbs. *Sci. Am. Spec.Iss. Mysteries of Mind*, Vol **7**,84
Miller GA (1991) *The science of words*. Sci Am Library, N.York
Mountcastle VB (1990) The construction of reality, in J.C. Eccles, O. Creutzfeldt (eds) *'The principles of design and operation of the brain'*. Springer, Berlin.
Penrose R (1994) *Shadows of the mind. A search for the missing science of consciousness.* Oxford Univ Press, N.York
Piaget J (1948) *Psychologie der Intelligenz*. Rascher, Zürich
Plomin R (1990) The role of inheritance in behavior. *Science* **248** 183
Pöppel E (1984) *Grenzen des Bewusstsein*. Deutsche Verlags-Anstalt, Stuttgart
Popper K (1983) *Realism and the aim of science*. Hutchinson, London
Pulvirenti L, Koob GF (1996) Die Neurobiologie der Kokainabhängigkeit. *Spekt Wiss* **2** 48
Rose S, Kamin LJ, Lewontin RC (1984) *Not in our genes biology, ideology, and human nature*. Panthenon, N.York
Sagan C (1996) *The demon-haunted world*. Random House, N.York
Savage-Rumbaugh S (1991) Multi-tasking: the Pan-human rubicon, *XX Semin.Neur* **3** 417
Schacter DL (1995) Introduction: Consciousness, in Gazzaniga MS (edit) *'The cognitive neurosciences'*. MIT Press, Cambridge, Mass
Schultz W et al (1997) A neural substrate of prediction and reward. *Science* **275** 1593
Simon HA (1969) *The sciences of the artificial.* (sec edit) MIT Press, Cambridge, Mass.
Smith SB (1983) *The great mental calculators*. Columbia Univ Press, N.York
Snyder SH (1986) *Drugs and the brain*. Sci Am Library, N.York
Sperry R (1983) *Science and moral priority*. Columbia Univ Press, N.York
Vogel G (1997) Scientists probe feelings behind decision-making. *Science* **275** 1269
Wald G (1965) Determinacy, individuality and the problem of free will, in J.R.Platt (edit) *'New views of the nature of man'* Univ Chicago Press, Chicago
Wald G (1988) *Cosmology of life and mind*. Los Alamos Sci Fell Coll, Los Alamos N.M.
Winson J (1990) The meaning of dreams. *Sci Am* **263** 5 42
Winson J (1997) The meaning of dreams. *Sci. Am. Spec.Iss. Mysteries of Mind*, Vol **7**,58
Zwicky F (1971) *Jeder ein Genie*. Lang, Bern

CHAPTER 9
ARTIFICIAL INTELLIGENCE

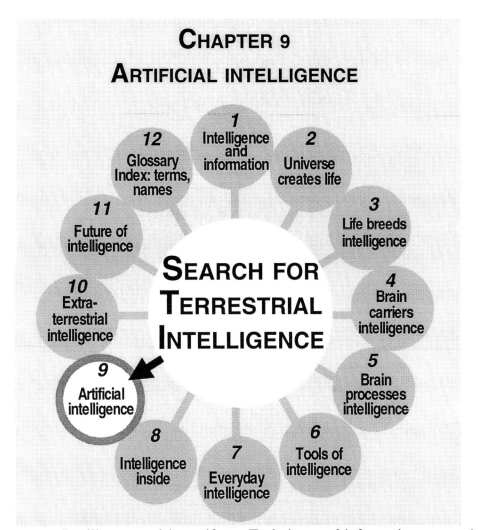

Intelligence and its artifacts. Techniques of information processing. Brain and computer: similarity and differences. Computers faultlessness. Fuzzy, connectial and other computers.

The real world and puzzles. Virtual reality. Artificial reflex and instinct: robots, chess player. Artificial discent: experts, learning robots. Artificial language translators.

Computational brain. Artificial intelligence, chance and reality. Artificial consciousness?

9 ARTIFICIAL INTELLIGENCE 336
9.1 INTELLIGENCE AND ITS ARTIFACTS 336
9.1.1 Artifacts; what are they? *336*
9.1.2 The brain and the computer: similarity? *338*
9.1.3 Stages of computer evolution *341*
9.2 COMPUTERS AND INTELLIGENCE; FAULTLESSNESS AND COMPLETENESS 342
9.2.1 Computers faultlessness? *342*
9.2.2 Fuzzy, connectial, and other computers *343*
9.2.3 The real world and jigsaw puzzles *344*
9.2.4 Virtual reality *345*
9.2.5 Semantics and information processing *345*
9.3 ARTIFICIAL REFLEX, INSTINCT, AND DISCENT 346
9.3.1 Artificial reflex; robots *346*
9.3.2 Artificial instinct; robots *347*
9.3.3 Artificial instinct; chess player *348*
9.3.4 Is communication with animals possible? *349*
9.3.5 Artificial discent; experts *350*
9.3.6 Artificial discent; learning robots *351*
9.3.7 Artificial language translators *351*
9.4 ARTIFICIAL INTELLIGENCE, COMPETITOR OR SLAVE 353
9.4.1 Computational brain *353*
9.4.2 Artificial intelligence; chance and reality *355*
9.4.3 Artificial consciousness? *358*
9.5 BIBLIOGRAPHY 360

9 ARTIFICIAL INTELLIGENCE

9.1 INTELLIGENCE AND ITS ARTIFACTS

9.1.1 *Artifacts; what are they?*

One of the most significant attributes of intelligent beings is the ability to generate a cultural system, based on numerous artifacts. An artifact is a material object made by a living being, especially that deliberately constructed by an intelligent being, which cannot emerge spontaneously. The difference between naturally occurring and man-made objects can sometimes be so small that it is impossible to distinguish between them. The origin of the oldest stone tools is a clear example of such uncertainty. However, the evolution of human civilization steadily brought more and more sophisticated devices which can be clearly classified as artifacts.

We must note that artifacts are also prepared by a number of animals, beginning with the rather less-developed, such as insects. The beehive and the termite mound are good examples, but bird's nests and beaver dams are also excellent illustrations. Some animals create tools in an ad hoc fashion, such as chimpanzees for hunting ants. An example of the material culture of chimpanzees is the nut-cracking behaviour which involves selecting 'hammers' of the appropriate weight to crack nuts and transporting both hammer and nuts to suitable anvils. Interestingly, Tai chimpanzees high on the trees, employ smaller stones to hammer nuts than they do on the ground floor. In Lope', where the chimpanzees don't crack nuts, wooden and stone hammers, and root and stone anvils are available as well as nuts themselves. (Nota bene: bonobos, the 'intelligent' apes which are 'able' to learn language, have not yet been observed to use tools). It is clear that stone tools were produced long before the species Homo reached the stage of full intelligence. One and a half megayears ago, at the level of so-called 'superdiscent' of Homo habilis and later at the level of 'proto-intelligence' of Homo erectus artifacts steadily developed and achieved more and more an 'artificial' character. At the level of 'sub-intelligence', represented according to this book by Homo sapiens neanderthalensis (Table 3-4), artifacts have primarily the nature of tools. It must be mentioned that at the lower levels of intelligence artifacts were not limited to those artifacts to be directly used, but also to tools required for preparing other tools—'tools for tool making'. Also, the use of 'artificially preserved' fire is a highly developed product of deliberate action and, in particular, tools and materials for 'making fire' must be thought of as a high-level artifact (Table 7-2).

Only with the emergence of Homo sapiens sapiens and the establishment of intelligence did the character of artifacts change radically. The emergence of the first ornamentation of tools and, later on, drawings and paintings on cave walls can be thought of as the beginning of the new era in artifact production; that of 'information artifacts'. A unique role in the development of artifacts can be attributed to the development of the means of processing (mostly storage) of human speech: the tools and materials for storage of writing and painting. Probably the oldest artifacts in this area are at least 6 thousand years old, with development passing through the construction of the printing machine 500 years ago, and the typing machine 100 years ago. A really revolutionary jump comes with the construction of the calculating machine—the computer.

A computer is a machine, produced at a large industrial scale, used for information processing. But what is information? Remember the anecdote about the ironsmith in the Iron Age who was asked 'what is iron?' (Fig. 1.1).

There is not an easy task to define the computer, as it is to see from following citation:

P.S. Churchland, T.J. Sejnowski (1992): "The definition of computation is no more given to us than were the definition of light, temperature, or force field. .. There is a lot we do not yet know about computation. The .. development .. of computer and software was accompanied by theoretical inquiry into what a sort of business computation is. We can consider a physical system as a computational system when its physical states can be seen as representing states of some other system, where transitions between its states can be explained as operations on the representations."

In our definition intelligence is the information processing about information processing (biotic and technical).

What is a computer? Rather, what is a computerised information processing system, which at present contains the following components:

–Human factors: computer user, program author and interpreter of results, computer's designer and builder,

–Software: programs for computers (that is, human-made instructions for guiding computer activities internally), and data-bases (input data prepared for information processing),

–Hardware: man-made electronic and mechanical construction, with the appropriate input and output devices for communicating with the human user and programmer?

A following caution is of significance:

C. Shirky (1995): "Often computers merely magnify existing inefficiencies."

9.1.2 *The brain and the computer: similarity?*

The two information processing systems—the biotic human brain and the technical, artificial computer—are often compared, but not always correctly. Let us begin a comparison with a review of the structural materials employed, even if this is not the best method. The structural materials of a brain have been discussed in Chapters 4 and 5. The computer is at present constructed mostly of electronic components, the most significant of which are called 'chips'. Only a small part of a computer (mainly the input and output units) is of a mechanical nature. However, we must remember that a machine able to compute could, in principle, be built of other components, such as hydraulic or pneumatic tubes, valves, and pumps. These 'computers' would be slower and much larger, but they could do basically the same job as electronic computers (Table 9-1).

G.E. Hinton (1992): "The brain is a remarkable computer. It interprets imprecise information from the senses at an incredibly rapid rate... Most impressive of all, the brain learns—without any explicit instructions."

A.R.Damasio (1994): "(O)ur minds would not be the way they are if it were not for the interplay of body and brain during evolution, during individual development, and at the current moment. The mind had to be first about the body, or it could not have been."

Of course, with time it seems that it will be possible, or even desirable, to construct at least some types of computer containing bio-chemical devices, with specially synthesized 'organic' sub-components, including memory, 'wiring', etc. It seems that having different structural materials is not the crucial reason for differences between the brain and the computer, even if some authors have written that a 'Human is jelly' (H.Moravec, 1988).

By contrast, the most efficient silicon technology currently requires about 0.1 microjoule per operation (multiply, add, etc.), that is 10 to 100 million times more than a neuron. There are some ideas about a quantum computer, which could use very small amounts of energy.

Table 9-1 Brain and body relation

Stage of information processing	Biotic reflex	Biotic instinct	Biotic discent	Biotic intelligence	Artificial intelligence
Animals	Sponges	Insects	Dogs	Humans	'Artificial thinker'
Brain	No	Small brain	Brain	Very large	Very large
Body	Large	Large	Large	Large	No
Principle	Body without brain	Brain serves the body	Brain serves the body	Body serves the brain	Brain without body

A direct consequence of the high energy efficiency of neurons is that brain can perform many more operations per second than even the newest supercomputers. The fastest digital computers in 2004 will be capable of around hundred-teraflops (teraflops = 10^{12} floating point operations per second). A group of computer designers begin laying the groundwork for a petaflops supercomputer that could be built around 2010.

Considering that the unit of information processing are not the neurons and synapses but the microtubules the calculation must be improved. There are 10^{11} neurons, each with 10^4 microtubules per neuron and the switching rate is of the order of one nanosecond (that is 10^9 per second). The rate of brain's information processing achieves a value of 10^{24} 'operations' per second. In the opinion of some scientists (R.Penrose) the source of emergence of consciousness is connected with this phenomenon.

The present state computers are mostly built of solid state components. D. Bradley (1993) put the question: will future computers be all wet? Could be—if visionaries in the growing field of supramolecular chemistry continue to have their way with molecules that self-assemble in solution. Chemists are beginning to exploit the properties of some supermolecules to make tiny devices: molecular switches, diodes, and transistors, as well as molecular wires to join them together. All these diminutive devices could be packed into 'wetware', it might one day be possible to construct an entire computer based on supermolecules interacting on thin film supports. J.M. Lehn, Nobel laureate 1987 in chemistry, calls supramolecular chemistry the designed chemistry of intermolecular bonds. Another scientist claims that if we use an external input, such as a light source or an applied voltage, then some supramolecules could be used to store information. Of course, this new branch of chemistry is not immediately intended to produce commercially viable techniques. Instead, it is working out general principles.

J.M. Lehn (1992): "The next, and very important, difference between the brain and the computer is in the relationship of each with its own system of inputs (sensors) and outputs (motors). The brain is inherently united with such a system—its own body. On the other hand the computer has a very rudimentary set of inputs and outputs, which are directed by a human user. It is not allowed to 'think'. Such an improvement of input and output in a present-day computer would give it the properties of a real brain."

Table 9-2 Human brain and computer/robot

Components	Computer	Robot	Human brain
Structural material	Silicon (germanium) chips, magnetic disks, metallic, plastic, glass		Neuron cells, glial cells, extracellular fluid, blood
Hardware	Chips (mostly silicon) and other components		Human brain from hemispheres up to synapses and neurotransmitters
Firmware BIOS	Hard-wired processor. When 'power on'—search for operating system on the hard disk		'Wired' connections and 'plasticity'
Operating system	On the hard disk (magnetic, optic)		Genetic + epigenetic
Memory	Hard disk, diskette (magnetic) Discrete erosion is possible		Dissipated in different parts.
Software	Programs which could be added or withdrawn on the diskette		Program = Instruction could be 'added' but not withdrawn
Inputs(sensors) Output(actuators)	Keyboard, sensors. Screen	Sensors, manipulators, actuators	Multiple differentiated sensors. Multiple differentiated muscles
Kind of connections	Serial/parallel	Digital/analogue	'Connectionial' connections
Energy source	Stationary and mobile electricity, hundreds watts	Some kilowatt	Autonomic metabolic energy for brain: about 20 watt (metabolic)
Environment	Human being able to use the PC	'Jigsaw Puzzle World' (simple, well defined environment)	Real world including other human beings

Last but not least the difference between brain and computer lies in the ability to discriminate the quality of information received; that is, the influence of the bonus-malus system. We should not forget that the bonus-malus system is the result of a very long biotic evolution and of the possession of an own body and cannot be simply and cheaply simulated by a computer (Table 9-2).

All these arguments allow us to ascertain that the modern computer and the human brain are very different systems. The computer, at the present time, is not an 'artificial brain'. But what happens in the future, or the very distant future?

9.1.3 Stages of computer evolution

In an arbitrary way, we will try to describe the development of the computer up to the present day and into the near future. The simplest way is to use our previously developed classification of biotic information processing in order to find out the present stage of evolution of the modern so-called supercomputer. Here it is appropriate to repeat the question of one scientist, who has asked what our descendants will do. Because we have applied the word 'super' to such a large number of things that they will be unable to apply the word correctly in the future.

We talk about 'supernova', 'supermarket', 'superunified force', 'superconducting supercollider', 'superconductivity', and, last but not least, 'supercomputer'. It is obviously an apt observation, because the 'supercomputer' of the present day is surely not the highest possible level of development. What name will we, or our descendants, give to the next higher stage of computer? Super-duper-computer? (Section 11.6.1). Information processing machines—computers—do those things well which humans find hard, such as rapid calculations, while doing poorly some things which are relatively easy for us, such as the problems of so-called 'common sense', or 'every-day problems'.

A typical computer at the end of the twentieth century is a sequential (serial) computer. It carries out step-by-step processing (that is one sequence after the other) and the processed information is also formulated in the form of sequences of 'bits'. Now researchers hope to build computers whose performance is measured in 'tera-ops'—that is, tera-operations per second, or 1000 times more powerful than today's computers, which have calculation rates measured in 'giga-ops' (giga = billion = 1000,000,000, 'ops' = operation per second). Early in the next decade (2000-2010) the semiconductor industry plans to sell gigabit chips now in the laboratory, with a transistor electrical channel, or gate, that measures between 0.1 to 0.2 micrometre in length. It is about the width of a DNA coil, or a thousandth the width of human hair which is in the order of magnitude of the size of a neuron.

This new effort aims to take a further step towards 'neuronal computers' by building them with the same architecture as neurons have in the human brain. These neuronal computers will be able to handle imprecise information, such as human speech, and should not need programming; they will be able to learn new tasks by trial and error, as humans do. But building such machines will need enormous technical breakthroughs in parallel processing. Recently parallel processors have been constructed which promise very high 'efficiency' for some special problems, such as language processing or picture recognition.

The increasing complexity of the circuits now etched on silicon will increase the difficulty of detecting errors in design and manufacture. In a closely related field, it is well-known that complex software programs are unlikely ever to be tested in all the circumstances in which they are meant to function.

Devices fabricated from biological molecules promise compact sizes and faster data storage. They lend themselves to be used in parallel-processing computers, three-dimensional memories and neuronal networks. Some of the candidates for this purpose is the molecule of bacteriorhodopsin. Rhodopsin in the human eye, when it absorbs a photon, changes its structure and releases energy that serves as an electrical signal able to convey visual information to the brain (R.R. Birge, 1995).

9.2 COMPUTERS AND INTELLIGENCE; FAULTLESSNESS AND COMPLETENESS

9.2.1 *Computers faultlessness?*

Information processing theory at present is concerned with information quantity (measured by binary units: bits) only, and is still incapable to deal with information quality, measured by means of bonus-malus systems (Section 1.2.5). Only a computer with a developed bonus-malus system coupled with semantics (appropriate representation of the real world) would have the ability to learn at a higher level, and therefore make self-correction of its own false programming.

It is trivial to say that information processing can produce not only irrelevant but also false results. The reason is, in most cases, because the program (the software) is false (containing errors). In rare cases, the erroneous output results in failure of the hardware, which is simpler to diagnose.

J.Horgan (1996): "N. Oreskes et al (1994) observed that numerical models were becoming increasingly influential in debates over global warming, the depletion of oil reserves, the suitability of nuclear-waste sites, and other issues. Their paper was meant to serve as a warning that verification and validation of numerical models of the natural system is impossible. The only propositions that can be verified—that is, proved true—are those dealing in pure logic or mathematics. Such systems are closed, in that all their components are.. true by definition. ...Natural systems are always open,.. our knowledge of them is always incomplete, approximate, at best, and we can never be sure we are not overlooking some relevant factors. In other words, our models are always idealizations, approximations, guesses. Moreover, it is always possible that other models, based on different assumptions, could yield the same results."

9.2.2 *Fuzzy, connectial, and other computers*

In an early phase of development today is the 'connectial' computer, which is based on the parallel processing of information and in which information is not organized exactly in a sequential manner. 'Highly parallel distributed processing' (also called a 'connectionist model' or 'interconnection network') is based on the idea of a more dimensional hypercube, which directly connects each processor to all others, but is irrelevant to the purely computational aspects of the process. Any function which can be computed on a parallel machine can also be computed on a 'classical' serial machine. Indeed, because parallel computers are still relatively rare, connectionist programs are usually run on 'classical' serial hardware.

On the other hand, parallel computers (called also 'artificial neural network') have some properties which drastically differ from those of a classical serial (sequential) machine. Massive parallelism means that the system is fault-tolerant and functionally persistent; the loss of a few connections, even quite a few, has a negligible effect on the character of the overall transformation performed by the surviving network. A parallel system stores large amounts of information in a distributed fashion, any part of which can be accessed in milliseconds. Information is stored, not in special memory devices, but in the specific configurations of synaptic-like connections with differentiated strengths, as shaped by past learning.

The most common design of such 'neural networks' has three layers, with each 'neuron' linked to those in adjacent layers. Those in the outer layers are also linked to appropriate input and output devices. The internal, 'hidden' layer is connected to the layer of output units (G.E. Hinton, 1992). The activity of the input units represents the raw information that is fed into the network. The activity and the behaviour of the output units, depends on the activity of the hidden units and the weights between the hidden and output units. We have now experience with hundreds of massively parallel computers in thousands of scientific applications. These computers require sometimes special programming languages.

The splendid image of computers has been somewhat tarnished by the existence of 'computer viruses'. This plague of the late 20th-century is well known. Can there ever be an all-purpose vaccine against this plague? The answer seems to be no. According to well-informed opinion, the existence of computer viruses is "an inevitable consequence of fundamental properties of any computing domain". There are two types of computer virus. The first simply reproduces; it is a program whose output is always a copy of itself. The second is a program that infects and alters the operating system of a computer itself. In practical terms this means that new computer viruses will continue to appear and new antiviral vaccines will be needed. New viruses are of course written by humans, and all that must be done is to find a clever or more patient human to defeat the virus. The analogy to biotic viruses appearing in the present world is striking.

9.2.3 The real world and jigsaw puzzles

Among numerous ambiguities in the development of the computer based 'Artificial Intelligence' is the neglecting of the principal difference between the real world and the so-called 'realm of jigsaw puzzles'.

Under the notion 'puzzle' (according to Webster's dictionary: a question, problem, toy, or contrivance designed for testing ingenuity) we understand a system of objects and their direct environment which is clearly defined, mostly artificially designed, which does not include any uncertainty, and in which all actions are well defined and can be expressed in the form of algorithms. Puzzles (jigsaw puzzles) have assured solutions. The number of possible courses of action is limited, even if the number of combinations is practically unlimited. The only source of uncertainty is a human opponent, if the puzzle is a game, such as chess (Table 9-3).

In the real world, all these boundary conditions are often not defined, unclear. The system, the objects, and the environment cannot be exactly defined and limited, and the representation or actions cannot be transformed simply and easily into algorithms. Real world problems have no sure solutions. And the solutions must be proved to be false or true or remain undecided.

Uncertainty of the future state results from both lack of knowledge of the nature of the phenomena and the so-called 'free will' of some actors.

Table 9-3 Realm of puzzles and real world

	Realm of puzzles	Real world	
Stages of artificial information processing	'Jigsaw puzzles', games, technology	Formal knowledge only	Full knowledge, semantics, ethics
Artificial reflex	Simple robots		
Artificial instinct	Chess player		
Artificial discent		Expert	
Artificial intelligence			Thinker

9.2.4   *Virtual reality*

At the end of 20th century the phenomenon of 'virtual reality' achieved an unexpected success, among others because of its unclear and fuzzy definition. Virtual reality is a series of pictures which mimic our own physical environment, and allows us to feel as being and acting inside this environment. But at present we are rather far from this stage.

Experts estimate that for each frame of a 'virtual reality' to appear photorealistic, it must contain 80 million 'polygons' (polygons are called three-dimensional objects which are composed of many two-dimensional figures bounded by straight lines). At least 10 frames per second are needed to sustain the illusion of continuous motion. So any 'virtual reality' system that aspires to visual realism must be able to compute and draw at least 800 million polygons per second. The best modern graphics supercomputers can render 2 million polygon per second. The next generation will achieve the rate of 30 million polygons per second. Useful virtual reality applications need more than just pretty moving pictures. Virtual objects must also mimic the behaviour of their real counterpartners, which means making millions of additional calculations each second to ensure that they act like massive solids rather than massless surfaces. Add a sense of touch, as many programs strive to, and the workload again increases dramatically, since textures must be updated hundreds of times a second to feel lifelike. Cognitive studies have shown that separating the sight, sound and touch (and what about smell?) of an event by a few tens of milliseconds can cause confusion.

9.2.5   *Semantics and information processing*

Computer programs are formal—that is, they are purely syntactic. There exists no correspondence between the symbols used in a program and the real world. The correspondence is given due to the mediation of a human being as the result of his or her ability to transform formal results from the computer into a meaningful representation of reality. This correspondence we will call 'semantics', even if this concept is usually used in the much narrower sense of 'the study of meanings', and not the 'correspondence between symbols and their meaning'.

Of course, human language (the natural and especially the scientific, of which mathematics is the extreme case) is also syntactic. Semantics is a product of the mutual influence of formal information processing with the bonus-malus system of pleasure and pain of the animals, that is with the real world.(Section 1.3.2).

Experts in the field of 'artificial intelligence' are conscious of the necessity of relating formal programs with the emotionally coloured bonus-malus system.

H.Moravec (1988): "The software (for robots) would receive two kinds of messages from anywhere within the robot, one telling of success, the other of trouble. I'm going to call the success messages 'pleasure' and the danger messages 'pain'. Pain would tend to interrupt the activity in progress, while pleasure would increase its probability of continuing. Pain could grow into an incapacitating phobia and pleasure into an equally incapacitating addiction."

Knowing this, the future development of artificial information processing systems, such as more highly developed robots, depends on discovering how to build semantics into these robots.

Is it possible to create a fast-learning process for implementing semantics, instead of passing through the long-lasting 'normal learning' processes? The answer may be in the affirmative at the stage of reflex and instinct, but such a process will be very difficult to achieve, if at all possible, at the stage of discent, not to mention that of intelligence.

9.3 ARTIFICIAL REFLEX, INSTINCT, AND DISCENT

9.3.1 *Artificial reflex; robots*

Using once more our classification of biotic information processing, we will try to order the development of artificial information processing into four stages: reflex, instinct, discent, intelligence. Arbitrarily we have chosen four representatives of artificial information processing for the sake of illustrating this development.

The lowest stage of biotic information processing is that of reflex (Section 3.4.1). A modern pocket electronic calculator corresponds to the stage of reflex, or more precisely to the nerve strings of reflex. However, a robot possesses more or less sophisticated sensors, such as artificial eyes, and artificial executive organs, such as artificial hands or legs.

A robot is more or less specialized for some functions which are continuously repeated in the same way—such as throughout the entire production process in an industrial environment. Yet it must be clear that, under these conditions, the objects and the environment of the robot are well defined and can be described by simple algorithms. This is a typical case belonging to the realm of puzzles. The level of information processing corresponds exactly to the stage of reflex. In this situation, the robot can be as effective as a human being, or even more so.

H.A. Simon (1983): "What distinguishes robotics from other areas of AI (artificial intelligence) is that the robot is embedded in an external environment that it can sense and act upon. This has two consequences. First, it imposes a severe performance test: the robot's performance is evaluated by its behaviour in its external environment, not by its self-imagined behaviour in a toy problem space. Second, the robot can adapt the complexity and other properties of its internal problem space to its own computational capabilities, depending upon feedback to eliminate the discrepancies between expectation and reality."

At the beginning of the nineties, there were in existence around 100 thousand robots in the whole world, for less than 20 years and used only in industry. This compares with 100 million of computers (existing only for the past 20 years) almost 1000 million privately owned cars (in existence for around 100 years), and 500 million television sets (existing for 50 years), used in private households.

At the beginning of the 21st century, a new generation of robots will probably appear on the scene. Some of these will be in the form of four-legged, two-handed 'smart' machines to work in hazardous occupations, but it is obvious that making a robot for use in hospitals or homes with direct contact with people is a much more difficult task. There are many mechanical and electrical components which influence the further development of robots: automotive energy sources (e.g. electric batteries), very fast 'portable' computers, better electrical motors, and improved sensors.

Among other factors, the size of such a robot must be determined. Should it be larger than, equal in size to, or smaller than, a person? At the beginning, the size of robots was made the same as that of man, because a robot should be a direct substitute for a human being. However, in the future it seems to be more reasonable for robots to be much smaller, or much larger, than man. Then they would be able to do their specific tasks more efficiently.

A.Asimov (1983): "First, a robot need not be as complex and as intelligent as a human being. Second, even if robots are very stupid at first, they may well become rapidly more intelligent as newer models are developed in rapid succession. Third, it is not absolutely essential that a robot carries its own brain, any more than it is necessary for each television to incorporate its own broadcasting station."

9.3.2 *Artificial instinct; robots.*

The next higher stage of biotic information processing is instinct (Section 3.4.2).

H.Moravec (1988): "I think the spider's nervous system is an excellent match for robot programs possible today.(We passed the bacterial stage in the 1950s with light-seeking electronic turtles). By comparison, the best of today's machines (robots) have minds more like those of insects than humans. .. Robotic equivalents of nervous systems exist, but they are comparable in complexity to the nervous systems of worms." (Note in this book: Worms correspond to the stage of 'reflex' and insects to the stage of 'instinct'.).

In 1990, a small thirty-centimetre-long six-legged robot was constructed which actually imitated a cockroach fairly well. It skittered across the floor and gamely climbed over books that were stacked in its path. The constructor of this 'artificial insect' decided that its 'brain' should be based entirely on simple reflexes, without making internal models of the world. Its 'brain' was organized around coherent behaviour patterns such as avoiding obstacles, wandering around, or exploring. Even this amount of information processing required a computer weighing some kilograms, obviously having to be located outside the 'artificial insect' and connected with it by means of wiring.

H.Moravec (1988) assured that by the middle of the next century, robots will be as intelligent as humans and will essentially take over the economy and will have all the important jobs. (Nota bene: H.Moravec omits the question if robots will be also professors for robotics engineering at Carnegie Mellon University, at the same institute which at present is Moravec by himself). Moravec claims that the machines will expand into outer space and will fan out through the Universe, converting raw matter into information-processing devices.

9.3.3 *Artificial instinct; chess player*

Computer's instinct is here illustrated by a computer chess player. That this case corresponds strongly to the realm of puzzles is obvious. A chess player can have input and outputs limited to operations only on the basis of simple abstract information (current position of the chess pieces). However, the computer must be relatively highly developed, very fast and with large enough memory.

It must be repeated that the most important difference between the biotic stage of instinct and artificial instinct discussed here is, on one hand the absence of the lower stage of reflex and, on the other hand the absence of the influence of the bonus-malus system, linked together through semantics.

Chess-playing computers were based, until the nineties, on the sequential processing of information concerning the current position of chess pieces and the enormous number of 'rules of thumbs' prepared and programmed by the best human chess players. During the game each next step is chosen from thousands of different, permitted steps, after calculation of the highest value of merit.

The constructors of the chess-playing computer (F.H. Hsu at al, 1990) reasoned that their machine could incorporate relatively little knowledge of chess and yet outplay excellent human players. However, one must remember that the computer does not mimic human thought—it reaches the same ends by different means. Computer sees far but notices little, remembers everything but learns nothing, rising above its normal strength.

May 1997, the day in which the grandmaster G. Kasparov lost a chess play with 'Deep Blue' (a computer chess program developed at IBM, based on a computer system consisting of several hundred special purpose chess chips tuning in tandem, enabling to search an average of 200 million positions per second) is the beginning of a new model of computer simulation of the human art of mental games. It must not be forgotten that in the end this is a play between a human player (Kasparov) and an other human program creator (Hsu et al, 1990). Deep Blue had two key advantages. First, the grandmaster could not study its 'track record': although he played and beat the computer 15 months ago, it has changed substantially since then. Secondly, he was facing a psychological obstacle. After nearly half of Kasparov's moves, the computer was able to respond immediately, having worked through 200 million possible moves every second while its opponent was still thinking. This is 20 times faster than Kasparov last faced Deep Blue. "It is miraculous he kept his sanity at all" says one specialist.

Even so, it sometimes produces insights that are overlooked by even top grandmasters. These developers are planning the construction of a chess-playing program which will examine more than a giga (billion) positions per second; enough to plan more than a dozen movements, in most cases. It seems that in the future a real chance to achieve the world-master grade is to build a chess-computer with a minimum of 'bonus-malus' for each situation, and with one simple 'will' for gain. May be the chess-bonus-malus must be connected with some 'aesthetic feeling'.

9.3.4 Is communication with animals possible?

At the end of the eighties a simple 'robot' was constructed which imitated an individual bee (the biotic level of instinct). This remotely controlled robot, of the size and appearance of a bee, was able to perform a 'bee-dance'—which generates information for other bees about the direction and distance to a rich nutrient source. The robot excreted a suitable quantity of sweet smelling and tasting drops, containing the appropriate pheromones, which were then tasted by its 'sister-bees'. It even emitted the appropriate acoustic signal of around 260 hertz, which corresponds to bee's wing movements. The communication of the 'robot-bee' with its sisters was relatively successful, and some acoustic replies from the sister-bees were detected and analysed by the robot's operator.

In the future, it may be possible for a human being, to communicate with some higher animals—for example dogs (biotic information stage: discent). As has been mentioned in Section 6.3.2, direct 'person-to-person' human-animal communication with chimpanzees has failed. But another way can be chosen for communication with a dog. With enough knowledge about the behaviour of dogs, it must be possible to construct an artificial dog which can move its body, legs, and tail. It must be able to be active acoustically and even emit some specific smells from the appropriate parts of its body. This artificial dog will be connected to a suitably programmed computer system, representing the behaviour of dogs in given situations and armed with numerous sensors—visual, olfactory, and even tactile.

Communication can be achieved by computer analysis of the signals received from a living dog and sending the appropriate signals to the artificial dog. A human being can send the desired commands through the computer via the artificial dog, and receive the answers given by the living dog.

Such a rather expensive way of communicating with one's own dog seems to be possible, but is very speculative and not very useful. It can only be a semi-scientific experiment. It must be clear that the hypothetical experiment described is a kind of translation of human speech, or human thought, from the stage of intelligence into communication at the stage of descent, and in the reverse direction. In the paper 'What is it like to be a bat' Th. Nigel (1973) claims that no matter how much one might know about the neuroanatomy of a bat's brain and the neurophysiology of its sensory activity, one would still not know 'what it is like' to have the bat's sensory experiences.

9.3.5 Artificial discent; experts

The third highest stage of biotic information processing is discent (Section 3.4.3). At a corresponding level of artificial information processing we will propose the 'expert'. To characterise the peculiarities of an expert (human and artificial), let us look at what some authors have written:

F.R.Wright (1956): "An expert is one who does not have to think. He knows."

M.Minsky (1986): "To be considered an 'expert', one needs a large amount of knowledge of only a relatively few varieties. In contrast, an ordinary person's 'common sense' involves a much larger variety of different types of knowledge—and this requires more complicated management systems."

L.Hunter (1982): "Most expert systems are based on a set of rules: If this, then that. But expertise does not come from following the rules. If you think about it, real experts learn what they know from experience and are able to recognize when a case is an exception to a rule. An expert has to learn to deal with the unexpected."

The present state of the practical use of computerized 'experts' is both encouraging and discouraging. Recently (1995) a 'parallel computer' achieved a rather good result in classifying the types of galaxies (e.g. spiral, elliptical) from photographs made by Hubble's telescope.

Discent is biotic information processing based on a large learning ability. How can machines learn? An anecdote from an expert in learning machines is apposite. When D.Lenat developed 'Eurisko', a learning system, he thought that Eurisko could generate its own rules for solving problems, and in doing this it surpassed itself. Eurisko was so clever that the first golden rule it generated was that all machine-generated instructions were rubbish! (according to E.Galea, 1990).

Linking 'artificial intelligence' and robotics, researchers are constructing an infinite automaton to learn how the body influences cognition (J.Travis, 1994).

This robot will have two hands with three fingers and will run on wheels. Its visual system will see black and white only. Its huge brain is placed in an other room, and includes a network of 64 sophisticated microprocessors. Hopefully it will be able to mimic the mental capacity of an infant in the period before language could be learned. Probably the robot will have the ability to establish 'contact' with a nearby person (recognizable 'mother') but at the same time pay attention to a distant 'conversation'. Cross-cultural cues, such as smiles, frowns, soft encouraging tones, and nodding heads, will also be hardwired. The critics of this project claim, that "there is so little known about the early stages of cognition (of babies) that it is kind of silly to spend hundreds of thousands a year (of scientists) to simulate what we don't know. It's a waste of time."

9.3.6 Artificial discent; learning robots

There is no easy way to teach a computer, that is a robot, all these things which humans take for granted, like the fact that one person cannot be in two places at once. This and a lot of other analogue problem are included in the phenomenon called 'common sense' (Section 7.1). There are some efforts to construct such a computer. The computer called 'Cyc' has been build on the basis of a project devoted to 'Large common sense knowledge basis' (Section 6.6.4). Cyc is organized a little like a library where knowledge exists in books and databases. The heuristics is here the deciding factor. Cyc must include around two million rules, appropriate handling time and space, causality, beliefs, emotions etc. At the end of the project Cyc will begin to read and learn itself. (C.Davidson, 1994) Some critics are of the opinion that the project Cyc is the last defence of the Artificial Intelligence dream to construct 'machine intelligence'. The activity is going further but nobody of the Cyc team is speaking about 'machine intelligence'.

9.3.7 Artificial language translators

There is a well-known anecdote about the artificial translator which translated the English proverb, 'The spirit is willing, but the flesh is weak', into, 'The vodka is good, but the meat is rotten', in Russian.

In spite of the fact that speech, in verbal or written form, is the expression of the highest stage of biotic information processing—intelligence—the kind of computerized translation which is done at present, corresponds in the best case to the stage of discent.

A computerized translation program is able to perform the simplest translation from one language to another, but cannot do it at the desired level, because the processing of language must be accompanied by such activities as 'understanding', meaning and context interpretation, semantics, etc. All these qualities could be managed by information processing at the highest stage—intelligence.

Table 9-4 Artificial information processing

Stage of information processing	Detectors (input) Effectors (output)	Environment	Artificial reflex	Artificial instinct	Artificial discent	Artificial intelligence, AI
Robots	camera manipulator	strong defined simple environment	simple algorithms			
Computer chess player	keyboard + monitor	strong defined environment (puzzle)		simple algorithms large number of combinations, large memory		
Translator	keyboard + monitor			complex algorithm (syntax) large memory		
Expert on Internet	keyboard + monitor				complex algorithm large memory, network	
Thinker	keyboard + monitor					extreme complex program extreme large memory
Artificial intelligence	artificial eye, artificial ear, sophisticated action					extreme complex program; extreme large memory, emotions awareness?

Modern computers, even so-called 'supercomputers', are a long way from reaching this stage. As we have seen, both partners—the speaker and the listener—must have the same, or at least very similar, world knowledge, similar life experience, and similar common sense, including the knowledge of mutual deceptions and lies (Section 7.4.4). The hopes of building a really effective computerized language translator must be postponed and new, more sophisticated knowledge of human language must be developed in the meantime.

J.J. Hopfield (1994): "Neurons are simple computers... The brain may be regarded as a composite computer made up of a network of neurons. Network computation with high connectivity between analogue elements is the means by which large brains gain an intelligence lacking in small nervous systems."

May be in a not too distant future we will use a small (and low-price) 'network (Internet) computer' 'NC' connected with a large central computer and appropriate program, which will allow the 'immediate' transformation of speech to a written and printed form, including translation into foreign languages, at least the most used.

9.4 ARTIFICIAL INTELLIGENCE, COMPETITOR OR SLAVE

9.4.1 *Computational brain*

Computer models are helpful in 'understanding' the principles whereby the human brain performs its manifold tasks. These models are, however, only a first step. They have not addressed a range of critically important factors such as attention and arousal, the role of emotions such as fear, hope, and joy, the role of moods, desires, expectations, long- and short-range plans (generally the human bonus-malus system). The constructive problems in the 'computational brain' concern how to go from simulating the neuronal processes in a computer to making synthetic retinas, spinal cords, parts of cerebral cortex—in short, how to make synthetic brains. The 'computational brain' science (neuronal information processing) is based on following principles:

–Testing the strengths and weaknesses of a model, seeing how it interacts with the actual world. Real time, for example, is an unforgiving constraint.

–With regard to input and output, a modeler can either simulate the world, or use the world (or only a puzzle-world). Simulating the world may turn out to be at least as difficult as simulating the brain.

–Often we do not know precisely what parameters in the world are relevant. A simulation of the world embodies assumptions about what in the world is likely to be relevant to an organism.

–Models eventually need to encompass not merely a single level of organization, but several levels. Consider again the many levels of brain organization, ranging from the whole brain, down through circuits, to neurons, to molecules, and from intelligence, through discent, instinct to reflex (P.S.Churchland, T.J.Sejnowski, 1992).

A.Prince and P.Smolensky (1997): "Can concepts from the theory of neuronal computation contribute to formal theories of the mind? Recent research has explored the implications of one principle of neural computation, optimization, for the theory of grammar. Optimization over symbolic linguistic structures provides the core of a new grammatical architecture....What is it that the grammars of all languages share, and how may they differ?"

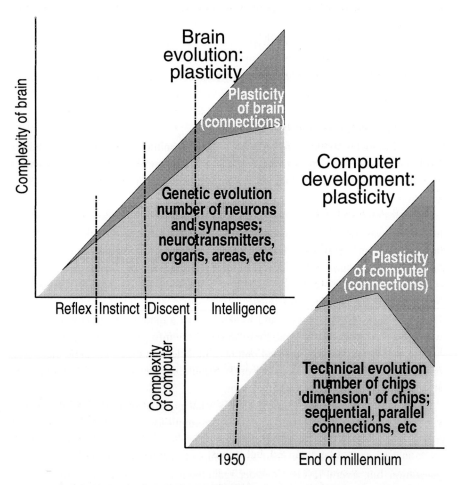

Figure 9-1 Plasticity of brain and computer

9.4.2 *Artificial intelligence; chance and reality*

The greatest challenge to human intelligence in the realm of artifacts is doubtless the realization of 'artificial intelligence'. In the past two decades, large efforts in this field probably have brought more disillusionment than success. There are a number of reasons for this. One of them is the underestimation of the crucial significance of the formulation of the notion of intelligence—human intelligence. Another reason is the fuzzy definition of the targets of the whole new branch of investigation which we call Artificial Intelligence (AI). Also, there has been pressure to utilize the ideas of artificial intelligence in industrial and, not least, military environments.

T M.Minsky defines artificial intelligence as the task of making machines perform tasks which, if done by humans, would require intelligence. In the early 1980s, hopes for the emerging ultra-sophisticated, artificially intelligent computers probably reached their highest level. A lot of theoretical study (rather unrigorous) and practical effort, such as the writing of suitable programs (rather inefficient) and even the construction of a special type of hardware (prepared rather for public relation effects), was carried out. But the breakthrough has not happened. Some people involved in AI said, at the beginning of 90s, that 'about half of the money spent on AI so far has produced nothing but very expensive conference papers (at a very shallow level, in our opinion) (Fig. 9-1).

R.A.Brooks (1991) argued that in order to really test ideas of artificial intelligence it is important to build complete agents which operate in dynamic environments (real world?) using real sensors. Internal world models that are complete representations of the external environment, are not at all necessary for agents to act in a competent manner. Many of the actions of an agent are quite separable—coherent intelligence can emerge from independent subcomponents interacting in the world.

The beginning of a real Artificial Intelligence will come only then when AI will show following properties: faulty reasoning, forgetfulness and repetition, the ability to detect deception, some level of emotion and some feeling of beauty, the real human characteristics.

The most sophisticated AI programs promise to help in situations where people do not know how to solve a problem, but in the opinion of prominent experts in this field these programs are bound to fail. The possibility of building functional AI cannot be excluded, but it must be postponed, to await the future development of new concepts in information science.

Scientific American (1990): "Attempts to produce thinking machines have met during the past 35 years with a curious mix of progress and failure. Computers have mastered intellectual tasks such as chess and integral calculus, but they have yet to attain the skills of a lobster in dealing with the real world. Some outside the AI field have argued that the quest is bound to fail: computers by their nature are incapable of true cognition."

D.B.Lenat (1995): "The goal of a general artificial intelligence is in sight, and the 21st century world will be radically changed as a result."

The development of Artificial Intelligence, as a branch of science, must evoke a strong reaction from philosophers:

J.R.Searle (1987): "This is the very common equation in the literature: mind/brain = program/hardware. ...According to `strong AI` the appropriately programmed computer with the right inputs and outputs literally has a mind in exactly the same sense that you and I do. It has the consequence that there is nothing essentially biological about the human mind. It so happens that the programs which are constitutive of minds are run in the wetware that we have in our biological machine, our biological computer in the head. That is, if you have the right program with the right inputs and the right outputs, then any system running the program, regardless of its chemical structure (whether it is made out of old beer cans or silicon chips or any other substance) must have thoughts and feelings in exactly the same way you and I do. In order that there be total clarity, I will state a set of `axioms` and derive the relevant conclusions: Axiom I. Brains cause minds; Axiom II. Syntax is not sufficient for semantics; Axiom III. Minds have contents; specifically, they have intentional or semantic contents. Conclusion: Instantiating a program by itself is never sufficient for having a mind."

Some researchers in AI believe that, by designing the right programs with the right inputs and outputs, they are literally creating minds. They believe, furthermore, that they have a scientific test for determining success or failure—the 'Turing test', devised by Alan M.Turing, one of the founding fathers of artificial intelligence (A.K.Dewdney, 1992).

P.M.Churchland (1995): "Criticisms of the Turing Test are legion. The background worry is that convincing verbal behaviour over a teletype link is something that might be produced from variety of different causes, none of which has anything essential to do with real conscious intelligence. .. Turing test is too exclusive, since intelligent creatures without linguistic competence are doomed to fail this linguafocal test. This includes prelinguistic children, conscious human adults with localised aphasia, most of higher animals on the planet (according to definition in this book of intelligence the apes are not on the stage of intelligence), and all intelligent aliens who don't communicate by human language. The Turing Test is a test precisely for people who have no adequate theory of what intelligence is, or no theory beyond the humble framework of our prescientific folk concepts.. The thing to do, plainly, is to develop a theory of cognitive activity and conscious intelligence that is genuinely adequate to the phenomena before us."

Nota bene: Turing test is a kind of sophisticated deception based on the conversional ability of this program. The proband types his problem on a keyboard and sees on the monitor a typed question concerning the mentioned problem. The proband types the next answer and sees the next question. The proband is not informed whether in the other room is a computer or a living person. The computer program has finished the test positively if the proband thinks he has to do with a real person. A 'perfect' Turing program must be sufficiently 'clever' to lie, to cheat, and dissemble. Some students think that even Turing did not regard the imitation game as a definition of machine intelligence. The pioneer of Artificial Intelligence J.McCarthy writes: "Anyway, I never regard the (Turing's) imitation game as a useful goal."

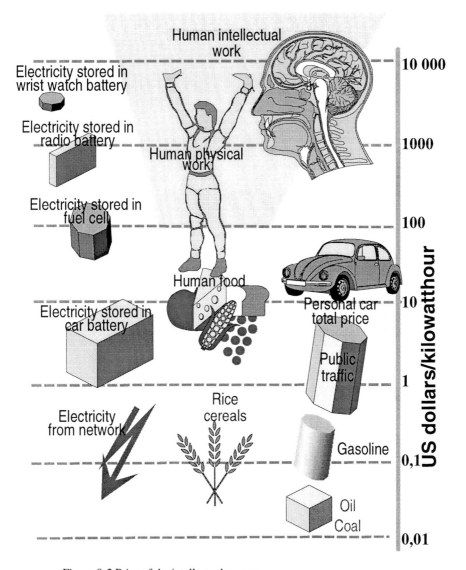

Figure 9-2 Price of the intellectual energy

This approach is what is called 'strong AI' by J.R. Searle, and is often summarized in the following way: "The mind is to the brain as the program is to the hardware". A more cautious approach is to think of computer models as being useful in studying the mind in the same way that they are useful in studying, let us say, molecular biology. This approach has been called 'weak AI'. Traditional artificial intelligence attempts to copy the conscious mental processes of human beings doing particular tasks. Its limitation is that the most powerful aspects of thought are unconscious, inaccessible to mental introspection and thus difficult to set down formally. One of the most important unconscious processes is the qualifying of each piece of information with the label of 'good-bad'—the bonus-malus system

M. Minsky (1986): "Artificial intelligence: the field of research concerned with making machines do things that people consider to require intelligence. There is no clear boundary between psychology and Artificial Intelligence because the brain itself is a kind of machine."

G.E.Hinton (1992): "Although investigators have devised some powerful learning algorithms that are of great practical value, we still do not know which representations and learning procedures are actually used in the brain. But sooner or later computational studies of learning in artificial neural networks will converge on the methods discovered by evolution. When that happens, a lot of diverse empirical data about the brain will suddenly make sense, and many new applications of artificial networks will become feasible."

But can we speak about intelligence without taking into account some, if not all, of the human attributes such as fantasy, intuition, strong emotions, motivation, ethics, faith and aesthetics? We are a long way from even hoping to find these features in a computer program, even using the largest and fastest computer to be built within the next two decades.

To expand this concept I wonder if we will ever be interested in creating not only artificial intelligence but also artificial stupidity? Or artificial humour?. Can artificial intelligence improve itself?

9.4.3   *Artificial consciousness?*

Some scientists claim that animal behaviour, including that of insects, can be explained 'more economically' in terms of consciousness: an internal model of the self. To this, the following remarks are necessary. Firstly, the information processing level of insects corresponds in our terminology to the stage of instinct, which is a relatively low stage. Secondly, the term 'more economically' is concerned not with the insect itself, but only with the way of thinking of entomologists (scientists concerned with insects). Because some people are of the opinion that modern robots, from the point of view of their information processing, have reached the level of insects (that of instinct), the consequence must be that modern robots are conscious.

M.Eigen (1990): "I do not believe that we ever posses a computer that even approximates the human brain in all its capacities, but a connected brain and computer will demonstrate 'superhuman' capabilities."

R.Penrose (1989): "At some time in the future a successful theory of consciousness might be developed—successful in the sense that it is a coherent and appropriate physical theory, .. and that this theory might indeed have implications regarding the putative consciousness of our computer. One might even envisage a 'consciousness detector', built according to the principles of this theory, which is completely reliable with regard to human subjects, but which gives results at variance with those of a Turing test in the case of a computer. .. One of the claims of AI (Artificial Intelligence) is that it provides a route towards some sort of understanding of mental qualities, such as happiness, pain, hunger... Perhaps human intelligence can indeed be very accurately simulated by electronic computers—essentially the computers of today. Perhaps, even, these devices will actually be intelligent; perhaps they will think, feel, and have minds. Or perhaps they will not, and some new principle is needed."

It is worth to mention that M.Minsky, one of the fathers of artificial intelligence, and here so often cited, is recently (1994) working on a new book entitled 'The emotion machine' said: "That's a person".

C.E.Shannon, the father of information theory, said during an interview (1990: "I'm a machine and you're a machine, and we both think, don't we?."

This book is based on principles which consider such a statement to be false.

9.5 BIBLIOGRAPHY

Anderson JR (1984) Cognitive psychology. *Artif Intell* **23** 1
Asimow I (1981) *A choice of catastrophes.* Fawcett Columbine, N.York
Birge RR (1995) Protein-based computers. *Sci Am* **272** 66
Bradley D (1993) Will future computers be all wet? *Science* **259** 890
Braitenberg V (1986) *Künstliche Wesen. Verhalten kybernetischer Vehikel.* Vieweg, Braunschweig
Churchland PM, Churchland PS (1990) Could a machine think? *Sci Am* **262** 1 26
Churchland PS, Sejnowski TJ (1992) *The computational brain.* MIT Press, Cambridge, Mass.
Davidson C (1994) Common sense & the computer. *New Sci* 4, 30
Dewdney AK (1992) Turing test. *Sci Am* **66** 1 14
Haugeland J (1986) *Artificial intelligence, the very idea.* MIT Press, Cambridge, Mass.
Hopfield JJ (1994) Neurons, dynamics and computation. *Phys Tod* **2** 40
Hsu FH, et al (1990) A grandmaster chess machine. *Sci Am* **263** 4 18
Johnson-Laird PN (1988) *The computer and the mind.* Harvard Univ Press, Cambridge, Mass.
Lenat DB (1984) Computer software for intelligent systems. *Sci .Am* **251** 3 152
Lenat DB (1995) Artificial intelligence. *Sci Am* **273** 3 62
Moravec H (1988) *Mind children, the future of robot and human intelligence.* Harvard Univ Press, Cambridge, Mass.
Searle JR (1990) Is the brain's mind a computer program? *Sci Am* **262** 1 20
Simon HA (1983) Search and reasoning in problem solving. *Art Intell* **21** 7
Tank DW, Hopfield JJ (1987) Collective computation. *Sci Am* **257** 62
Travis J (1994) Building a baby brain in a robot. *Science* **264** 123

CHAPTER 10
EXTRATERRESTRIAL INTELLIGENCE

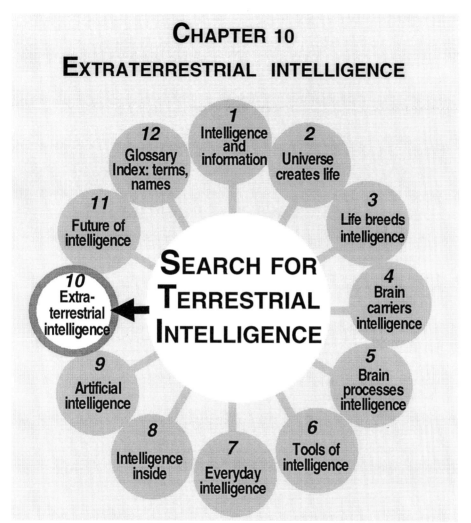

Extraterrestrial intelligence: some assumptions. Extraterrestrial intelligence could be different and much older.

Interstellar communication. How to communicate with aliens? Aliens: friend or foe?

Extraterrestrial artificial intelligence?

10 EXTRATERRESTRIAL INTELLIGENCE 362
10.1 EXTRATERRESTRIAL INTELLIGENCE? 362
10.1.1 Extraterrestrial intelligence, some assumptions 362
10.1.2 Extraterrestrial intelligence could be different and much older 364
10.2 EXTRATERRESTRIALS COMMUNICATE WITH US ? 367
10.2.1 Extraterrestrial communication? 367
10.2.2 How to communicate with aliens? 367
10.2.3 Extraterrestrial intelligence; friend or foe? 369
10.2.4 Extraterrestrial artificial intelligence 370
10.3 BIBLIOGRAPHY 370

10 EXTRATERRESTRIAL INTELLIGENCE

10.1 EXTRATERRESTRIAL INTELLIGENCE?

10.1.1 *Extraterrestrial intelligence, some assumptions*

At present we know one and only one example of intelligence, the terrestrial.

Does extraterrestrial intelligence exist? What form would extraterrestrial intelligence have? It is obvious that aliens must not be exactly like us. Extraterrestrial intelligence is probably, and must be, multifarious. Cosmic uniformity of life and intelligence is an extremely simplified assumption. As we have mentioned before, the detailed structure of intelligence is obviously a 'local' property; that is, significantly depends of specific Earth properties and historical contingencies. However, a too extreme diversity is also not justifiable, because our assumption is that intelligence has some important properties which are probably of universal character.

What boundary conditions for the emergence and development of extraterrestrial life can be formulated? In the background there still exists the question of the nature of the only kind of intelligence we know—our own. Without knowing this, all further questions remain meaningless.

Our first assumption is therefore that we are able to formulate a universal definition of intelligence which has not only local, terrestrial value, but also more universal, cosmic meaning. The second assumption is that intelligence emerges spontaneously within a long-term, self-developing biosphere. A further assumption is that our definition of life is sufficiently universal and is valid on a large, even galactic or cosmic, scale.

Table 10-1 Planets bearing intelligence
(Remark: all data are chosen arbitrarily, only for the sake of illustration)

10 civilizations in whole Galaxy in average distance of 10,000 light-years		
	10 planets with technical civilization and interest for interplanetary communication	Too 'old' civilization, without interests to interstellar communications
Without technical civilization	100 planets with **high technical civilization**	Australian Aborigines only
Without intelligent beings	1000 planets with **intelligent beings**	'Ape's planet'
Without animals		Plants only.
Without multicellular organisms	With **multicellular** animals some hundreds of planets	Unicellular organisms only
No 'spontaneously' emergence of cells	**Cellular** structures emerge contingentally, 'spontaneously'	
Without 'organic' matter from come-tary impacts etc.	With **abiotic 'organic'** matter from comets, meteorites	
Without moon: without flood/ ebb	With large **moon** causing flood/ebb (with changing coastline)	
Without open water surface	10 kilo of planets with **continents** and large oceans	Too much water, no continents, oceans surface only
Without large neighbour planet; impacts of comets cause mass extinction	30 kilo of **planets** with right mass about 10^{25} kg (right gravitational attraction) allowing existence of **atmosphere** and **hydrosphere**	Neighbour large planet as Jupiter being a 'comet scavenger' in the Solar system
Too small mass (gravitation attraction small), without atmosphere and hydrosphere		Too large mass; atmosphere to thick and therefore opaque for star's light. Jupiter-like
With planets, but too near to star, with high light flux, temperature	100 kilo of **planets** on right **distance** from central star	Too far with to low light influx; too low surface temperature
Without planets	1 mega of **stars** of 3rd generation (Sun-like) with right mass, **age,**	
With Supernova of 1st generation, devoid of radio-nuclides	10 mega of **stars** with heavy long-time **radio-nuclides** from supernova 2nd generation	
Multiple systems (double), prohibitive for stable orbits	100 mega of **stars** solitary, enabling stable **planetary orbits**	
Too young < 2 gigayears: young main sequence	1 giga of **stars** with right **age,** about some gigayears	Too old >10 gigayears: red giant, white dwarf
Too small mass, therefore too slow burning; brown dwarfs	10 giga of **stars** with right mass about ½ to 2 solar masses	Too large mass, too fast burning, evolving to supernova
400 giga of stars of all types in our Galaxy (100,000 light years diameter)		

But some doubts remain: it is possible that we will not be able to define the nature of extraterrestrial intelligence. It may be so unlike ours that even the richest fantasy and speculation will be inadequate to define it.

If our way of thinking is principally correct then an extraterrestrial intelligence must be linked with the development of life, and therefore carried by the electromagnetic elementary force and by the lightest chemical elements, in particular hydrogen, oxygen, and carbon. Also, the source of free energy needed for the emergence and development of life is a Sun-like star (Section 2.3.1 and Fig. 2-4).

The local conditions have a significant influence on the properties of living and especially of intelligent beings. This opinion is far from being broadly accepted in he academic world. Here is an example of a more exotic opinion which, from our point of view, is unacceptable:

O.H. Moravec (1988): "If you found life on a neutron star and wished to make a field trip, you might devise a way to build a robot there of neutron stuff, then transmit your mind to it. Since nuclear reactions are about a million times quicker than chemical ones, the neutron-you might be able to think a million times faster."

Of course, neutron stars are significant components of the realm of the stars, being the very late phase of their evolution. However, because neutronic matter is, to a first approximation, governed not by the Electromagnetic but by the Strong Nuclear force (Table 2-1), the only structure which could exist is a trivial sphere and not a complex structure, which is the only conceivable carrier of life and intelligence.

10.1.2 *Extraterrestrial intelligence could be different and much older*

Even if our assumption is valid that extraterrestrial intelligences are in some way similar to ours, we cannot forget that our civilization is very young—only some thousands or at most some tens of thousands of years old. Alien intelligences could be millions, or even hundreds of million, (but not gigayears, because of stars evolution restriction) of years older than ours. And this factor alone gives us enormous difficulties in coming to any conclusion. However, we cannot stress enough how important contact with extraterrestrial intelligence could be for the further development of our species. Also a lack of contact in the future will be of enormous significance for humankind, because it stresses the universal value of the intelligent being.

We began this discussion with the question of possible locations of extraterrestrial intelligences. Can intelligence evolve in the void of cosmic space? Can intelligence exist in dust/gas clouds, or on meteorites or comets? Is the existence of highly organized systems, and especially intelligent structures, conceivable on the surface of a star, assuming that a star has a temperature of at least some hundreds of degrees Celsius on its surface? The older highly developed civilizations could exists on the internal surface of artificially constructed solid spheres around central stars (so-called Dyson's sphere).

The only reasonable answer seems to be that intelligence, together with a cultural and technical environment, can emerge on the surface of a planet. The later existence of intelligence on artificial quasi-planetary colonies in cosmic space we will discuss below. Recently almost a dozen planetary systems in a radius of ten light-years has been observed (S.V.W. Beckwith, A.I. Sargent, 1996). In all these cases the planets have masses about ten times larger than Jupiter, this means are covered by a very thick, opaque atmosphere. Also the distance from the central star seems to be far from optimal for emergence of life. The future will bring clearer data concerning the nature of these planetary systems. The discussion about the traces of life in the Martian meteorite are still going; the pros and contras increase.

How many planets in our Galaxy (including 400 giga of stars) may carry intelligent beings? Must an intelligence-bearing planet include oceans and continents? We are continental animals. What are the conditions for the existence of marine creatures with a brain which is able to carry intelligence? We can agree that such animals are as highly evolved as terrestrial mammals; that is, they are warm-blooded (sufficient transport of oxygen to a large brain) and therefore air-breathing. Consequently they are forced to live near the surface of the ocean, in order to be able to breathe fresh air containing free oxygen. We can not forget that free oxygen exists only as 'by-product' of water photolysis of mostly continental plants. Large artificial structures (buildings, large machines, telescopes), being a significant component of developed technical civilization and scientific activity, need a solid base and cannot be built in oceans, while waves, storms, etc. are very destructive for underwater constructions near the surface of the sea. On the other hand, the marine environment is uniform, both at a small and large scale, and therefore is rather 'uninteresting' and cannot encourage brain development in the same way as a continental jungle does. But we must remember that dolphins and whales—the most highly developed terrestrial marine mammals—have their origin in continental mammals which went back to the sea about 50 megayears ago.

Could birds or large flying mammals be intelligent? From what we know about the mass of a brain which is able to carry intelligence, it seems to be impossible for a creature which is able to fly long distances to have such a large brain. In addition, such a flying intelligent being must have six, and not four, extremities—two legs, two wings and two hands for subtle manipulation This is not impossible, but rather improbable (Fig. 3-5).

However, why do we not also consider the possible existence of another type of central nervous system: brain? The answer is rather simple. Independent of the microscopic and macroscopic structural principles, the function of this system must be analogous to that of the human brain. This function is to process signals (electromagnetic phenomena) into sets of representations of the real world. Ultimately the function of every central information processing system which emerges on an evolutionary path is probably very similar and of universal nature.

Table 10-2 Is there intelligent life on the Earth

Two questions: **F. Drake's** question: Is there intelligent life on the Earth? **E. Fermi's** question: Where are they (aliens)? (EI = Extraterrestrial Intelligence = Aliens)							
Does an extraterrestrial intelligence exist at all?							
EI does not exist: we are alone for ever	**Extraterrestrial intelligence exists**						
EI has never existed	Only in the past	Aliens have contact with the Earth		Aliens have no contact with the Earth			
		We cannot observe them	No 'UFO's	Aliens are too young; not enough developed	Aliens are mature		
				Aliens are not interested in us; they are too intelligent to want this	Intelligence is very rare and remains locally isolated	Aliens are able and willing to make contact with us	
						Aliens will isolate us: We are in 'Zoo' or in 'Prison'	We will receive their message in the future, we must also reply
							1) We will co-operate on a cosmic scale
							2) We will die before having contact with aliens

10.2 EXTRATERRESTRIALS COMMUNICATE WITH US ?

10.2.1 *Extraterrestrial communication?*

If our assumptions are valid, then we can try to estimate how many planets may exist which could be locations of life and biotic information processing systems: that is, intelligent beings.

Many answers can be given to this question, depending on the estimation of one parameter or another. At present we have only weak evidence for the existence of extrasolar planetary systems.

From a non-localized viewpoint, a possible systematic analysis is shown in Table 10-1. Of course, all the parameters used can be given other probabilities, etc., and the numbers given in this figure are arbitrary. More 'optimistic' estimations depend only upon the subjective opinions of those making such calculations (M.Taube, 1985,1988).

10.2.2 *How to communicate with aliens?*

The problem of how to communicate with extraterrestrial intelligent beings is important. Probably the best way to start a mutual communication will be a message including series of prime numbers. A prime number is an integer that is not divisible by any number except itself. The numbers 2,3,5,7,....41,43,47, et cetera, are all prime. The Greeks knew that there are infinitely many prime numbers. The assumptions is that each intelligent being elsewhere in the beginning of the scientific and technical development 'discover' the peculiar property of prime numbers. (Nota bene: a rat can learn to turn left at every second fork in a maze, but not at every fork corresponding to a prime number).

Much more complicated will be to begin an interstellar talk with a written or spoken terrestrial language. The answer depends strongly on the theory of the origin and nature of speech. Starting from Chomsky's theory of 'universal grammar' (Section 6.5.4), it must be clear that there is little likelihood that our terrestrial genetic basis of speech will be even remotely similar to that of an alien. But Chomsky does give a way to tackle this problem. He thinks that, at the beginning, linguists (and we can add information scientists) must compare the 'speech' of aliens and terrestrials, and then, with known physical and mathematical methods, investigate the type of alien coding and logic. This task will be more like a physical investigation—by using hypotheses, formulas, and experiments—than a 'translation', and will probably be similar to the investigation of the information content of DNA. Contact between alien intelligences will not be easy, but it must in principle be possible, because intelligence is a universal phenomenon.

Communication with aliens is primarily difficult, not because of the complexities of 'translation', but because of the very small probability of the existence of intelligent societies in our Galaxy. If the number of such civilizations is of the order of tens or even hundreds, and they are distributed more or less randomly throughout the galaxy, the mean distance between them will then be of the order of thousands of light-years. Is our social and political order able to reach global agreement on the text of a message, to resolve all possible doubts and, last but not least, to pay for the equipment needed to send an answer to a calling alien? And what profit can be expected when the returning message only arrives some thousands of years later? Assuming that an alien civilisation is one thousand light-years away from us, and we send our signal today, then we will receive an answer in 2 thousand years. We will be two thousand years older, the answer of the aliens would one thousand years old, and both question and answer would be rather out of date!

It is an established opinion that within a radius of tens of light-years there is no planet carrying intelligence, or more exactly a technical developed civilization which use electromagnetic waves for communication. Also, the probability that alien intelligence exists within a radius of even a hundreds light-years is very low.

Since some decades a systematic search for extraterrestrial intelligence (SETI), by detection of radio transmission coming from different directions of Universe has been realized. It is obvious that such search has costs. Now a 100 million US dollar programme (BETA = Billion-channel Extra Terrestrial Assay) is being realized, including a detection and analysis system with 15 million channels simultaneously: 800 Sun-like stars in a radius of 80 light-years.

Table 10-3 Alien artificial intelligence

	Biotic intelligence	**Artificial intelligence**
On this planet	Human terrestrial intelligence, e.g. Homo sapiens sapiens without intermediate stages(Neanderthals, extinct 50,000 years ago)	Artificial intelligence based on computers with semiconductors, with algorithmic programmes, but without connection with quality of information
On a distant extra-solar planet	Human-like intelligence, but without Gödel's incompleteness theorem limitation; existing parallel to sub-intelligence beings	Artificial intelligence based on organic components with parallel networks, with inherent bonus/malus system including morals, ethics

After five years of continuous sky survey (23,000 observations of 209 Sun-like stars) there have been found a handful of candidate radio signals but none of these sources repeats, and in science nonrepeating data are usually not worth much. All other signals were identified; they were terrestrial. But it is well known that absence of evidence is not evidence of absence. For the next couple of years the search will include further 1000 Sun-like stars. In case the Galaxy is not so thickly populated by garrulous aliens the SETI plans to extend the search even further.

F. Drake (1996): "All we know for sure is that the sky is not littered with powerful microwave transmitters."

But one evidence we have: we find no traces of activity of extraterrestrial intelligence on our planet. A simple piece with long-living, artificially produced technetium, would suffice for such evidence.

10.2.3 *Extraterrestrial intelligence; friend or foe?*

Much more important is the question of interest in communication which must be faced before mutual dialogue can begin. Are we interested in contact with aliens? And what motivation would aliens have in contacting us? How could they react to our efforts to contact them? What can we expect from such contact? Is it possible that aliens can help us in the solution of our current painful problems—which, as we know, are mostly of social, political and moral nature.

Of course, we are very interested in making progress in science, but are we ready to do this on the basis of a gift from a stranger? Assuming that Isaac Newton obtained Einstein's solution to gravitation problems, could we assume that the development of terrestrial civilisation would have been positively influenced? Would this help us or, on the contrary, would it only bring complications and difficulties, direct or indirect, foreseeable or hidden? Is contact with extraterrestrial intelligence in any way at all a desirable event? Of course, the simple knowledge that another intelligent society exists in our cosmic neighbourhood is very important. It is at present somewhat disturbing, even alarming, that we cannot find any trace of the existence of extraterrestrial intelligence—in this Solar System, in our Galaxy, or even in the whole Universe. Are we alone? With time and with improving technical means, we will feel more and more uneasy if the 'Great Silence' continues.

F Drake, D. Sobel (1992): "I don't believe the human brain is limited in any fundamental way. I think it can emulate the power of any intelligence we may find in the universe. And I expect the discovery of extraterrestrial life to bear me out on this account soon."

The problem of contact with alien intelligence is part of a more general problem of differences and similarities between terrestrial and extraterrestrial intelligences. It is also partially a problem of future development, of the possibility of divergent or convergent evolution (Section 11.6).

One solution to this dilemma could be the 'hypothesis of the zoo'. Another intelligence, from a more highly developed extraterrestrial civilisation, observes us as a wild, dangerous, even self-destructive society, and therefore does everything possible to keep us in isolation, without any contact. Who could be interested in receiving a message from us and learning of our terrestrial experience?

As children of the last decade of the twentieth century we all think of any alien intelligence as an invader, having only one aim—to conquer or destroy mankind.

10.2.4 *Extraterrestrial artificial intelligence*

One option of more purely academic interest is that we can try to consider extraterrestrial intelligence which is not natural (biotic, based on living beings), but which is an artificial superintelligence, and perhaps exists somewhere in our Galaxy as the survivor of a former 'biotic civilisation' which has disappeared.

We are not even able to speculate more or less reasonably about the possibilities of another type of artificial intelligence, such as one based on biotic components (hydrogen, oxygen, carbon) instead of our present inorganic ones (silicon). But the most important obstacle is our uncertainty concerning the moral principles of an alien intelligence or artificially intelligent robots—if such things exist.

10.3 BIBLIOGRAPHY

Asimow A (1983) *The roving mind.* Prometheus, Buffalo
Beckwith SVW, Sargent AI (1996) Circumstellar disks and the search for neighbouring planetary systems. *Nature* **383** 139
Davoust E (1991) *The cosmic water hole.* MIT Press, Cambridge, Mass
Drake F, Sobel D (1992) *Is anyone out here? The scientific search for extraterrestrial intelligence.* Delacorte Press, N.York
Hart MH (1979) Habitable zones about Main Sequence stars. *Icarus* **37** 351
Sagan C (1994) The search for extraterrestrial life. *Sci Am* **271** 4 70
Shklowskii IS, Sagan C (1966) *Intelligence life in the Universe.* Holden-Day, S.Francisco
Taube M (1985) *Evolution of matter and energy on cosmic and planetary scale.* Springer, N.York
Taube M (1988) Materie, *Energie und die Zukunft des Menschen.* Hirzel, Stuttgart
Von Hoerner S (1978) Where is everybody? *Naturwiss* **6** 553

CHAPTER 11
FUTURE OF INTELLIGENCE

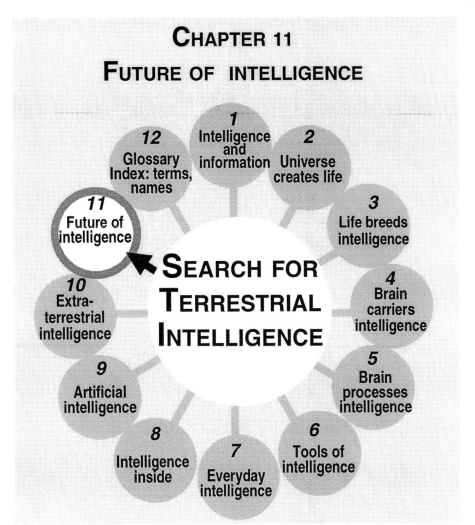

Human brain; is further evolution possible? Why terrestrial intelligence emerged only once? Distant future of the Universe. Galaxy, Sun and Earth in a distant future. A catalogue of natural and man-made future catastrophes. How stable is our planetary and cultural environment?

Probability of emergence of superintelligence. Are there limits to the growth of intelligence. Doubts concerning the future. Artificial intelligence in the future, with or without selfconsciousness? Will intelligence exist for 'ever'? Is science for ever? Mind and brain: an excellent unification theory?

11 FUTURE OF INTELLIGENCE	372
11.1 THE SURVIVAL OF TERRESTRIAL INTELLIGENCE	372
11.1.1 Can terrestrial intelligence survive?	372
11.1.2 Why has terrestrial intelligence emerged only once?	373
11.1.3 Human brain; further evolution?	373
11.2 HOW STABLE IS OUR COSMIC ENVIRONMENT?	376
11.2.1 Totality is always growing	376
11.2.2 The distant future of the Universe and the Milky Way	376
11.2.3 Our Sun will remain unchanged for 5 gigayears	378
11.3 HOW STABLE IS OUR PLANETARY ENVIRONMENT?	380
11.3.1 Earth, distant future	380
11.3.2 A catalogue of natural catastrophes	*381*
11.3.3 Natural biotic catastrophes	*384*
11.3.4 Biosphere, distant future	*384*
11.4 HOW STABLE IS OUR CULTURAL ENVIRONMENT?	385
11.4.1 Man-made catastrophes	*385*
11.4.2 New catastrophes: a superdrug and quasi-reality	*386*
11.5 SUPERINTELLIGENCE, THE PROBABILITY OF ITS EMERGENCE	387
11.5.1 A thousand future generations around one table	*387*
11.5.2 Are there limits to the growth of intelligence?	*389*
11.5.3 Superintelligence, the future highest level?	*390*
11.5.4 Doubts concerning the 'brave future'; cloning	*392*
11.5.5 Can an artificial thinker be selfconscious?	*395*
11.6 THE VERY DISTANT FUTURE OF INTELLIGENCE	396
11.6.1 Will intelligence exist for ever?	*396*
11.6.2 Is science for ever?	*397*
11.6.3 Mind and brain—an excellent unification theory	*399*
11.7 FUTURE OF THE INTELLIGENCE; SOME DOUBTS	401
11.8 BIBLIOGRAPHY	402

11 FUTURE OF INTELLIGENCE

11.1 THE SURVIVAL OF TERRESTRIAL INTELLIGENCE

11.1.1 *Can terrestrial intelligence survive?*

Each view of the future reflects wishful thinking of the authors. Because of some social resonance of printed words, it has also a self-fulfilling prediction. And sometimes, or even mostly, a pessimistic point of view is not very helpful.

What can human intelligence say about its own future? What fate awaits the intelligent beings on this planet? The species Homo sapiens sapiens has only existed for 100 thousand years, and must be considered as rather young.

On average, so far a species survives for a megayear. However, the end of a species is not abrupt, but continues for hundreds of thousands of years. Thus, one thousandth of the number of individuals will still be living after 10 megayears, if our assumption is correct

In this book we will contest the argument for the decline of the human species. There exists a number of species, from snails and frogs to crocodiles, which have existed on this planet for more than two hundred megayears without being threatened with extinction, even at the present time.

Why would a species which has the highest stage of information processing—the intelligent species—be deprived of such longevity? Why can there not be a long-term future for Homo sapiens on the Earth?

11.1.2 *Why has terrestrial intelligence emerged only once?*

The following question cannot be ignored: Why, during the long history of the terrestrial biosphere, has fully developed intelligence emerged only once? Why have other species not produced the highest stage of biotic information processing, parallel to Homo sapiens sapiens? Among some millions of existing species, and tens of millions of extinct rather high evolved species, only our species has been the bearer of intelligence. Of course, our species has had the possibility of violently eliminating competitors, such as Homo sapiens neanderthalensis; and there are suspicions that he did this!

But what of the future? Is it impossible to expect that another terrestrial species, over million or giga of years, could achieve the same level of intelligence? (Especially if Homo sapiens sapiens cannot control his own development in the desirable direction). The idea behind the film 'Planet of the Apes' was perhaps not so naive.

However, before such 'terrible' scenarios are depicted, we should discuss the more realistic problems of the near future.

11.1.3 *Human brain; further evolution?*

If one postulates that humans will survive for, say, millions or even giga of years, the question of the development of the biotic basis of intelligence—the human brain—must arise. As a first approximation, albeit a very naive one, we could postulate that the development of the human brain will progress and will be accompanied by an increase of mass. The present brain mass is around 1.4 kg, and it consumes some 20 per cent of the total bodily energy consumption. Assuming a future increase to around 2 kg, with an energy consumption of 30 per cent of the total, the problem of adequate brain cooling emerges.

This direction of development seems to be impossible, because of the disproportionate use of energy by the brain. At present, the act of birth is strongly influenced by the baby's brain diameter, because the size of the mother's pelvis is the limiting factor. A larger brain will result in premature birth; that is, in a shorter pregnancy and, therefore, increased physical weakness of the new-born. Of course, a highly developed technical civilization could cope with this situation, but naturally it poses difficulties.

Another way for the human brain to develop could be by further lateralization and further specialization of some of its functional areas. Extensive lateralization, however, causes loss of redundancy of both brain hemispheres. This fact has a rather negative influence on the resistance of the organism to injury. We are not referring to possible deliberate and conscious control of the brain's development, but only possible spontaneous evolutionary factors which influence natural selection. We cannot postulate that, natural selection in the future will be eliminated, but only that its harsh methods will have to be softened.

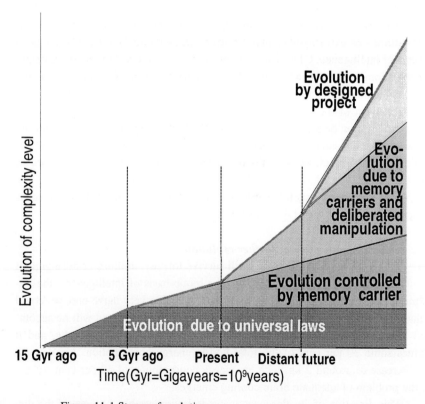

Figure 11-1 Stages of evolution

Another possible trend results from the phenomenon of the continuous death of neurons and synapses (Section 4.3). Perhaps neurons' death, being the expression of spontaneous selective elimination, can be controlled in such a way that the quality of the brain's functions increases (even if, at present, it is difficult to define what a higher brain quality should be). Another possibility is an increase of the number of different neurotransmitters and other chemical agents influencing the brain's functions.

It may be that the human brain currently contains all the components, areas, modules, neurons, synapses, and neurotransmitters which are needed to achieve the highest possible level of intelligence, even on a cosmic scale. Perhaps we have enough neurons to be able to understand the very nature of the human mind, and must only make good use of these neurons. Exploiting the future potential may lie not in an increase of the biotic basis of brain components, but more in optimization of its cultural conditioning.

New species emerge due to more-or-less stepwise environmental pressure, but always in isolated populations. The future of mankind will be influenced by deep-going globalization in all fields of activity. Therefore isolation of some groups, territorially, nationally, racially, culturally and socially, seems to be improbably and even undesired. A global quasi-one-step genetically improvement of Homo sapiens sapiens is economically and logistically almost impossible, ethically suspicious and includes a number of unknown threats, especially during the transition phase.

The only way is the controlled influence by means of a number of parallel activities: elimination of hunger, improvement of medical care, general education, global cultural exchange and last but not least some neuro-active substances improving the neuronal processes in human brain. The most probable direction of this controlled 'self-evolution' seems to be the social and moral behaviour and only in a second step the intellectual abilities. Such scenario results not in 'creation' of new species, but in some controlled 'domestication' of Homo sapiens by Homo sapiens.

Genetic self-manipulation of the human brain may be the way to further evolution in distant future. For example beginning with elimination of pathological genes. However, the complex mutual influence of some regulation genes, with their not always clearly defined topological connections, may have as a consequence that the elimination of some pathological genes could lead to unpredictable complications. (Topology, is part of mathematics, and relates to the study of those properties that an object retains under deformation; here topology is used in a more general sense as knowledge of the property of complex systems multiply internally connected).

Only in very distant future, based on deeper knowledge of the human genome and of the self-knowledge of our selfconsciousness will it be possible to jump to a new species, or even new genus called here 'Homo humanus' (Fig. 11-1).

We must not forget that the level of intelligence is based, not only on the biotic base of brain mass, modules, neurons, etc., but also on the cultural base of communication between individuals, social environment, and reasoning aids such as computers.

11.2 HOW STABLE IS OUR COSMIC ENVIRONMENT?

11.2.1 *Totality is always growing*

Totality includes all things and produces new things, new entities. Therefore it can be said that due to evolution of galactic and stellar matter (evolution of chemical elements) and evolution of terrestrial (and probably of extraterrestrial) life including increase of number of biotic species, and enormous number of man-made true and false concepts, the totality of all things grows continuously and uninterruptedly, and that the number of new entities, which complexity grows steadily, is always increasing. Even if this growth is periodically interrupted by catastrophic extinction periods. Since the emergence of intelligent beings, the number of new entities, especially material and mental artifacts, has increased rapidly.

The complexity of new entities also increases. Totality is not only richer, but also more complex (in spite of this that the definition of 'complexity' is yet only in first stage of formulation. It is important to recognize that intelligence is able to comprehend Totality including its growing complexity. It is also of greatest importance to see that a large proportion of the emerging new entities consists mostly of artifacts (man-made material and ideal entities); that is, the products of intelligence. Intelligence is not only able to understand Totality but is also the co-creator of it.

11.2.2 *The distant future of the Universe and the Milky Way*

One of the most important of the inherent abilities of human intelligence is being able to look into the future. Most of our thoughts, emotions, and motivations are connected with the future: of course, primarily with the near future, often measured only in seconds, but sometimes people try to think about the long-term future—even about eternity (Nota bene: 'eternity' is an excellent example of a human abstract concept). The nature of intelligence will dramatically change at the moment when we are forced to admit that 'doomsday' is unavoidable within, let us say, a thousand years.

It is one of the 'duties' of intelligence to do its best in examining the future fate of Totality. This is the reason for our remarks concerning the future of the Universe, our Galaxy, the Sun, and the Earth (as the seat of life and intelligence).

There are at least two hypotheses concerning the future fate of the Universe. Which of these hypotheses better corresponds to the taste of the reader must remain an open question. Two extremely different possibilities are given:

–The Universe is 'open' and evolves and expands without limit towards 'eternity'. After some period of time (giga times giga larger than the present age of the Universe) atomic matter will have exploited all the resources of nuclear energy (Strong and Weak nuclear forces (Section 2.2.2)), and almost all matter will either fall into black holes or remain in the form of photons and neutrinos. Later, even black holes decay, due to ultimate radiation processes. It is difficult (but not impossible) to imagine the existence of intelligence under these extremely unfriendly conditions; of course, in non-biotic form (F.J.Dyson, 1979).

–The Universe is 'closed.' This means that, after around 50 gigayears, universal expansion will stop and then turn into contraction. After around 120 gigayears from the present, the Universe will end in a Big Crunch—the point-like symmetrical reverse of the Big Bang. Assuming that the hypothesis of the Big Crunch turns out to be the most probable, then there exist two questions:

–After 50 gigayears the Universe will begin to contract. Some scientists doubt if intelligence can exist in such a Universe (S.Hawking, in 1988 published this opinion, but recently he has changed his mind). Assuming that intelligence can exist further, the question to be answered is whether the principle of the steady growth of totality remains valid. Of course, contraction of the radius of the Universe does not prohibit the further increase of complexity and differentiation, especially due to the creation of new artifacts.

–A more extreme question is: What will happen as the Big Crunch comes nearer and Doomsday unavoidably approaches? This must be the end of intelligence! But is this not too pessimistic a point of view, which could weaken the motivation for the long-term existence of intelligence? Is there a 'possibility to take control over the fate of the Universe before the Big Crunch comes, to try and prolong the existence of the Universe, and intelligence itself? In a contracting Universe, it must be possible to send signals to the 'limits' of the Universe and do something to weaken the contraction. In this case, we come back to the scenario of the 'Open Universe', as it has been described above.

Can intelligent beings, billions of years in the future, influence the fate of the Universe? Is this question so stupid that we must immediately forget it? Is it outside the realm of physics? Or is it only a long way from our present state of knowledge? In any case, the surprising fact is that such a question can be formulated at all.

Assuming that the Universe is very long-lived, what do we know about the longevity of galaxies? It seems to be proved that the longevity of our Galaxy — the Milky Way—is similar to that of the Universe.

F. Dyson (1979) in his paper 'Time without end: physics and biology in an open Universe', sought to show that in an open, eternally expanding Universe intelligence could persist forever—perhaps in the form of a cloud of charged particles, or even neutrinos and photons.

Table 11-1 Past and distant future of the Sun

Time, gigayears	Radius	Surface temperature Kelvin	Luminosity, relative	Internal processes Type of star
4.6 gigayears ago	Smaller than present	< 5,000 K	Luminosity 0.75 of present	Burning of hydrogen into helium **Young Main Sequence star**
Present	600,000 km	5 800 K	Luminosity: $4 \cdot 10^{26}$ watt	Half of disposable hydrogen (10%) has been burned **Main Sequence star**
1.1 gigayears in future	600,000 km	> 6,000 K	Luminosity increases 10%	**Main Sequence star**
4.5 gigayears in future	Fast expansion of outer regions, hundred times of present radius	< 3,000 K	Luminosity thousand time higher	All disposable hydrogen will be burned out. Contraction of inner mass and rapid expansion of outer regions: **Red Giant star**
After 4.5 gigayears,	Fast contraction of radius	>10,000 K	Luminosity high during tens of gigayears	No nuclear reaction, long time cooling **White Dwarf**
After hundreds gyrs	Further contraction	< 1000 K	Luminosity decreases	**Black Dwarf**

11.2.3 *Our Sun will remain unchanged for 5 gigayears*

The future history of individual stars is much more complex than the evolution of galaxies including hundreds giga of different stars. Here is not the place to discuss the evolution of all types of stars, but the fate of Sun-like stars is doubtless of the greatest interest.

Stars with a mass near to that of the Sun, which are also stars of the third generation (Section 2.3.1), have a total longevity of around 10 gigayears. After this relatively stable and quiet phase, this type of star begins a new, very dramatic phase of evolution. In the very short time of a few hundred thousands years, they undergo transformation into so-called Red Giants, with a thousand times larger radius and therefore greater luminosity.

The Sun is around 4.5 gigayears old, and is in the middle of its quiet, stable phase of evolution. Solar eruptions, sunspots, and other activities on its surface are rather benign. After another 5 gigayears or so, the Sun will probably transform into a Red Giant. Its relative luminosity (measured on the Earth as the so-called 'solar constant') will increase so far that temperature on the Earth will increase to some hundreds of degrees Celsius. The oceans will boil, the biosphere will be destroyed, and terrestrial civilization will die out, but 'civilization' may continue elsewhere.

It seems that one of the most important tasks of terrestrial intelligence will be to investigate the future evolution of the Sun and to prepare measures to avoid the consequences of the Red Giant catastrophe.

Discussing the future evolution of our central star, scenarios have been suggested which influence the Sun and/or its planetary system in such a way that favours the development and survival of intelligence. And we must repeat once more that it is not the physical aspect of the problem that is so astonishing, it is the fact that human intelligence is able to formulate such questions at all and think about their solutions.

11.3 HOW STABLE IS OUR PLANETARY ENVIRONMENT?

11.3.1 Earth, distant future

Our future is also influenced by the evolution of the Earth as a planetary body. The Sun–Earth system could alter in dangerous directions; for example, the Earth could approach, or recede from, the Sun. In almost all models of the further evolution of Sun, it will increase its luminosity by 10 per cent in 1.1 gigayears. The solar constant will increase and the greenhouse effect will achieve fatal limits.

The Earth–Moon system could also evolve in undesirable ways.

The Earth changes continuously. The continents are drifting, and after hundreds of million of years the configuration of the Earth's continents, oceans, mountains, and rivers will be very different from that seen today. The amount of carbon dioxide will decrease because of chemical reaction with the minerals of the Earth crust. Concentration of atmospheric oxygen is also rather variable. Even the mass of terrestrial water decreases due to photolysis and escape of hydrogen into cosmic space. All these numerous effects (not least human activity) will dramatically influence the global climate.

In spite of the fact that our study of the Earth, as a whole, is only in its initial phase, it seems probable that in the future the influence of humans on their own planet will become larger and larger. It also seems that, in spite of mankind's past experiences (the destruction of the natural environment; wars, with and without nuclear weapons; unsolved social and moral problems), intelligence is the only force with the potential to guide the future fate of the Earth in a better direction. But what is 'better'? Is intelligence at all able to understand and formulate the necessary answers?

Table 11-2 Future of the Earth
Solar constant S in kW/m² **Earth surface temper.** T in Kelvin

Time, gigayears	S /kW/m² T/K	Hour per day	Continent	Water	Oxygen	Biosphere
4 gigayears ago	S<1 T <290 K	10		Continuous increase	O is chemically bonded	Large bombardment sterilized the Earth
2 gigayears ago	S = 1.2 T<290 K	18	Supercontinent	Oceans increase	Emergence of free O_2	Emergence of prokaryotes
Present	S = 1.34 T=290 K	24	Five continents	Large world ocean	0.2 bar partial pressure	Full developed biosphere
2-3 gigayears	S >1.4 T > 300 K	27	Supercontinents	Decrease loss of H_2	Oxygen bonded by iron	Biosphere more rich or less?
4.5 gigayears	S > 450 T>1100 K	>30		Oceans boiled out	Oxygen bonded	Sun transform to a Red Giant*)
*) 'Possible' scenario: 3.5 gigayears: shifting of Earth from the threatening Red Giant towards Jupiter during a one gigayear voyage (M.Taube, 1982)							

11.3.2 *A catalogue of natural catastrophes*

Present mankind is a descent of numerous and very dramatic catastrophes. Even Big Bang could be considered as a catastrophic begin of the Universe. And primordial annihilation of matter and antimatter. The origin of part of the chemical elements in the pot of supernova explosion about 4.5 gigayears also represents a 'catastrophe'. The impact of a meteorite 65 megayears ago dramatically changed the terrestrial biosphere. The spectacular almost catastrophic change of the global climate during Ice Age and the not less spectacular end of this period are further examples. Mankind and it ancestors have a rather large 'experience' in this respect.

However, in the future we must expect further 'catastrophes' and we must be prepared for survival.

Our Galaxy, as we have mentioned, develops in a rather tranquil manner, but from time to time enormous explosions do occur within it. The influence of these events upon the climate and biosphere of the Earth seems to be rather small. The reason for this is that the Sun circulates around the galactic centre at the rather large distance of around 25 thousand light years, and we are also living in a quiet sector of the Milky Way on the Sun-street number 3rd. Additionally, the Sun itself will probably evolve over the next five gigayears without any unpleasant surprises.

A star with a mass around ten times greater than that of the Sun evolves very rapidly, and after tens of millions of years terminates its quiet life in an enormous explosion, called a supernova. The energy flux emitted during a supernova explosion is of similar magnitude to the energy flux of the whole Galaxy.

Recently in neighbour galaxies two supernova explosions have been observed: supernova 1987A in Large Maghellanic Cloud and supernova 1993J in galaxy M81. If a supernova would occur nearer to us than some tens of light-years, its effect on the Earth's biosphere, (and especially to the humans in it) would be catastrophic. In the long history of the Earth we cannot find any unequivocal trace of such a catastrophe. Recently the hypothesis that a supernova in a distance of 30 light-years will have a fatal impact on the biosphere (extinction of more than 90 per cent of species) on the Earth by destroying the ozone layer has been published. Such a supernova in the mentioned distance from the Earth occurs in the Milky Way each 240 megayears. The evolution of a neighbouring massive star in the direction towards a supernova could be probably recognized thousands of years before it occurred. It is not clear what countermeasures could be taken, but a passive, resigned approach must not be adopted.

Another situation exists considering the behaviour of the smaller bodies in our planetary system. A number of comets, meteorites, and even planetoids come into our vicinity, and from time to time collide with the Earth. A dangerous collision could probably occur once in every twenty-five megayears or so. If the Solar System had not contained giant planets (Jupiter and Saturn, with respectively 318 and 95 times the mass of the Earth) intelligent life might never have arisen on Earth, because the Solar System would be full of comets striking the Earth so frequently that they would impede the development of advanced life forms. Such collisions with the Earth caused wide-spread catastrophes and the mass extinction of living creatures. The last large catastrophe has been caused by a fall of a cosmic body of about 10 km diameter 65 megayears ago.

Do humans recognize such an approaching catastrophe and will they undertake some countermeasures? Our answer is, 'Yes'. Even today, mankind could predict such a threatening event. If the time period is in the order of one or two years, it seems possible that the trajectory of a small cosmic body could be altered in such a way that impact with the Earth could be avoided, by means of rockets with thermonuclear warheads.

Just 50 years ago an approaching large meteorite, would have destroyed our civilization, even if its course had been detected one or two years before a collision. However, such an event within the next decade could be prevented by a united mankind (neglecting possible religious and philosophical differences about the role of doomsday) using existing technology. This speculation is designed to illustrate that human intelligence has reached a critical point—a point of no return of the first kind—with the ability to control various catastrophes, even on a planetary scale.

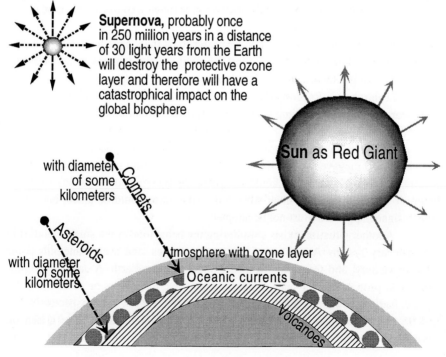

Figure 11-2 The spectrum of natural catastrophes

The meaning of the term 'point of no return', as used here, is the following. Before human intelligence and civilisation reached this critical point, the danger existed that a severe catastrophe could more or less completely destroy a large part of the human race, and therefore push back civilisation and culture. This would result in a significant decrease in the level of intelligence. We must remember that intelligence is the product not only of genetics and 'natural catastrophes' but also of culture. At the moment of crossing the 'point of no return', civilisation has the ability to foresee an approaching catastrophe and also either counter it or survive it. The reason for the additional term 'of the first kind' will be clear from the following section.

11.3.3 Natural biotic catastrophes

The biosphere, in spite of the increasing negative impact of technical civilization and explosion of global population still develops. New species are emerging. Among them also the simplest. The emergence of a virus which is lethal for all terrestrial living creatures, or only for human beings, cannot be excluded and cannot be foreseen. Of course, we can imagine some possible countermeasures.

Let us repeat; our intelligence has reached the stage where it can be hoped (but no more) that any future development in the biosphere, including unexpected deadly threats, can be controlled. It must be stressed that mankind has only reached such a level of bio-technology within recent years. We can call this situation the 'point of no return of the second kind'. The term 'of the second kind' refers to the 'man-biosphere' relationship. Of course, such an opinion is at present scientifically not well enough supported but cannot a priori be excluded (Table 11-3).

11.3.4 Biosphere, distant future

For some giga of years, the biosphere has been increasing—both quantitatively (amount of mass) and qualitatively (number of families and species). To date, human activity has a negative influence on the biosphere.

The present number of species equals some millions (catalogued number: 1.5 million species). The number of exterminated species during recent tens thousand years is of the order of ten of thousands, and the end of this man-made extinction is not yet in sight.

Only in the last decade has human knowledge gained the technical ability to change (hopefully in the right direction) the further development of the genetic pool. There is no reason to exclude the possibility that the deliberate and responsible action of human intelligence could give a significant impulse to the further qualitative and quantitative development of the whole terrestrial biosphere.

The past development of our biosphere has been based on the general principle of overabundant production of species and selective elimination—the Darwinian principle of the 'survival of the fittest'. In the future, this principle can be strongly influenced by deliberate human intervention. Such influence can be positive or, unfortunately, very negative.

Table 11-3 Points of no return

Number	Name	Content
First	**Cosmic** catastrophe	Collision with small cosmic body (meteorites etc.) Nearby Supernova explosion. Sun evolves to Red Giant (in about 5 gigayears).
Second	**Biotic** catastrophe	Deadly super-virus, degeneration of genetic code.
Third	Man-made **environmental** catastrophe	Destruction of environment (mass extinction of species). Irreversible impact on the climate (greenhouse effect etc.). Global nuclear war (nuclear winter) or/and biological war. Social conflicts and self-destruction of civilization.
Fourth	Man-made somatic catastrophe: **superdrug**	Superdrug which 'fulfilled' all human needs even the need for the superdrug. Phantomatics instead of real world.
Fifth	Intelligence **degeneration**	Decline of human intelligence because of genetic and/or social and moral degeneration
Sixth	**Extraterrestrial aggression**	Invasion of Earth by aggressive aliens. Interplanetary war.

There exists a hypothetical possibility that, at some time in the future, extraterrestrial genes could be important as the result of contact with extraterrestrial intelligence; or, at least, information could be gained concerning the structure and manipulation of extraterrestrial genes. Remember that genes are always a local phenomenon (for example, terrestrial), though containing some elements which are universally valid. It seems that there cannot be any great advantage in importing extraterrestrial genetic information.

11.4 HOW STABLE IS OUR CULTURAL ENVIRONMENT?

11.4.1 *Man-made catastrophes*

With each year our civilization is going to be more complex. H. Reeves (1996) claims that: 1) Nature generates complexity; 2) complexity generates efficiency; 3) efficiency threatens the future of complexity.

The various crises faced by humans through the recent decades call into question the very validity of this complexity and raise fundamental issues: Is complexity not doomed to destroy itself after reaching a certain level, precisely the level it has reached today?

Not only natural catastrophes threaten our future. The greatest threat to mankind is man himself. We are a long way from the position where we could say that we have reached the 'point of no return of the third kind'. Catastrophes considered here cover all social and political conflicts, including, among other things, global nuclear war.

A phenomenon of our day is the more or less accepted conviction that the population explosion and the accelerated destruction of the natural environment on this planet are the most significant threats to our future existence. It is astonishing that the consciousness of these dangers (formulated 200 years ago by T.R. Malthus) has emerged in our day, when it seems that our civilization in principle, could solve the problems, even if only with the greatest effort and at a very high cost. This critical situation will perhaps be solved in the next half century. However, it is obvious that the possibility of crossing the 'point of no return of the third kind' is still remote.

We will later (Section 11.4.3) discuss the problems of higher levels of intelligence, including so-called 'superintelligence' and self-control. The negative aspects of self-controlled intelligence can be found in A. Huxley´s 'Brave New World' (1936) and in numerous later science-fiction books (G. Turner, 1992)

S.Lem (1981), in his book 'Summa technologiae', includes the following chapters, whose titles are almost self-explanatory: 'Cosmic civilisation', 'Intel(l)ectronics', 'Prolegomena of omnipotence', and 'Phantomology'. The final chapter, 'Creation of worlds', has the sections: 'Breeding of information', 'Engineering of transcendence', and 'Cosmogenic engineering'.

Lem describes 'Phantomatics' in the following way: We construct quasi-reality, which, for intelligent beings, is indistinguishable from the real world, but apparently governed by other rules that those which we know from physics. The problem is not in the construction of such a quasi-world, but in the organization of direct inputs to our brain, without the mediation of the human senses. Lem knows that a civilisation achieving this level of 'phantomatics' (at present we are speaking about 'virtual reality') can no longer exist, because its contact with the real world would be almost negligible.

11.4.2 *New catastrophes: a superdrug and quasi-reality*

The emergence of a higher stage of information processing, of improved intelligence (though what does 'improved' mean?), or even of superintelligence can and probably will, be accelerated by parallel activity in different areas:

–Influence on mental processes by means of physical, chemical, or biotic agents, such as drugs, alcohol, etc., added from outside,

–Assistance by means of extrasomatic information processing devices (the highest stage could be artificial intelligence),

–Controlling the brain's activity and structure by means of genetic manipulation.

The most important (and probably the most likely) of all man-made catastrophes, and therefore the most dangerous, would be the 'superdrug doomsday'. The science-fiction idea of Lem concerning so-called 'crypto-chemo-cracy'—a society ruled by continuously distributing drugs which make life on a plundered and devastated world bearable, and even attractive. Chemo-virtual reality?

If we could influence the bonus so far that the satisfaction of success is achieved, and the malus is not increased when the superdrug is absent, then the decoupling of the internal and real worlds would be perfect. The motivation for additional individual activity would be lost, and the intake of more drug would substitute for other feelings. The following quotation sums this up well.

M.Minsky (1986): "If we could deliberately seize control of pleasure systems, we could reproduce the pleasure of success without the need for any actual accomplishment. And that would be the end of everything."

Would the collective intelligence of the world population ever relinquish its own self-determinism to a superdrug? Given the past history of mankind's susceptibility to alcohol, opiates, and nicotine, we cannot say that the human race is not capable of consciously destroying itself by taking a superdrug.

We can only hope that by bringing to our consciousness the whole scale of a 'doomsday superdrug'—our point of no return of the fourth kind—we will able to prevent all its consequences. But how? (Section 11.4.2). We can conclude that the creation of a utopian world by means of 'phantomatics', or with the help of a 'superdrug', cannot be considered desirable. This is another example of the 'point of no return of the fourth kind'.

Criticizing an 'idea' that psychoactive agents users should receive the appropriate genes from opium poppy or the coca bush which would allow them to synthesize their narcotics endogenously, M.F.Perutz (1996) makes the remark that there is a simpler way; we already posses the genes for making enkephalin, which has a similar effect.

Future of intelligence 387

11.5 SUPERINTELLIGENCE, THE PROBABILITY OF ITS EMERGENCE

11.5.1 *A thousand future generations around one table*

We have discussed the topic of communication with animals (Section 9.3.4), but what can we say about communication with our own descendants, with future humans. We proposed a 'thought-experiment' consisting of a meeting of representatives of former generations around one table (Fig. 6-10). We concluded that a conversation (even if not direct) between a present-day individual and his ancestors (even over a time span of 25 thousand years) is principally possible, but only covering very limited areas of mutual interest, corresponding to mutual world knowledge (Fig. 11-3).

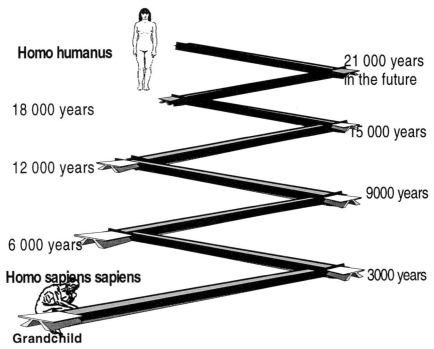

Figure 11-3 A thousand future generations at one table

From the point of view of the terms used in this book, we have postulated that our ancestors, too, at least over this period of time, had an intelligence level corresponding to our own. No qualitative jump should be postulated. However, if one goes back to the more distant past, and invite ancestors living not only 25 thousand years ago, but let us say 100 thousand years ago (individuals of Homo sapiens neanderthalensis), or even 500 thousand years ago (individuals of Homo erectus), then our conclusion will be different. We will be forced to assume that such distant ancestors were at a lower stage of information processing, called 'sub-intelligence' for Neanderthals and 'proto-intelligence' for the Homo erectus individual. Communication would then be almost impossible.

An analogous situation would (probably) be possible in reverse, if our long table were filled with representatives of future generations. Conversation, even if not direct, ought to be possible. Of course, as before, we could discuss topics covering a very limited area of mutual interest and world knowledge. But a question remains: Would our descendants have roughly the same level of intelligence and the same language as we? Or would they have attained a level we could label as 'superintelligence'?

11.5.2 *Are there limits to the growth of intelligence?*

New species emerge due to 'punctuated equilibrium' (according to S.Gould) or more stepwise but always in isolated populations (according to E. Mayr). The future of mankind is possible only in deep-going globalization. Isolation of some groups, territorially, nationally, racially and socially, seems to be improbable and undesired. A global quasi-one-step genetical improvement of Homo sapiens sapiens is ethically suspicious and includes a number of unknown threats, especially during the transition phase.

We have mentioned the possibility of the further development of the material basis of human intelligence—the human brain. Now we will consider the future of the development of intelligence itself. The crucial point is the possibility of the emergence of a new, higher stage of biotic information processing; qualitatively higher than the present level and so far different that it needs a new term to describe it, such as 'superintelligence'.

Although it is an affliction of our times to prefix many things with the adjective 'super' (Section 9.2.2), we cannot find a more appropriate term for a higher level of biotic information processing other than 'superintelligence'.

In other words, we can ask if the existing level of our intelligence is the highest possible, and we will therefore meet, in the future, clear limits to its further growth? Or could it be transformed laterally into the higher level of superintelligence? Some scientists have attributed very high qualities to the present level of intelligence:

P-S. de Laplace (1810): "An intelligence that at a given instant was acquainted with all the forces by which nature is animated and with the state of the bodies of which it is composed would—if it were vast enough to submit these data to analysis—embrace in the same formula the movements of the largest bodies in the Universe and those of the lightest atoms. Nothing would be uncertain for such an intelligence, and the future, like the past, would be present to its eyes."

This extreme opinion has, of course, been criticized throughout the past two centuries, and is still criticized today.

D.Layzer (1990): "Laplace's Intelligence is an idiot savant. It knows the position and velocity of every particle in the Universe; but because this vast fund of knowledge is complete in itself, there is no room in it for information about the stars, galaxies, plants, animals, or states of mind. .. We humans must watch the film unwind, but Laplace's Intelligence sees it whole."

I.Prigogine (1980): " (Laplacian idea can be expressed in a short form, as follows): God is reduced to a mere archivist, turning the pages of a cosmic history book that is already written, including the final chapter. Here is no chance for an intelligent being with free will, even if this freedom remains very limited."

But even if superintelligence would not have the capabilities of Laplace´s intelligence, it could have other, today unforeseen, abilities.

11.5.3 Superintelligence, the future highest level?

It sounds paradoxical, but the sole description of beings with higher than human intelligence can be found only in science-fiction writings. The best of this type of speculation has been formulated by S. Lem (1967).

However, it is not clear if our level of information processing allows us to define the properties of a qualitatively higher stage of information processing which we could call superintelligence. There are no explicitly formulated limits which prevent us from discussing the properties and features of superintelligence. On theother hand, it is not clear how to begin to characterize it, beginning from our definition of intelligence.

Biotic development, spontaneous or under human control, cannot be the only way to achieve a qualitatively higher stage of information processing. Another way could be to intensify the symbiosis between the human brain and artifacts such as the computer, for better information processing.

Here the field for speculation is very large, but we will give some lines of possible development in this direction. Not only science-fiction writers believe in a future 'synthesis' of the human and the computer. The extreme case of this proposed 'symbiosis' is called the 'cyborg' (cybernetic organism). One such scenario:

H. Moravec (1988): "Successive generations of human beings could be designed by mathematics, computer simulations, and experimentation, like aeroplanes, computers, and robots are now. They could have better brains and improved metabolism that would still be made of protein, and their brains would be made of neurons. At that time, a genetically engineered superhuman would be just a second-rate kind of robot, designed under the handicap that its construction can only be by DNA-guided protein synthesis. Only in the eyes of human chauvinists would it have an advantage—because it retains more of the original human limitations than other robots."

Another, more conservative, picture is still speculative enough to provoke an avalanche of other more or less credible speculations. In this picture, future development proceeds along parallel paths. On one hand, the biotic basis of the human brain is improved by manipulating its macro- and micro-architecture. The brain's hardware, and software could be manipulated by means of genetic and somatic operations.

It must be hoped that these changes would lead to a significant growth of the very human qualities of selfconsciousness, and the feelings of 'Myself' and 'Ourselves'. Alongside this, would also proceed the development of the artificial (that is technical) components of superintelligence. Intimate communication with our 'own personal supercomputer', partially or fully implemented in the human body, could then further improve human intelligence.

On the stage of intelligence as we have mentioned (Section 8.2.1) the so-called sub-consciousness is not directly achievable. Will the superintelligence be able to have direct contact (direct communication) with subconsciousness? This would mean the end of subconsciousness. Superintelligence will be conscious and will control all desires, all urges and impulses. Intelligence is in some sense coupled with the ability to lie. Will superintelligence able to super-lie and capable of super-deception? And what does this mean?

The highest level of biotic information processing, the superintelligence, could directly be linked to a global super-Internet and cause an emergence of a new feeling of 'We, ourselves', for the whole mankind, with a higher level of moral and ethical altruism. This sounds unrelated but the notion of intelligence cannot be separated from the bonus-malus system (that is, from morals and ethics).

In the past, the strategy in all spontaneous natural processes was the overabundant production of species, types, etc., and subsequent selective elimination. In the future a new strategy will arise resulting from the activity of intelligent beings as the creators of artifacts. Instead of overabundant production there will be purposeful activity. Instead of selective elimination in the real world, through the random action of the environment, there will be deliberate elimination of activity at the level of planning and mental preparation.

This new strategy can be applied to the further development of human intelligence itself. Neurons and synapses are already generated in overabundance. One possibility is to control and to steer the elimination (that is, the death) of neurons and synapses, in such a direction that desirable areas of the human brain will evolve (larger, more complex, better 'wired'), and the undesirable areas and organs, which influence 'bad' properties (if we are able to define such properties) can be contracted or better controlled (Table 11-4).

L.Wolpert (1995): "We could, in principle, try to device a genetic program which would generate...an angel. The problem of the angel is to provide both an extra pair of wings as well as angelic temperament. To get extra pair of feathered limbs would require considerable ingenuity but we know enough about the patterning of the body plan, and the development of wings and feathers, it would be plausible. One would be making use of known genes from birds and mammals. It is most unlikely we would know what neuronal connections to generate to get an angelic temperament, but one could probably devise a selection procedure given enough time."

Table 11-4 Human brain: future evolution

		Global direct brain-brain network. Implementing in human central nervous system specialized bio-chips enabling 'direct' communication between individuals.
	Hard chip-brain manipulation. Implementing in nervous system some specialized bio-chips for improvment of selected intellectual activities.	
	Soft genetic manipulation Genetic manipulation for improving the neurons connections, the action of neurotransmitters, synapses, and for elimination of large brain's failures.	
Directed cultural evolution 'Soft' influence on the plasticity of neurons connections and neurotransmitters action due to 'psychoactive agents'. Danger of spontaneous genetic degeneration.		
Spontaneous undirected evolution. Cultural impact due to learning, training.		
in 10–100 years	in 100–1000 years	in 1000–10,000 years

Superintelligence does not mean 'superman' described by F. Nietzsche (1887) in 'The Genealogy of Moral', as an individual free from moral laws.

11.5.4 *Doubts concerning the 'brave future'; cloning*

A new species emerges: Homo sapiens humanus or better Homo humanus? Will the modern Homo sapiens and the future Homo humanus speak with mutually understandable language? Will they be able (only as an experiment of thought) to mutual sexual breeding? Will the Homo humanus emerge stepwise only in a part (selected or random) or will he include the total global population more or less homogeneously. Will the superintelligence be able to solve the Gödel´s paradox and other logical paradoxes (Section 7.4.2) with which our thinking system is burdened? Do we need superintelligence to solve the problem of unification of some physical ideas, such as quantum theory and general relativity theory, which are at the present stage of evolution of intelligence non-unifiable. Nota bene: the searched theory is at present called: 'superunified theory' or Theory of Everything.

In the 'optimistic' case a number of questions discussed in previous chapters, such as some paradoxes in thinking, some crucial problems as Gödel´s incompleteness, unification of the principles of physical description of the world, the deeper meanings of Laws of Nature, the origin of Natural Constants, and last but not least the deeper mechanism of intelligence itself, will remain on the agenda of future generations of scientists.

Optimism or pessimism are concepts which should not be directly coupled with individual bonus-malus systems. There are people in whom the 'good feeling', the bonus labelled state is caused by confirmation of their pessimistic opinion, and other people being in a bad mood, malus situation, are looking for ideas which could be optimistic.

An intelligent being is obliged not only to ask questions, but also to formulate doubts about all things and all concepts, about their definition, meaning, validity and correspondence with the real world, whether or not the conclusion will be called optimistic or pessimistic. Even some doubts about the 'existence of reality' is obligatory.

The development of the human brain will in the future, probably be influenced by 'genetic engineering': the artificial manipulation, modification, and recombination of DNA molecules in order to modify an organism or population of organisms. Cloning (clone is a genetically identical organism derived by transfer of DNA molecules, taken out from a somatic cell of the adult or even died 'parent') is a popular object of speculations. There seems to be a lot of overestimation of the function of the 'naked' DNA, and a clear underestimation of the impact of the egg´s

Future of intelligence

Table 11-1 Somatic and embryonic cloning and the human				
Individual	**Individual, natural**	**Identical twins (mono-zygous)**		**Somatic cloned individuals**
		natural, spontaneous	**in vitro, deliberate**	
Genetic **paternal DNA**	DNA is around 25 years (one generation) older than those of the father and therefore includes some spontaneous mutations (impact of cosmic and technical radiation, chemical mutagens, spontaneous methylation)			
	DNA from the father's sexual cells(gamete)			**DNA** from the **somatic** cells of the 'unique' **adult** parent (e.g. male)
Genetic **maternal DNA**	DNA is included in the maternal egg			Maternal **DNA** is removed
Egg (**always maternal**)	Egg with complete organelles including cell membrane, cytoplasm, tens-hundreds of: mitochondria (the powerhouse of the cell) with genuine **mitochondrial-DNA,** ribosomes (the site of protein synthesis), lysosomes (protecting the other components of the cell from random destruction), peroxisomes (organelles that are the site of metabolic processes involving hydrogen peroxide), centrioles (acts at a precise time in the cell division cycle during DNA replication), and other components			
Fertilization of egg results in **zygote**	Natural sexual act: spontaneous fusion of sperm and egg		In vitro fertilization: also by mean of electro-impulse	
Number of embryo and foetus	One embryo	Two embryos and two foetuses (deliberate or accidental): **identical twins** || One embryo	Deliberate quintuple
Uterus	'Primary' maternal uterine environment: supplier of blood (including all nourishing but also pathological components: drugs, alcohol, nicotine, etc.) and microorganisms (benign or harmful).			One generation later with clear different properties	
Prenatal development of brain (uterine environment)	Human brain produces during the embryo-foetus phase around 100 giga (100 thousand million) of neurons. About 50 to 70 per cent of these 'new-born' neurons die during the development during baby, childhood and infancy			
Postnatal development of brain familiar environment	Human brain produces during the embryo-foetus phase around 10 tera (10 million million) of synapses. Further number of synapses (10 times more) are born and die during the development			Different 'familiar environment': one generation later	
Social-cultural environment; one **generation** later	Identical climatic and social-cultural environment			Different climatic and changed social-cultural environment	
Child, adolecent development of brain social, cultural environment	Human brain does not produce new neurons. Life experience, especially emotional, influences emergence and destroying of synaptic connections and other mechanisms of long-term memory (including bonus-malus system)			
Cloned 'Elvis Presley' identical pop singers				Five 'Elvins'

'out-side-DNA-elements', cytoplasm, plasma membrane, organelles: mitochondria, lysosomes, etc., during first phases of egg development. The impact of non-genetic mechanisms going on in the uterus is significant. It strongly influences the development of embryo and foetus (including many symbiotic micro-organisms, drugs etc. transmitted from uterine environment). Later the birth shock on the new-born, and the impact of the environment are further examples of epigenetic influence on the gifts (like singing and other inclinations).

Intelligence is information processing about the information processing (biotic and technical) and it cannot be excluded that intelligence is not interested at all in the knowledge what it is itself. Maybe in this case knowledge constitute more 'danger' than ignorance. In other words, maybe the atoms in our brain are not able to investigate how they act when they investigate the atoms in the Universe.

The physical unfeasibility of such a kind of super-duper-computer is obvious because of the limitation of the maximum velocity of light.

Based on our definition of biotic information processing, we must assume that any future superintelligence, including artificial superintelligence, will comprise not only very highly developed hardware and appropriate software. The software for artificial superintelligence should be prepared by the superintelligent machine itself.

It is an open question whether it will be possible to introduce into a supercomputer all the necessary qualia (Section 6.6.3) (including bonus-malus labelling), by means of a single 'injection' or if it will require longer-lasting 'educational training', beginning from the baby stage. In each case, a machine with artificial superintelligence must not be deprived of the faculty known as common sense.

11.5.5 *Can an artificial thinker be selfconscious?*

A current question, which absorbs a considerable number of people, not only experts and scientists, is whether or not a supercomputer (in our terminology, a supercomputing 'thinker') could have consciousness? (Fig. 11-4).

A pragmatic question is whether we need such a machine? Will we deliberately try to construct such a machine or, on the contrary, will we be unable to?

A second question is if, in the future, a supercomputer would be able to control a super-robot, which could construct by itself a selfconscious super-robot?

Question three assumes that a supercomputer could act for example as a physician and establish a diagnosis in a human patient. Would it conclude that it itself is deprived of selfconsciousness, and therefore cannot feel what its patient feels, particularly when faced with fatal illness?

A fourth question is if an 'intelligent robot' could feel tired and unwilling to do its job? Could it simulate stupidity? What could we do in this situation? Can 'artificial stupidity' also arise from 'artificial intelligence'? Could an 'intelligent computer' have enough social motivation for not wanting to be called 'stupid' and not use deception and lies for its own profit?

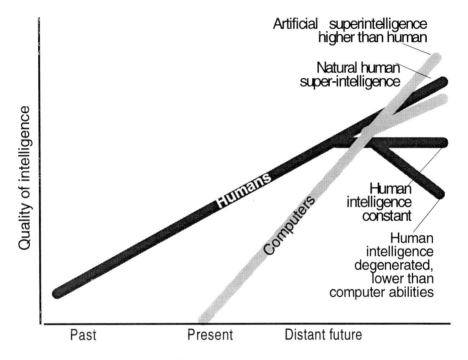

Figure 11-4 Human intelligence and artificial impact

Fifthly a selfconscious supercomputer may be creative in the field of the arts—in literature, poetry, or music. (The term 'creative' must be taken in the fullest sense, as it is used for, say Dante and Mozart). Though why should the field of sciences be excluded? Why not an artificial Newton or Einstein? Of course, we are not speaking about the serial production of a brilliant supercomputer, but why could a brilliant one not emerge by chance, by 'mutation' ?

A final question is, if a supercomputer could be as creative as Goethe or Gödel, why could it not be as 'good' as Lincoln or deGaulle? Are we interested in being governed by such an 'honourable' supercomputer, or electing one as President of a United States of the World? At a lower level, should such an intelligent machine be deprived of the usual human rights, including the democratic right to vote?

11.6 THE VERY DISTANT FUTURE OF INTELLIGENCE

11.6.1 *Will intelligence exist for ever?*

For some people this question is, at best, ridiculous or stupid. In the worse case, such a question is forbidden. There are a number of reasons for these attitudes. Firstly, it is not clear what we understand by the term 'eternity'. Here, we will consider eternity to be the lifetime of our Universe; perhaps a little shorter, but not by very much. Therefore, we will refer to 'eternity' and mean 'as long as the Universe exists'. This means that, in the case of the hypothesis of the 'closed Universe', we are talking about around a hundred gigayears; or even longer, depending upon the eventual fate of the Universe, which can still only be roughly predicted.

The next question is if it is possible for a biotic species to exist for giga of years. Now, one cannot ignore the fact that Homo sapiens sapiens is not an average biological species. He can control his own genetic base and it is not a priori excluded that he will be able to preserve his own species for a very long time. It is, of course, another question whether this will be worth doing, or not.

The existence and relative stability of the Universe does not automatically ensure the physical stability of the direct environment of future mankind. (This point has already been considered at the beginning of this chapter).

Different possible catastrophes, natural and man-made, can be surmounted, under some pre-conditions. Eternal existence does not seem to be prohibited, a priori, but some doubts still exist. We must ask ourselves about the motivation for such a long existence, and we should try to understand what role mankind could play in the Universe.

11.6.2 *Is science for ever?*

What is our brain that all mathematicians, in all times, in all parts of the world, are speaking the same formal language, using the same logic and producing an internal coherent mathematics? (Nota bene: And what about 'communication' with extra-terrestrial mathematicians? Do they speak also understandable 'mathematicalese'?) And how it is that mathematics is so effective and useful in the development of physics, the most abstract representation of the real world. How does it happen that a good part of our concepts concerning the world and our brains, is so successful in practical activity of the humankind on this planet?

The prophesies of the end of sciences at the end of our century and millennium are numerous. The scientific and public acceptance of all these scientists is high.

The following titles of the books, mostly published 1995-1997 are self-explaining:

    S.Hawking (1980) "Is the end of theoretical physics in sight".

    F. Fukuyama (1992) "The end of history and the last man".

    J.Horgan (1996) "The end of science. Facing the limits of knowledge in the twilight of the Scientific Age".

    J.Gimpel (1995) "The end of future. The waning of the High-tech world".

    D.Lindley (1993) "The end of physics —The myth of a Unified Theory".

    J.Leslie (1996) "The end of the world. The science and ethics of human extinction".

    J.Rifkin (1996) "The end of work: the decline of the global force and the dawn of the post-market era".

Number of books with titles: 'End of arts', 'Painting after the end of painting' is almost endless.

Some remark: F.J.Dyson (1996) in criticism of J.Leslie's "The end of the world" writes: "Half of the book is devoted to application ..(of the Bayes rule) to the question of human survival. This discussion is worthless, being based on invalid use of the rule. The other half of the book is a recital of many real dangers to which our species is exposed, but (Leslie) does not examine them critically. A mistaken philosophical argument may have serious consequences in the real world"

The list of books concerned with the end of history, of theoretical physics, of physics, of science, of world even of future is large and still increasing. Who will try to write a book entitled 'The end of endology'. It seems that a better solution will be to write 'The future of futurology'. And not one book per generation but a large number of it. In a large spectrum of opinions, beliefs, hopes and fears some questions could be helpful to anticipate coming difficulties and to prepare solutions.

    R.P.Feynman (1985): "We are lucky to live in an age in which we are still making discoveries. (This) is the age in which we are discovering the fundamental laws of nature, and that day will never come again."

However, the following citation is a good counterargument and does not need explanation:

    M.Planck (1924): "When I began my physical studies in Munich in 1874 and sought advice from my venerable teacher Philipp von Jolly...he portrayed to me physics as a highly developed, almost fully matured science....Possibly in one or another nook there would perhaps be a dust particle or a small bubble to be examined and classified, but the system as a whole stood there fairly secured, and theoretical physics approached visibly that degree of perfection which, for example, geometry has had already for centuries."

Some general remarks concerning the future of science are needed here. The development of natural sciences has occurred with some breaks since more than two millennia, and in an astonishingly explosive kind during the recent three centuries. The scientific objects include the Universe and the elementary particles, the geometrical figures and the human intelligence. However, it must be said, that in the future the continuation of scientific activity is not for all scientists self-evident. It is not clear for them if mankind in the future will continue this heroic effort to investigate the intelligence, consciousness, mechanism of free will. May be other priorities (social, ethical, mystical-irrational) will be of higher significance. We are convinced, that scientific activity is a genuine attribute of human intelligence, also in next and in distant future.

J.Horgan (1996): "(Some scientists) find it hard to accept that pure science, the great quest for knowledge, is finite, let alone already over. But the faith that science will continue forever is just that, a faith, one that stems from our inborn vanity. We cannot help but believe that we are actors in an epic drama dreamed up by some cosmic playwright, one with a taste for suspense, tragedy, comedy, and—ultimately, we hope—happy endings. The happiest ending would be no ending."

The authors of this book are convinced that human intelligence, in spite of being a product of an 'uncommon sense' and sometimes finishing in tragedy, includes as indivisible component the search for the 'Truth', even if an individual's motivation leads to social profit. Scientific activity could be thought as a 'drive', as an 'instinct' of higher stage, inborn in some individuals, coupled with a high, even very high, label in the human bonus system.

S.Hawking (1996): "However, if we do discover a complete theory, it should in time be understandable in broad principle by everyone, not just a few scientists. Then we shall all, philosophers, scientists, and just ordinary people, be able to take part in the discussion of the question of why it is that we and universe exist. If we find the answer to that, it would be the ultimate triumph of human reason—for then we would know the mind of God."

A.R.Damasio (1994): "I made my view clear on the limits of science: I am sceptical of science's presumption of objectivity and definitiveness. I have a difficult time seeing scientific results, especially in neurobiology, as anything but provisional approximations, to be enjoyed for a while and discarded as soon as better accounts become available. .. Perhaps the complexity of the human mind is such that the solution to the problem can never be known because of our inherent limitations."

11.6.3 *Mind and brain—an excellent unification theory*

Real progress in the study of human intelligence depends upon the overall development of all the natural sciences. Knowledge of the nature of the Universe, of elementary particles and forces, of genetics, of the emergence of species, and of neuroscience are all mutually and indivisibly related. If we want to have a better understanding of the very nature of Homo sapiens sapiens, we must be ready to

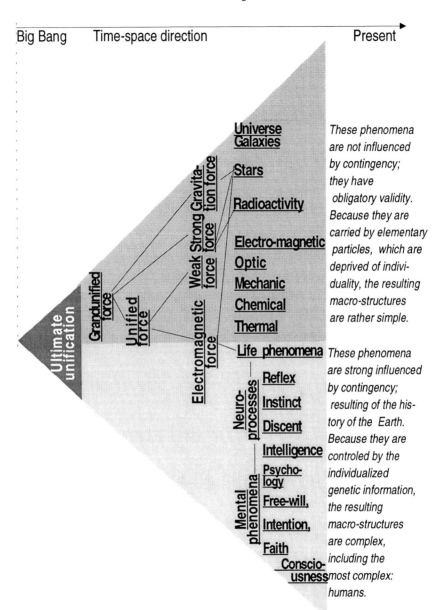

Figure 11-5 Excellent unification: mind-matter

expand efforts in all these branches of science. Our progress must be towards all horizons, even if this is a costly process.

In this way, we will achieve, step by step, a unified picture of all natural phenomena, including our own intelligence and minds.

D.J.Chalmers (1997): "Thus a complete theory will have two components: physical laws, telling us about the behaviour of physical systems from the infinitesimal to the cosmological, and what we might call psycho-physical laws, telling us how some of those systems are associated with conscious experience. These two components will constitute a true theory of everything."

Mankind has come a long way from separate descriptions of mechanical, thermal, chemical, magnetic, electrical, biological, astrophysical, and even psychological phenomena; we have reached a point where most, if not all, physical phenomena can be (or will be, in the not-too-distant future) described by one physical theory—the so-called 'Theory of Everything' (Fig. 11-5). The term 'everything' is connected to these phenomena which have universal validity, that is are self-controlling processes. Some examples: Big Bang, emergence of time and space, emergence of so-called Natural Constants, emergence of forces and particles, origin of chemical elements, origin of galaxies and stars.

There are other phenomena which in a significant way depend on local contingencies, on local 'history'. This includes the typical 'terrestrial' processes, such as origin of life, and later the evolution of terrestrial biosphere, including the numerous species. They all are a product of 'accidental' mass extinction, due to impact of meteorites, of global volcanic activity and other contingencies. The origin and evolution of biotic information processing (reflex, instinct, discent, and intelligence) is of the same class of phenomena. The essence of the biotic information processes seems to be comprehensible on the basis of an amalgam of concepts from different branches of the natural sciences including the 'Darwinian' evolution theory. Of course, some phenomena such as feelings, emotions and beliefs, are still far from being investigated completely.

11.7 FUTURE OF THE INTELLIGENCE; SOME DOUBTS

Does human intelligence include innate elements which leads to self-destruction? Is intelligence not sentenced to degeneration and final self-extinction? Is the investigation of the future not too difficult and an almost forbidden target for the present state of development of intelligence? Has an intelligent being not the 'duty' first to improve itself and then......But what means 'improvement' of intelligence? Has mankind not numerous other problems which must be solved before we begin to think about the future of intelligence?

Plautus (270 B.C): *"Homo homini lupus"* (A human being appears to another human being as a wolf)

P.Connes (1979): *"Intelligens intelligenti homo"* (An intelligent being appears to another intelligent being as a human)

11.8 BIBLIOGRAPHY

Damasio AR (1994) *Descartes' error. Emotion, reason, and the human brain.* Putnam, N.York
Dyson FJ (1979) Time without end: physics and biology in an open Universe. *Rev Mod Phys* **51** 3 447
Dyson FJ (1996) Reality bites. *Nature* **380** 296.
Gimpel J (1995) *The end of future. The waning of the High-tech world.* Praeger, Westport, Conn.
Gott JR (1993) Implications of the Copernican principle for our future prospects. *Nature* **376** 315
Hawking S (1980) *Is the end of theoretical physics in sight.* An inaugural lecture on the University Cambridge
Hawking S (1996) *The illustrated A brief history of time,* Bantam Press, London
Hecht J (1991) Will we catch a falling star? *New Sci* 7.8,.48
Horgan J (1996) *The end of science. Facing the limits of knowledge in the twilight of the Scientific Age.* Addison-Wesley, Reading, Mass.
Islam JN (1979)The long term future of the Universe. *Vistas Astron* **2** 3 265
Kates RW (1994) Sustaining life on the Earth. *Sci Am* **271** 4 92
Leslie J (1996) *The end of the world. The science and ethics of human extinction.* Routledge, London
Lindley D (1993) *The end of physics-The myth of a Unified Theory.* Basic Books, N.York
Moravec H (1988) *Mind children.* Harvard Univ Press, Cambridge, Mass.
Minsky M (1994) Will robots inherit the Earth? *Sci Am* **271** 4 86
Reeves H (1996), *Newsweek,* 5 Feb
Sagan C (1994) The search for extraterrestrial intelligence. *Sci Am* **271** 4 70
Taube M (1982) Future of terrestrial civilization over a period of billions of years: Red Giant and Earth shift. *J Brit Interplan Soc* **35** 219
Turner G (1992) *Brain child.* Avon Books, N.York
Wolpert L (1995) Development: is the egg computable or could we generate an angel or a dinosaur?, in MP Murphy, LAJ O'Neill (edit) *'What is life? The next fifty years'.* Cambridge Univ Press, Cambridge, UK

12
GLOSSARY, INDEX

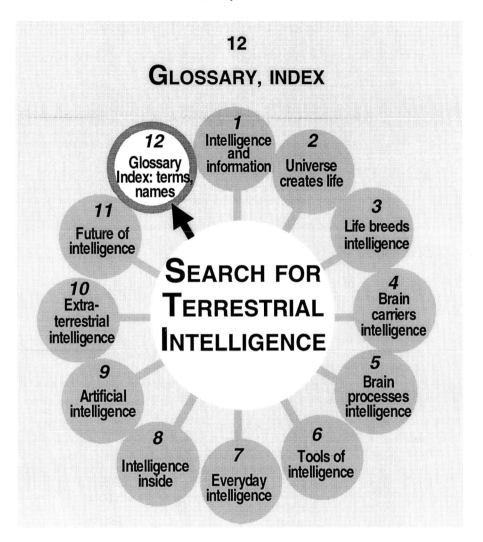

Glossary
Index of names and terms

12 GLOSSARY

Addiction: compulsive use of habit-forming *drugs*. There are probably persons who, for *genetic* reasons have lower than normal ability to produce or release these *'bonus-carrying'* substances. These persons, deprived of the natural opiates, such as endorphin, may become highly sensitive to pain and lapse into depression, and even are more predisposed to suicide. Addiction results not only from *genetic*, but also from *epigenetic*, causes. (Section 5.3.3)

Algorithm: (Arabian: after *al-Khuwarizmi*–Persian mathematician, 9th century) rule for solving a particular problem. In many cases the rule consists of applying a special set of steps; as opposed to *heuristics*. (Section 6.6.2)

Amygdala: (Latin *amygdala* almond) a small organ in the brain (5 gram), receives input from other brain organs, contains neurons among others responsible for perception of facial expression of other individuals and therefore carries the ability for appropriate social communication. It plays an essential role in many aspects of *emotiona*l information processing. (Section 4.1.5; Fig.**4-3**)

Artifacts: (Latin *artis* art + *facere* to make) something made by human effort or intervention. (Section 9.1.1)

Artificial consciousness: (see *Consciousness*) hypothetical state of *computer*ized machine possesses some understanding of *mental* qualities such as happiness, pain, hunger, humor. (Section 9.4.3, Tab.6-1)

Artificial discent: (see *Discent*) a *computer*ized machine, possessing inputs (keyboard or microphone) and executive organs (*effectors*, printer or loud speaker), and *information processing* on the level of *discent*. Examples: 'expert' or 'translator'. (Section 9.3.6, 9.3.7, Tab. 6-1)

Artificial instinct: (see *Instinct*) a *computer*ized machine, possessing artificial *sensors* and executive organs (*effectors*), and *information processing* on the level of *instinct*. Example: 'chess player'. (Section 9.3.3; Tab. 6-1)

Artificial intelligence (AI): (see *Intelligence*) the study of how to make computers behave intelligent or to 'model' the human brain mechanisms. (Section 9.4.2; Tab. 6-1)

Artificial reflex: (see *Reflex*) a *computer*ized machine, possessing artificial *sensors* (such as eye) and executive organs (*effectors*) such as hands and legs, and *information processing* on the lowest level of *reflex*. Example: robots. (Section 9.3.2; Tab. 6-1)

Australopithecus: (Latin *australis* southern; Greek *pithekos* monkey, ape) genus Australopithecines, small brained but with a fully upright posture, in East Africa between 4 to 3 megayears ago. Almost complete skeleton (Australopithecus afarensis, 3.25 megayears old) of a small female individual is called 'Lucy'. (Section 3.5.5; Fig. 3-9)

Axon: (Greek *axon* axis) a filament whereby a *neuron* exerts the most characteristic and unique component of neuronal cells, by means of which it sends *information nerve signals*) to its partner neuronal cells. (Section 4.2.3; Fig. 4-7)

Biotic: (Greek: *bios* life) pertaining to any aspect of living beings.

Biotic information processing: spontaneously evolved system of processing information in animals, including four stages of evolution from the lowest to the highest: see *Reflex, Instinct, Discent, Intelligence*. (Section **1.4.1** 3,4,1; Tab. **1-6**; Fig. **1-5, 3-7**, 3-8)

Glossary, index **405**

Bonus-malus system: (Latin *bonus* good + *malus* bad) every animal must have its own scale of 'good or bad'. In this book, we will speak about the 'bonus-malus scale'. All remembered information, old or new, has a quality rating which is measured on a continuous 'bonus-malus scale'. It is impossible to overestimate the significance of the bonus-malus system in biotic information processing, especially at the highest stage—that of human intelligence. We will often return to this value scale. The systems are more or less equivalent:–bonus/malus,–pain/pleasure,–reward/punishment,–appetite /aversion,–hedonic/anhedonic system. These involve the whole range of feelings, from pleasant to unpleasant. The bonus-malus system is partially carried by some neurotransmitters in the brain. (Section 1.3.2; Fig. **7-2, 8-7**)

Brain: the largest part of the Central Nervous System with a mass in humans of around 1.4 kg and around hundred subdivisions (organs). Human brain can be considered from different points of view, 1) macrocomponents; organs 2) origin and *phylogenetic* age; young and old 3) anterior-posterior differentiation; corresponding to a *motor* and *sensory* activity 4) hierarchy of activity—primary, secondary, tertiary 5) left and right asymmetry, *lateralization*; left and right hemispheres 6) function in mental activity; *speech*, vision etc. 7) *ontogenic* development; from child to elderly age 8) microcomponents; *grey* and *white matter*, *neurons, glia* etc. The energy flux of the human brain equals around 20 watt; that is, some 20 per cent of the consumption of the whole body (100 watt corresponds to a daily consumption of 8.64 million joules, or, in other units, around 2 100 kilocalories). (Section 3.3.2, 3.3.4, 4.1.2, **4.1.3**; Fig. 3-6, **4-3, 4-4**)

Central nervous system: part of nervous system which is condensed and centrally located: the *brain* and spinal cord. (Section 4.1.3; Fig. **4-1**)

Cerebellum: (Latin diminutive form of *cerebrum*) belongs to the oldest part of the *brain* and is larger in man than in other animals. The cerebellum lies at the bottom back of the brain (cerebrum). From outside, it resembles two balls of wool a few centimetres in diameter. Has a volume of 1/10 of the brain, however, includes almost half of all brain's neurons, that is approx. 30 billion ($30 \cdot 10^9$) neurons. The cerebellum seems to be responsible for 'automatic', actions, especially for co-ordination of the movements of hands, legs, etc. When one is learning a new skill, such as cycling, or the sophisticated movements of the trained typist or musician, one must initially think through each action in detail, and control is performed by the cortex. (Section 4.1.6; Fig. **4-3**)

Cerebral cortex: (Latin *cerebrum* brain) the part of folded neural tissue that envelopes the brain**,** see *Cortex.* (Section 4.1.4; Fig. 4-3)

Chess player, artificial: see *Artificial instinct.* (Section 9.3.3)

Cognitive operations: (Latin *cognoscere* to become acquainted with) the main objects of linguistic, cognitive psychology, and 'artificial intelligence', mostly without paying much attention to the nervous mechanisms. (Section 5.4.1)

Common sense: good judgement or prudence in estimating or managing affairs, especially when free from emotional bias and intellectual subtlety or special or technical knowledge. (Section 7.1.1; Fig. **7-1**)

Computer: the present fastest digital computers are capable of around teraflops (10^{12} = tera floating point operations per second); the brain of the common housefly, for example, performs about 0.1 tera operations per second when merely resting. Events in an electronic computer happen in the nanosecond (one billionth) range, whereas events in neurons happen in the millisecond (one thousandth of second) range. (Section 9.1.2, 9.1.3., 9.2; Tab. 9-2, 9-4; Fig. 9-1)

Concept: (Latin *conceptus* collection, gathering) mental object resultant of a generalizing (abstracting) mental operation; an idea comprehending the essential attributes of a class of objects. (Section 6.6.1, Fig. **6-11**)

Consciousness: (Latin *conscientia* knowledge) totality of sensations, perceptions, *concepts*, ideas, *thoughts*, and feelings (*emotions*); internal awareness of internal processes and external events. (John Locke: "the perception of what passes in a man's own mind"). (Section 8.1.2; Fig. 8-1, 8-2, 8-3)

Cortex: (Latin *cortex* bark) cerebral cortex in the human *brain*, has a surface area of 220,000 sq. mm, a thickness of around 1.5 mm, and a volume of 300,000 cu. mm that is 0.3 litre. The mass is 0.3 kg, and it is the largest organ in the human brain. The human cortex is three times thicker than that of the mouse. The area of the cortex of the chimpanzee and gorilla is about 50,000 sq. mm. The cortex areas which directly send executive signals to muscles are called 'primary motor areas'. More complex signals are prepared at a 'higher level', in the secondary motor area. The 'highest level' of processing is the tertiary motor area. (Section 4.1.4; Fig 4-3, **4-4**)

Dendrite: (Greek *dendron*, tree) protruding elements of the *neuron* resembling the branching of a leafless tree. They are receiving zones for information (*nerve signals*) from other neurons. A large number of appendages, about 0.1 micrometers long, about 1 micrometre in diameter, cover the dendrites. (Section 4.2.3; Fig. **4-7**)

Detectors: see *Sensors*. (Fig. **3-3**)

Discent: (Latin *discere* learning) The third stage of biotic information processing is called 'discent, (higher than *instinct* and lower than *intelligence*) The term 'discent' is not used in English or in any other language: it is proposed here. Discent occurred some 150 megayears ago and is associated with birds and, later, mammals. (Section 3.4.3; Tab. **1-6**, Fig. **1-5**, 6-1, 6-2)

DNA: Deoxyribonucleic Acid, a macromolecule being carrier of *genetic* information (see *Genome*). The characteristic structure of DNA, made of two intertwined, complementary long strands, is the so-called 'double helix'. The stability of DNA is very strong, because of the self-repair-mechanism of DNA which is enormously effective. The spontaneous decay of DNA is likely to be a major factor in mutagenesis, carcinogenesis and ageing. Some DNA sequences (*genes*) are responsible for 'genetic diseases'. (Tab. **4-2, 4-3**)

Effectors: (motors) organs specialized for transformation of chemical energy of living beings into mechanical movement (muscles) or tissue specialized to discharge chemical active substances (hormones) see also *detectors*. (Section 3.3.1; Fig. **3-3**)

Electroencephalography: in short **EEG**. The recording of brain electrical activity (electrical waves) by means of electrodes attached to the skull. (Section 4.4.4)

Elementary forces: the 'simplest' forces (interactions) which explain almost all natural and all technical phenomena: 1) gravitational force 2) electromagnetic force 3) weak nuclear force 4)strong nuclear force. (Section 2.2.2; Tab. **2-1**; Fig .**2-1**)

Elementary particles: the 'simplest' particles, which explain almost all natural and all technical phenomena: 1) leptons, a family of 6 species, including electron and neutrino 2) quarks, being constituents of proton and neutron. (Section 2.2.1; **Tab 2-2**, 2-3 Fig. 2-3)

Emotions: There is no universally agreed upon definition or theory of emotion that delimits emotional phenomena in a way that is useful for relating those phenomena to neural systems. In humans, emotions cause changes in heart rate, breathing, and stomach and intestine contractions. However, the dominant role of the brain in generating emotions both consciously and unconsciously, cannot be questioned. (Section 5.3.2, 5.3.3, 6.3.1)

Epigenetic: (Greek *epi* besides, outside; *genesis* birth) epigenetic mechanisms are the properties not strictly or directly determined by *genetic* factors, and are acquired in the course of development of individuals, and depend upon experience within the environment, the body itself, and the current state. Epigenetic mechanisms cannot produce new neurons, or even new synapses, but they can very effectively kill (that is, cause the death of) excessive *neurons* and *synapses*, and therefore influence the ultimate state of neuronal *networks*, modules, etc.

Evolution: all *phylogenetic* changes in organisms from generation to generation, carried by *genetic* mechanisms, *DNA*. (Chapter 3; Section 11.1.3; Tab. 3-1; Fig **3-2**; 3-4, 11-1)

Extraterrestrials: hypothetical intelligent beings on distant extrasolar planets, having a developed technical civilization with an ability to get in contact with us. (Section 10.1; 10.2.3; Tab. 10-1, 10-2, 10-3)

Free will: feeling that one is free to make personal choices. (Section 8.3.2, **8.3.3**, 8.3.4)

Ganglion: a group or concentration of neuron cell bodies. (Section **3.2.2**, 3.2.3)

Gene: (Greek *genos* descent, kin, sex) a macromolecular unit of hereditary material included in *DNA*. On average consists of some thousand base units, and controls a synthesis of mostly one protein macromolecule. (Section 3.1.2, **4,6**)

Genetics: (preserved from one generation to another) directly controlled mechanisms, which govern the proliferation, categorization, migration to the ultimate location, and formation of widespread and enormously excessive connections (axons, dendrites and synapses). Genetic mechanism are responsible only for a limited number of instructions. Genetics gives the general rules, chance results in a spectrum of possibilities. (Section **4.6.1**, 4.6.2, 4.6.3)

Genius: extraordinary native intellectual power, manifested in unusual creative capacity especially in the field of mental activities. (Section 7.2.4; **Fig. 7-3**)

Giga: = 10^9 = 1,000,000,000   (thousand million = billion).

Genome: total sum of the genes of an organism (included in *DNA*) The human genome, with about 3 giga base pairs (from which about 97 per cent is not directly carrying genetic information) is responsible for the human body, with some 10 thousand giga cells, including 100 giga neurons in the central nervous system. 80,000 genes are included in the human genome. That means that the human genome provides the blueprint for about 80,000 different proteins.

Genus: (Greek *genos* stock, kind) a group of related, similar *species*. Our own genus, *Homo*, branched off *australopithecine* stock the last 2 megayears. Brain size increased much more rapidly in Homo than in Australopithecus, going during 2 megayears from an initial 0.7 kg to its present average of 1.4 kg. The genus *Homo* includes some species, such as *Homo erectus, Homo habilis, Homo sapiens neanderthalensis, Homo sapiens sapiens*.

Glial cells: (nerve glue), neurons make up only 15 per cent of brain cells; the other 85 per cent are glial cells, but only 50% of mass, which were once thought to be little more than passive support elements. They hold the brain's neurons together, and nourish (by transporting nutrients from the blood and preparing glucose for the neurons), protect, and listen to neurons. They do not send out the long projections called axons that terminate in synapses on neurons or other cells. (Section **4.2.4**)

Gödel's incompleteness: mathematical and logical evidence, that human intelligence is not able to proof all mathematical and logical statements. (Section 7.4.2; Fig **7-7**)

Grey matter: the collection of neurons also in the cerebral cortex that constitutes the outer layer of the *cerebrum* and is responsible for integrating sensory impulses and for higher *intellectual* functions. (Section 4.2)

Gyr: gigayears (1,000,000,000 years).

Hedonic: (Greek *hedonikos* pleasure) relating to pleasure. *see Bonus-malus system*

Hemisphere: the *brain* is macroscopically and microscopically asymmetric. The left-right differences between the brain's hemispheres are measurable, and are related to differences in the higher activity of the brain. (Section 4.1.3; Fig. **4-3, 4-4**)

Hertz: cycles per second (abbreviated as Hz).

Heuristics: (Greek *heuriskein* to discover) general 'rule of thumbs' providing aid or direction in the solution of a problem, especially complex, unclear problems. Heuristic rules originate from previous experience, own or other member of society, as opposed to *algorithmics*. (Section 6.6.2)

Hetero- (Greek *heis* other, different) a prefix used to mean that two or more different conditions or properties are involved

Hippocampus: (Greek *hippokampos* sea horse) small region in the brain, with the mass of some grams. (Section 4.1.5; Fig. 4-3)

Hominids: family including fossil and recent genus of Homo. (Fig. 3-9)

Hominoidae: superfamily including family of great apes and family of hominids.

Homo: genus including species (hominidae) such as *Homo habilis, Homo erectus, Homo sapiens neanderthalensis, Homo sapiens sapiens.* (Section 3.5.2; Fig. **3-9**)

Homo erectus: (Latin *erectus* upright) species of *genus Homo* had a bigger brain and body than *Homo habilis.* He lived about 1.9 million–200 thousand years ago. Adults grew to 1.8 m and were at least as heavy as ourselves. Brain mass averaged 1 kg, more than that of species *Homo habilis*, though less than *modern man's*. Groups evolving in Africa probably spread to Europe, East Asia and South-east-Asia. Improved technology included standard toolkits, big- game hunting, the use of fire, and improved building methods, enabling this species to invade new habitats and climates. He killed and ate large animals, including boar, bison, deer and rhinoceros and used wooden spears with stone projectile points. (Section 3.5.5; Tab. **3-4**; Fig. **3-9**)

Homo habilis: (Latin *habilis* handy) species of genus stood no more than 1.5 m tall, with a weight of about 50 kg. His *brain* had a mass of around 0.7 kg, still only half the size of ours and his species lived about 2-1.5 megayears ago. *artifacts* found near its bones suggest that Homo habilis made basic stone tools, built simple shelters, gathered plant foods, scavenged big meaty limbs from the carcasses of animals killed by carnivores, and hunted small game. (Section 3.5.5; Tab. **3-4**; Fig. **3-9**)

Homo sapiens: (Latin *sapiens* wise) a species including two sub-species: *Homo sapiens neanderthalensis* and *Homo sapiens sapiens*. (Section 3.5.5; **Fig. 3-9**)

Homo sapiens neanderthalensis: sub-species of *species Homo sapiens*. He had a surprisingly large brain of about 1.5 kg (*modern man* has a brain of 1.4 kg). However, existence of fully-developed *lateralization* within his brain, the different features and architecture of his acoustic equipment and the different functions of the left and right *hemispheres* of his brain are still open questions. The Neanderthals were nearing their apogee 100,000 to 80,000 years ago. Their improved technology, especially fire-heated shelters, construction and clothing manufacture enabled them to endure the rigors of winter in the cold climate of the last glaciation period, 70,000-30,000 years ago. The last traces have been found around 40,000 years ago. (Section 3.5.6; Tab. **3-4**; Fig. 3-9)

Homo sapiens sapiens: species of *genus Homo*, the *modern man*. (Older examples: Cro-Magnon man) Is a sub-species of *Homo sapiens*, originated between 200,000 and 120,000 years ago in Africa and migrated to Europe and Asia by way of the Levant. If, as many molecular biologists believe, modern Homo sapiens arose as a discrete evolutionary event some 120-150 thousand years ago (the so-called hypothesis of 'mitochondrial Eve', the mother of us all, who lived in Africa), common origin of all human language is likely. He made extensive use of antlers, bone and ivory, not just for tools, but also for ornaments and objects of art. Some featured animals and other forms, mostly poorly drawn, on small, portable objects, and are 32,000 years old. Fertility symbolism could account for human figurines with exaggerated female features. Graves contained refined goods, such as necklaces. (Section **3.5.7**; Tab. **3-4**; Fig. **3-9**)

Illusion: (Latin *illusio* action of mocking) is an *unconscious* self-deception (mostly visual and acoustic) aimed to improve the momentary state of own *bonus-malus* level. (Section **5.4.3**; 8.2.3; Fig. **8-5**)

Information: (Latin *informare* to form) something acquired, received or transmitted including knowledge of external objects and events and their properties, or knowledge of internal state and emotion of intelligent being. Information can be measured by quantity and quality. (Section **1.2**; Fig. 1-1, **1-2, 1-3**; Tab. 1-3,1-4, **1-5**)

Information processing: storage, retrieval, creation, forgetting, or transformation of *information*. (Section **1.3**; Tab. 1-3, Fig. 1-2, **1-3**, 1-5, 3-7)

Instinct: (Latin *instinctus* drive, impuls) lower stage of biotic information processing (the *reflex* is the lowest stage and the *discent* is the higher stage). Is a complex and specific response on the part of an organism to environmental stimuli, that is, largely hereditary and unalterable. An instinct could be generated either when the animal (typically: worms, insects, spiders) becomes so congenitally predisposed to a stimulus that response cannot be improved through learning. Five hundreds *megayears* ago, the crustaceans emerge and, later, the insects. Some researchers feel that human behaviours such as aggression and territoriality may have instinctive components. (Section **3.4.2**; Tab. 1-6, Fig. 1-5)

Intellect: (Latin *intellectus* perception) higher cognitive activity of intelligent being, mostly free from emotions and prejudices, self-critical, see *Intelligence*.

Intelligence: (Latin *inter* among + *legere* to gather, choose) means the ability to *reason*, perceive, or understand; the ability to perceive relations, differences, etc., distinguished from *'will'*, and *'feeling'*. The term 'intelligence' to describe the whole system, including the stage *'intelligence'*, and the lower stages of biotic information processing: *reflex, instinct* and *discent*. The specific property of intelligence is creation of new, *semantically* significant, *information*, concerning the real world and last but not least theory of the intelligence itself . (Section **3.4.4**, 6.2; Tab. **1-6**, Fig. **1-5, 6-1**, 6-2, 6-3)

IQ: (Intelligence quotient): technique of measurement of human *intelligence* in relation to the average 'intelligence' at an appropriate age; partially misused for practical purposes. (Section 7.2.1, Fig. **7-3**)

Interneurons: neurons that form the connection between neurons that collect information from the *senses* and those that send signals to the *motors*. see *Neurons* (Section 4.2.2, 4.2.3; Fig. **4-6**)

Intuition: (Latin *intueri* to contemplate) act or process of coming to direct knowledge or certainty without reasoning, often without conscious knowledge of the reasoning chain, resulting from previous mental information processing. (Section **8.1.4**; Fig. 8-2)

Kyr: kiloyears (1 000 years).

Lateralization: (Latin *later* side) differentiation between left and right hemispheres. Modern *Homo sapiens sapiens* is a product of the lateralization of the brain's hemisphere. There are some demonstrations of the existence of behavioural asymmetry, in the form of right-handedness before birth. The left tertiary motor area has taken over the function of the speech production area the so-called *'Broca's area'*, and the left tertiary sensoric area, the function of speech reception and understanding, the so-called *'Wernicke's area'*. (Section 5.5.3; Fig. **5-10, 5-11**, Tab. **5-4**, 5-5)

Learning: process of acquisition and extinction of modification in existing knowledge, skills, habits. It is essential, to have mental concentration, with a planned strategy of action and subsequent evaluation and correction of errors in successive attempts. When well learned skills can be accomplished without voluntary attention, as in walking and swimming, cycling, skiing, and other similar activities where the performance of the skilled action has become automatic. The list of learned motor skills is enormous, and includes almost all kinds of sports, as well as artistic activities as ballet and music and, last but not least, the driving of automobiles. Cerebral *cortex* is the main organ involved in human learning, especially in the first phases, encompassing motivation, attention, evaluation, and correction. The performance of automatic motor activities is controlled mainly by the *cerebellum*, and the basal ganglia. Both belong to the oldest components of the human brain. Learning changes the pathways for performing the task of remembering. (Section **6.4**; Tab. **6-2, 6-3**; Fig. **6-4**)

Malus-bonus system: see *Bonus-malus system*

Mapping of brain: a series of computerised maps of the brain's anatomy, function, physiology, biochemistry, and molecular biology. The macro-architecture of the brain contains more than hundred different structures, such as the thalamus and amygdala. The mapping is realized by means of new imaging techniques such as positron emission tomography (*PET*), magnetic resonance imaging (*MRI*), computerised X-ray tomography and the magnetoencephalography, measuring the magnetic currents that are generated by electrical impulses in the brain. (Section **4.4.2**; Fig. **4-12**, 5-1)

Mega = 10^6 = 1,000,000 (one million).

Memory: storage of information. The short-time memory acting over tens of second. The working memory has been studied by *positron emission tomography* (PET). These results begin to uncover the circuitry of working memory system in humans. The long-term changes require something entirely new: the activation of genes, the expression of new proteins and the growth of new connections. There are other possibilities, based upon psychology, of classifying human memory. (Section **5.2**; Tab. **5-1, 5-2**; Fig. 5-3, 5-4)

Mental processes: relating intellectual and emotional activity.

Microtubule: very small tubes, with a diameter of 20 nanometres ($20 \cdot 10^{-9}$ m) and a length of 20 micrometre, consisting of the protein tubuline being inside the neurons. (Section 4.2.6; Fig. **4-8**)

Mind: another term for describing the totality of all information processing going on in a human intelligent being. (Section **11.6.3**, Fig. **8-6**)

Mitochondrial Eve: (Greek *mitos* thread + *chondrion* small grain) a hypothetical 'mother' of all *Homo sapiens sapiens* around 150-200 thousand years ago. This is a name for a small population of our species and not for one woman. (Section **6.5.3**)

Modern man: see *Homo sapiens sapiens*

Motors: see *Effectors*

MRI: (magnetic resonance imaging) technique for non-invasive measurement of structure and some activities of human brain. (Fig. **4-12**)

Myr. Megayears (1,000,000 years).

Natural selection: The process in the biosphere that brings about the evolution of new *species*, based on the so-called 'survival', the adaptation of the *genetic* information carrier, the *genes* (that is the genotype), to the environment, the 'niche'. (Fig. **3-10**)

Neanderthals: (German: name of river in Germany) see *Homo sapiens neanderthalensis*. (Section **3.5.6**; Fig. **3-9**)

Neocortex: see *Cortex* (Section **4.1.4**; Fig. **4-4**)

Neoteny: (Greek *teinein* to stretch) retention of juvenile characteristics produced by retardation of body development. Neoteny is a phenomenon among some salamanders, in which larvae of a large size, while still retaining gills and other larval features, become sexually mature, mate, and produce fertile eggs. Adult humans share numerous features found in baby chimpanzees, but which are lost as these grow up. Like humans, baby chimpanzees have a sparse covering of body hair, and a relatively large brain shielded by a bulbous cranium. Like ours, their skull bones are thin. (Section **3.5.4**)

Nerve: a strand of tissue comprising many nerve fibres (*axons*) plus supporting tissues.

Nerve signal: electric current in the *axons* propagating with velocity up to 100 meter per second. (Section 4.2.3)

Neural network: a *computational* device made of units that are like very oversimplified *neurons*. (Section 9.2.2)

Neuron: (Greek *neuron*; Latin *nervus* string) fundamental cellular units of the nervous system for transmission and processing of information. They transmit some thousand bits per second. The number of neurons in the human brain: 100 giga. Neurons are 'immortal', they rest alive during the whole existence of the human body. *Cerebral cortex* thus contains around 30 *giga* neurons. In cerebellum occur also 30 billion ($30 \cdot 10^9$) granule cells, 30 *million* Purkinje cells, and many others. (For comparison, the Milky Way consists of 100 billion stars, and the present number of human population on this planet is around 6 billion). Almost all neurons in the *cerebral cortex* are so-called 'interneurons'; that is, they connect one part of the cortex to another. Only 0.5 million neurons are '*sensory* neurons'—coming from sensors—and only 2-3 million are '*motor* neurons'—sending nerve signals to the muscles. Neurons are primarily electrochemical devices. They perform their work of receiving and transmitting *nerve signals* by means of electro-chemical energy. The energy is generated by the flow of charged chemical entities—ions—through the neuron's membrane. Neurons include *dendrites*, *axons* with *synapses*. (Section 4.2.2 , **4.2.3**; Fig. **4-5, 4-7**)

Neurotransmitter: substances (mostly small and rather simple chemical molecules) which are produced directly in neurons—more precisely in *synapses*, in the pre-synaptic part. There are other substances which have a direct effect on the activity of synapses, but are produced outside neurons, for example, some hormones. The neurotransmitters are packed in so-called synaptic vesicles (small number of around 100 small and rather simple molecules, built of hydrogen, oxygen, carbon, nitrogen, and, in some cases, sulphur. (Section 4.5.2; Table **4-1**)

Ontogeny: the life history (development from birth to death) of an individual, as opposed to *phylogeny*.

Percept: mental object, a product of the process of transformation of incoming nerve signals to the primary sensory area of brain. (Section 6.6.1; Fig. **6-10**)

Personality: the characteristic way in which a particular individual thinks, feels, and behaves. Personality embraces a person's moods, attitudes, and opinions and is most clearly expressed in interactions with other people. Personality contains those behavioural characteristics, both inherent and acquired, that distinguish each individual and is observable in the individual's relations with the environment and the social group. (Section 7.3.1; Tab. 6-7, 7-1, 7-5; Fig. 7-4)

PET (positron emission tomography): a technique for studying activity in the living brain by the use of radioactive short-lived substances emitting positrons (positively charged electrons). (Section 4.4.2; Fig. **4-12**)

Peta: $= 10^{15} = 1,000,000,000,000,000$ (one thousand million million).

Phylogeny: The evolutionary history of a group of organisms from 'lower' and less 'complex' species to 'higher' and 'more complex' species, as opposed to *ontogeny*.

Placebo: an inert substitute which in some cases effects the state of a patient. (Section 7.3.4; Fig. 7-6)

Protein: a macromolecule made by association of a large number (hundreds to thousands) of small molecules: amino acids. Synthesis of proteins is controlled by appropriate *genes*.

Psychoactive agent: includes following substances: drugs (synthetic, semi-synthetic, and vegetable natural substances) that affect the mind, and is used instead of: psychedelics, narcotics, *drugs*, also pharmacologically active drugs affecting the 'mind': sedatives, antidepressants, stimulants, neuroleptics, opiates, aphrodisiacs. Psychoactive agents include tobacco, coffee, (mental endurance), alcohol, and also myrrh, incense (used in most religious ceremonies). Neurotransmitters and other body own substances do not belong to psychoactive agents (endorphin = *endo*genous m*orphine*). (Section 5.3.3; Tab. **7-7**, 7-8; Fig. **5-6**)

Puzzle: a question or problem which is clearly defined, without any uncertainty, have assured solutions. (Section 9.2.3 Tab. 9-3)

Quantum mechanics: a branch of physics accurately describing the properties and transformation of some *elementary particles* and some *elementary forces*. Its basic ideas are not conform to everyday common sense.

Reason: reason is in opposition to sensation, perception, feeling, desire, as the faculty by which fundamental truths are intuitively apprehended. (Section 6.1.1)

Reflex: reflex (Latin *reflexus* reflected) is an involuntary action, when a stimulus is carried by an input nerve to a nerve centre, and the response is reflected along an output nerve to a muscle or gland. Reflex stage was reached around 700-800 megayears ago, when large multicellular marine organisms, with soft tissue, and without hard exoskeletons, emerged in the ocean. (Section 3.4.1; Tab. 1-6, Fig. 1-5)

Representation: reflection in human mind of the real world, see *Concepts* and *Percepts*. (Tab. **6-8**)

Glossary, index   413

Robot: (Czech *robota* forced labour) any automatically operated machine that replaces human effort, though it may not resemble human beings in appearance or perform functions in a humanlike manner. see *Artificial reflex*. (Section 9.3.1)

Schizophrenia: (Greek *schizein* split + *phren* mind) psychiatric disease, symptoms include: paranoia, delusions, social withdrawal, auditory and visual hallucinations, and disorganized thoughts—often does not surface until adulthood. In some cases is *genetically* influenced. (Section 7.2.7)

Self-consciousness: the quality or state of being self-conscious; self-conscious: aware of oneself as an individual that experiences, desires, thinks, and knows himself as thinking. (Section 8.1.2)

Semantics: (Greek *semaino* to mean, to signify) Is the philosophical and scientific study of meaning. (Section 1.3.4; 9.2.5)

Sensors: biotic specialized organs (senses) sensitive to incoming *signals*: optic (eye) including writings, acoustic (ear) including *speech* and music, volatile chemical (smell), soluble chemical (taste), heat or cold, mechanical contact; see also as opposed *effectors*. (Section 3.3.1; Fig 3-3)

Sexual selection: the choice of a mate on a basis of various attractive characters: as colour of skin, bird song, appearance of woman's face or breasts. (Fig. **3-10**)

Signal: material carrier of different physical properties which could be interpreted by special systems (*sensor*) and be transformed (processed) into an information. (Section 1.2.4)

Sleep: is characterized by synchronized events of billion of synaptically coupled neurons in thalamo-cortical systems. The rapid patterns characteristic of the aroused state are replaced by the low-frequency, synchronized rhythms of neuronal activity. Sleep oscillations are highly orchestrated and highly regulated. Coherent activity at a variety of frequencies, ranging from 1 Hz to over 40 Hz, but the extent of its spatial coherence is also quite variable. Human memory consolidation, active during sleep, is strongly dependent on REM (rapid eye movement) sleep. (Section 5.3.4)

Species: (Latin s*pecies* kind) a basic lower unit of classification of living beings, consisting of a population of closely related and similar individuals, capable of interbreeding freely each other but not with member of other species.

Speech: Speech sounds involve complex waves containing vibrations at a number of different frequencies, the lowest being the voice pitch of singing and intonation, produced by the vocal cords in voiced sounds. Every different configuration and movement of the vocal tract creates corresponding differences in the air vibrations that comprise and transmit sound. (Section 6.5.3; Tab **6-4**, 6-5, 6-6; Fig. **6-6**, 6-7, **6-8, 6-11**)

Supercomputer: Computers that have attained the maximum performance with the most advanced hardware and software technology available are called supercomputers. It consists of up to about a dozen Central Processing Units and is used for high-speed computation of large-scale scientific or engineering problems. (Section 9.1.3)

Superintelligence: see *Intelligence,* (Section 11.5.3; Tab. 11-4, Fig. 11-4)

Synapse: (Greek *synapsis* union, junction) A small cleft separates the nerve terminal from the other side, the postsynaptic membrane, belonging to another *neuron* and lying on a *dendrite* or neuronal cell body. The synaptic cleft is only a few millionths of a millimetre wide, and the whole synapse is just a few thousandths of a millimetre in size. The average number of synapses in the human *cortex* is around 6,000 per neuron, rising in some cases probably even to 20,000. The number of synapses in the cerebral cortex is something like 600 million per cubic millimetre, giving somewhat less than 1 thousand of billions (10^{15}) synapses in the human cortex. Information on synapses, in form of *neurotransmitters*, is transmitted in one direction only—from the *axon* terminals of the sending *neuron* (presynaptic neuron), across the synapse cleft, to the receptive surface (postsynaptic membrane) of the receiving, or postsynaptic neuron. (Section 4.2.5; Fig. **4-7**)

Synchronization: (Greek *syn* together) occurrence at the same time. Neurons in the visual cortex, activated by the same object in the world, tend to discharge rhythmically and in unison. The synchronized oscillations occur over less than half a second, and have a frequency of 40 *hertz*. These oscillations can be evoked in thin slices of *brain* that were kept alive in a nutrient bath. Similar oscillations have been seen in the olfactory cortex, which is involved in discrimination between different odours. (Section 5.4.5; Fig. 5-8)

Taboo: (Tongan *tabu*) prohibition instituted for the protection of a cultural group. All societies have a concept of incest, and all societies have a prohibition, or taboo, against it. Definitions of incest vary according to definitions of who is close kin. Reactions to violations of the taboo also vary from society to society. (Section 7.3.4; Fig. 7-6)

Tera: = 10^{12} = 1,000,000,000,000 (one million million).

Thalamus: (Greek *thalamos* inner room) is a large grouping of nuclei situated just anterior to the midbrain, and consists of two small ovoids. The thalamus is the final relay station for the major sensory systems that project to the cerebral *cortex*, and is called the gateway to the cortex, because the main inputs from all *sense*s (excepting smell) to the cortex have to pass through it. Each thalamic area (about two dozen regions) also receives massive connections from the cortical areas to which it sends information. Probably a specific thalamic nucleus exists for pain. (Section 4.1.4; Fig. 4-3)

Thinking: the highest cognitive processes occurring in human brain, based on numerous parallel going elementary processes in *neurons*, in all *brain* organs. Thinking is *conscious* or *unconscious*. (Section 7.2.3)

Unconsciousness: as opposed to *consciousness*. (Section 8.2.1)

Unification theories: present efforts of theoretical physicists to formulate one, only one, theory describing all natural and artificial phenomena. (Fig. **1-2, 11-5**)

Virtual reality: a form of computer simulation in which the user has the impression of being in an 'other' environment and action. (Section 9.2.4)

White matter: beneath the cerebral cortex is a mass of white matter, which is composed of nerve fibres projecting to and from the cerebral cortex. Commissural systems connect the two hemispheres via the corpus callosum and other systems. Association fibres connect different regions of a single hemisphere.

Wisdom: see *Intelligence*. (Section 7.1.4)

INDEX

Gls = Glossary

A

Abnormal movement pattern, chorea 272
Abstract 237
Abstraction 228
Accept 237
Acheulian culture 88
Action potential, neuron 117
Addiction **Gls** 142; 174
Adolphs R 106; 172; 279
Aesthetics 296
Ahissar E 184
AI, see artificial intelligence 353
Algorithm **Gls** 220
Algorithmics **Gls** 239
Alien
    brain of 364
    intelligence 364
    language 368
    material carrier 364
Altruism 328; 329; 332
    hominids 88
Alzheimer's disease 271
American Sign Language (ASL) 244
Amerind people 96
Amnesia 244
Amygdala **Gls** 105
    emotion 214
    bonus-malus system 172
Analogue processing, neurons 113
Anandamide, cacao 292
Andreasen NC 151; 273; 333
Anhedony 21
Animal, consciousness 312
    sign language 226
Anthropic principle 53
    intelligence 25
Anticipation 187
Antonini A 133
Anxiety 214
Aphasia 243
Aphrodisiac, psychoactive agent 174
Apoptosis 126
Appetite 19
Area, motor 109
    sensory 109
Arnett D 42

Artifact **Gls** 336
    beehive 336
    bird nest 336
    general 336
    ideal 289
    ideal, material 77; 207
    stone tools 336
    information 337
Artificial consciousness **Gls** 358
    discent **Gls** 350
    instinct **Gls** 348
Artificial intelligence **Gls** 344; 353
    expert 350
    extraterrestrial 370
    translator 351
    neural network 343
    reflex **Gls** 346
    stupidity 358
Artistic skill, depression 273
Asimov A 347
Association area 109
Astrocyte 121
Atmosphere, distant future 380
Attention 188
    intelligence 186
Attwell D 140
Aurignacian technology 95
Australopithecus 85
    anamensis **Gls** 85
Autism 274
Aversion 19
Axon **Gls** 114
    growth cones 124
    neuron 113
Ayala FJ 95

B

Babbling, deaf children, manual signed 235
Baddeley A 165
Bain A 142; 257
Barlow RB 25
Barondes SH 177
Barrow JD 53
Bar-Yosef O 89
Basal ganglia 103; 105
    damage 272

Beauty, symmetry 296
Bechara A 309
Beckwith SVW 365
Begley S 288
Beloff J 315
Beringia 96
Berk LE 235
Berridge MJ 121
Bible 228
Big Crunch 377
Binding problem 182; 184; 185
Biotic **Gls** 140
Biotic information processing **Gls** 77;140
Birge RR 342
Black DC 367
Blinds, reading 150
Blindsight 180
Bliss TVP 163
Blumenschein RJ 88; 329
Body language 222
Boesch Ch 290
Boesiger E 290; 297; 327
Bohr N (Nobel laureate, 1927, physics) 305; 324
Boltzmann L 65; 102
Bonus-malus **Gls** 19; 21; 310
    biochemical basis 141
    cocaine 142
    computer 358
    deception 21
    drive, feeling, emotion 171
    emotion 23
    free will 256
    intelligence 221
    intuition 172
    learning 220
    morality 327
    psychoactive agents 174
    semantics 21
    system 19;257
Bouchard TJ 279
Bradley D 339
Brain **Gls**
    asymmetry, function 189
    atlas 129
    blood barrier 144
    blood flux 133
    body surface 77
    chemical composition 137
    complexity 74
    computer 338
    damage, abnormality 273
    damage, speech 243
    development 122
    diseases 271
    distant future 373, 376
    domestication 73
    emergence 71
    energy flux 77; 131
    evolution 74; 101
    future 374
    genetics 101; 278
    gene manipulation 150
    lateralization 77
    limit of growth 77
    macrostructure 101
    mapping 157
    monkey 73
    neuron number 114
    phylogeny 102
    physics 128
    plasticity 74; 150
    rate of processing 339
    software, hardware 102
    sex difference 194
    size 72
    stem, Pakinson's disease 272
Bridgeman B 326
Brillouin LN 17; 256
Broca's aphasia 231
    area 110
    automatic speech 186
Brooks RA 355
Bystron JS 289

C

Calcium ion 121
Callaway E 135
Calvin WH 216; 226
Cann RL 95
Cantor G 242
Caramazza A 282
Carbon, origin 43
Card play, deception, deception 296
Catastrophe
    biosphere 384
    natural 381
Cavalli-Sforza LL 22; 227; 233; 265
Cavallo JA 88; 329
Cave paintings 96
    psychedelic drugs 290
Celibacy 280
Central nervous system **Gls** 366
Cephalization 71; 73
Cerebellum **Gls** 105; 107
    automatic activity 219

automatic memory 168
  coordination of movement 107
Cerebral cortex **Gls** 73
Chagall M 283
Chalmers DJ 12; 305; 311; 401
Chandrasekhar S (Nobel
  laureate,1983,physics) 297
Changeux J-P 20; 141; 142; 237; 329
Chaos 12
  epilepsy 136
Cheating 287
Chess 296
  computer 348
  'Deep Blue' computer 348
Children, lie 235
  questioning 242
Chimpanzee
  language ability 226
  material culture 337
Chomsky N 91; 191; 216; 297; 231; 232; 233;323;368
Chromosome
  asymmetry Y, X 189
  number 85
Chunk
  memory unit 165
  of time 156
Churchland PM 224; 304; 315; 356
Churchland PS 262; 337; 354
Ciba Found.Symp. 265
Clairvoyance 315
Cloning 393
Coca-Cola, cocaine 291
Coffee 174
Cognitive operation **Gls** 150
Cohen P 285
Collingridge GL 163
Comet
  catastrophic collisions 45; 383
  encounter 382
Commissurotomy
  personality 277
Common sense **Gls**
  computer 341
  definition 254
  general 205
  heuristics 255
  robot 351
  science as uncommon sense 262
Communication, evolution of 222
Complexity, brain 101
  definition 25
  human brain 74

information 385
Comprehensibility 3
Compression, information 9
Computation, definition 338
Computer
  bonus-malus 358
  brain 338
  common sense 341
  chess player 348
  development of 341
  energetics 339
  faultlessness 342
  fuzzy, connenctial 343
  learning ability 343
  network 353
  neural 342
  parallel processing 342
  programs 342
  reflex, instinct, discent, intelligence 341
  syntax 345
  virus 343
Concept **Gls** 237
  abstract 296
  imaging 186
Connes A 20
Connes P 401
Conscience 329
Consciousness **Gls** 304
  animal 312
  artificial **Gls** 358
  asymmetry 307
  definition 303
  detector 358
  gravitation and quantum mechanics 307
  intelligence 209
  microtubule 119;339
  neural basis 307
  neural correlate of 306
  nonconsciousness 314
  own death 332
  primary phenomenon 311
  supercomputer 395
  unconsciousness 314
  working memory 163
Cook ND 192
Copernicus M 297
Corpus callosum 103; 105
Corpus striatum 105
Cortex **Gls**
  structure 101
  volume 103
Cosmides L 287
Crane T 315

Creativity, schizophrenia 274
Creole, sub-language 224
Creutzfeldt O 224; 324
Crick F (Nobel laureate, 1968, medicine) 20;
 156; 178; 180; 184; 229; 240; 306; 307;
 316; 317; 324312
Crime 329
Criminal behaviour 331
Cro-Magnon man 95
Cromer A 262
Crosswords 296
Crypto-chemo-cracy 387
Culture
 intelligence 289
 psychoactive agents 177
Culture 289

D

Damasio AR 172; 212; 228
Damasio H 186; 228; 282
Darwin Ch 62; 93; 139; 328
Davidson C 351
Davies P 53; 261
Dawkins R, selfish gene 328
Death, selfcognition 331
Deception 229; 287
 bonus-malus system 21
 card 296
 intelligence 26
 self 287; 332
 self, free will 323
deCharms RC 184
Delighted-Terrible Scale 260
Dementia 271
Demosthenes 287
Dendrite, neuron Gls 114
Dennett DC 224
Deoxyribonucleic acid 9
Depression, artistic skills 273
Descartes 281
Desimone R 187
Detector, sensor Gls 13
Determinism neurogenetic 279
Devlin K 8
Dewdney AK 356
Diamond J 267
Diener E 221
Dill M 178
Dirac PAM (Nobel laureate ,1933, physics)
 298
Discent Gls 78
 artificial 346

birds 77
definition 80;210
evolution 101
genetics 147
learning 80
mammals 77; 81
occurrence 80
DNA Gls 393
 bacteria 147
 chemical stability 147
 human 145
 information processing 15
 mitochondrial 89
 worms 145
Dobzhansky T 290; 297
Dopamine, neurotransmitter 311
Doubfulness, intelligence 394
Douglas RJ 111
Down syndrom 271
Drawing 248
Dream
 meaning 317
 subconsciousness 316
 theory of 315
Drevets WC 131; 187
Drugs 140
 alcohol 145
 antidepressant 175
 cocaine 142; 175
 emotions 169
 morphine 140
 nicotine 175
 opiate 175
 psychedelic 174
 psychoactive agents 175
 stimulant 175
 tranquilizer 175
Dual symmetry 35
Dudai Y 163
Duve de Ch (Nobel laureate, 1974, medicine)
 62;66
Dyson FJ 262; 377; 398
Dyson's sphere 365
Dysphasia 244
Dysphoria 173

E

Earth
 distant future 380; 382
 future 381
 Jupiter influence 382
 -Moon system 380

origin 45
-Sun system 380
uniqueness 45
Eccles JC (Nobel laureate, 1963, medicine) 305
Ecstasy 292
Edelman GM (Nobel laureate, 1972, physiology) 5; 11; 126; 156; 160; 222; 232; 239; 282; 302
Edwards FA 140
EEG
   high-resolution 135
   electroencephalography 135
Effector **Gls**
Einstein A (Nobel laureate,1921, physics) 3; 7; 17; 33; 53; 220; 282; 298; 324
Electroencephalography **Gls**
Elementary forces **Gls** 4
Elementary particles **Gls** 4
Elimination, selective 18
Emotion **Gls** 212; 214; 310
   amygdala 105; 214
   bonus-malus 23
   information processing 169
   thalamus 214
Emotional memory 171
Encephalization, Homo 88
Endocrine system 142
Endorphin 140
Engram 167
Energy flux, brain 131
Engel AK 182
Enkephalin 172
   neurotransmitter 278
   see endorphin 140
Enquist M 296; 297
Epigenetic development **Gls** 128;147
Epilepsy 136; 168
ERP 135
Eternity, concept of 376
Ethics 327
Euphoria 173; 221
Eve mitochondrial 95; 229; 409
Evolution **Gls** 4
Evolvability 3
Excitatory message 124
Expert; artificial discent 350
Extinction, Cretaceous/Tertiary 45
Extrasensory perception, ESP 315
Extraterrestrial **Gls**
   communication 367
   friend-foe 369
   invasion 370

   intelligence 362
Extraterrestrial intelligence
   artificial 370
   material carrier 364
Eyeblink 180
Eysinck H 207

F

Faerman M 331
False intentional information 225
Falsifiability 285
Fantasy 267
Fear 213; 214
Feeling, missing limbs 317
Fermi E (Nobel laureate , 1938, physics) 369
Fertilization, personality emergence 312
Feynman RP 398
Fischbach GD 161
Fluoxetine, see Prozac 177
Fodor JA 101;283
Foelfsema PR 199
Force
   electromagnetic 38; 52
   grand unified 38
   gravitation 38
   nuclear strong 38
   psychic 325
   ultra-weak 37
   unified 38
Forgetfulness 222
   selective 164
Fowler WA (Nobel laureate , 1983, physics) 297
Fox PT 185
Fraudulence 287
Free will **Gls** 280; 306
   bonus-malus 256
   chaos, chance 320
   game 296
   myths 287
   mechanisms 320
   natural sciences 324; 323
   necessity, chance 319
   self-consciousness 323
   spontaneity 161
Freedom 221
Fremlin JH 323
Freud S (Founder of psychoanalysis) 139; 305; 315; 316
Frith U 274
Frontal lobe, pathology 271
Fukugita M 40

G

Galaxy
   explosions 382
   distant future 376; 378
Galea E 350
Ganglion **Gls** 179
Gardner H 207
Garthwaite J 140
Gazzaniga MS 277; 293; 303; 307
Gell-Mann M (Nobel laureate, physics, 1969) 15; 20; 25; 309
Gene **Gls**
   imprinting 149
   jumping 149
   methylation 149
Genetics **Gls** 147
   brain 278
   brain abnormality 272
   information, terrestriality 61
Genius **Gls** 267
Genocide 329
Gnome **Gls**
   junk 146
Genus **Gls** 90
Gibbons A 91
Gibbs WW 171; 199; 217; 250; 330
Gift 205
Gigachips 341
   errors in function 342
Gilbert CD 110
Gimpel J 398
Giurfa M 296
Glassman AH 291
Glia cells **Gls** 113; 116
Gödel K 260; 285
   incompleteness **Gls** 282; 284
Gold PW 144
Goldman D 213; 274
Gopnik M 244
Gottesmann II 273
Gould SJ 63; 85; 207; 263; 389
Grammar, universal 232
Gravitation
   quantum mechanics, consciousness 306
Gray CM 184
Grayson L 288
Grey matter, brain **Gls** 116
Grice P 16
Grosof DH 180
Guilford JP 208
Gundersen HJ 196

H

Haefner K 11; 16
Hallucination 273
Hallucinogens 293
   plants, monkey 171
Hameroff SR 240; 306; 307
Handedness 198
   Cro-magnon man 198
   human foetus 190
Happiness 221; 260
Hardcastle VG 185
Hari R 180
Hauser MD 190
Hawking S 20; 35; 34; 261; 377; 399
Hedonic **Gls**
Heinze HJ 187
Heisenberg W (Nobel laureate, 1932, physics) 20; 297; 305
   unpredictability 286
Hemminger H 330
Hemisphere **Gls**
Herrnstein RJ 264
Heuristics **Gls** 239
   common sense 255
   wisdom 258
Hilbert D 260
Hinton GE 338; 343; 358
Hippocampus **Gls**
   pathology 271
Hirokawa N 119
Hobson JA 317
Hofstadter DR 208; 303
Homicide 329
Hominids **Gls** 84
   chimpanzee and gorilla 85
   sexual division of labor 88
Hominoidae **Gls** 85
Homo **Gls**
   chromosomes 85
   genus 83
Homo erectus **Gls** 83; 88; 217; 389
Homo habilis **Gls** 83; 87; 217
   rudimentary speech 229
Homo humanus 392
Homo sapiens **Gls**
   distant future 373
   genome 145
   mendax 288
   neanderthalensis **Gls** 89; 217; 389
   neoteny 87
   out of Africa 89

Glossary, index   **421**

psychedelic drugs 290
sapiens **Gls** 83; 95; 373
Hopfield JJ 353
Horgan J 25; 299; 329; 398; 399
Hormone, endocrine system 142
Hoyle F 5
Hsu FH 348
Hubel D (Nobel laureate, 1981. Medicine) 179; 180
Humour 287; 289
Hunch 172
Hunger strike 280
Hunter L 350
Huntington's disease 272
Hutchison JB 144
Huxley A 386
Hydrogen, origin 43
Hydrosphere, distant future 380
Hypnosis 281
Hypothalamus 106

I

Idea imaging 186
Idiot savant 268
Illusion visual **Gls** 180
  feeling missing limbs 317
  self-deception 180
Immune system, mind 144
Imprinting 149
Incompleteness
  Gödel K 284
Information **Gls** 7
  artifacts 337
  artificial 10
  artificial 15
  biotic, genetic 10; 15
  carrier DNA 12
  complexity 385
  compression 9
  creation by intelligence 81
  decomposition 180
  definition 12
  direct 9
  emergence 12
  indirect 9
  inputs 71
  meaning 14
  overabundance 18
  processing, carrier 63
  processing, locomotion 75
  processing, neuron 9;111
  processing, purpose 67

quality 16; 19
quantity 16
semantics 345
selective elimination 18
theory 16
vagueness 317
value system 256
Information Age 8; 337
Inhibitory message 124
Inquisitiveness 242
Instinct **Gls** 75; 78
  arthropods 77
  artificial 346
  chess player 348
  evolution 101; 210
  fish 77
  genetics 147
  insects 77
  reptile 77
  scheme 82
Intellect **Gls**
  activity, energetics 134
Intellectual energy, price 355
Intelligence **Gls** 78
  algorithmus 29
  anthropomorphism 29
  animal 215; 225
  animal, Kluge Hans 215
  artificial 353
  consciousness 209
  creator of information 81
  culture 289
  deception 26
  definition 29; 205
  discent, instinct, reflex 205
  distant future 372; 396
  doomsday 376
  doubtfulness 394
  emotion, motivation 213
  evolution 101; 210
  extraterrestrial 5; 362
  faultlessness 342
  game, play 295
  genius 267
  inheritance 278
  language 231
  learning ability 217
  military intelligence 29
  morality 326
  occurrence 83
  quotient, IQ **Gls** 263
  related terms 205
  riddle 296

school 280
space-time perception, representation 155
speech 222
terrestrial 5; 25
theory of 205
universal definition 207
universality 25
weakness, strength 7
Intentional false information 225
Internet 352; 391
Interneurons **Gls** 116
Interplanetary communication 368
Introspection 194
Intuition **Gls** 172; 308; 309
   bonus-malus 172
   working memory 240
Ion channel 121
IQ, and lie, swindle, cheat 265
IQ, Intelligence quotient 263
Israelshwili J 60

J
Jamison KR 267
Japanese writing, Kana, Kanji 247
Jonides J 165

K
Kalin NH 214
Kamin LJ 263; 265
Kanzi, pygmy chimpanzee 226
Karni A 178
Kasparov G 348; 349
Khalfa J 211
Kiedrowski G 65
Kimura D 196; 198
Kinomura S 187
Kinsbourne M 128
Kleinnman A 274
Kluge Hans, animal intelligence 215
Knowledge, of real world 240
Koch Ch 19; 97; 111; 113; 156; 184; 240; 312
Kolb B 266
König M 171
König P 182
Koob GF 142; 291; 294
Korenman SG 175
Kosslyn SM 282
Kramer DA 258

L
Landry DW 295
Language 222
   chimpanzee 225
   deaf-and-dumb 244
   extrasomatic memory 224
   honey-bee 225
   instinct 231
   intelligence 231
Laplace de P-S 389
Lateralization **Gls**
   human cortex 190
   monkeys 190
   Neanderthals 89
   personality 265
   redundancy 77
   sex difference 198
Laugher 289
Laws of Nature 319; 393
Layzer D 10; 390
Learning **Gls**
   ability of 217
   memory 163; 217
   sleep 178
Lederman LM (Nobel laureate, 1985, physics) 38
LeDoux JE 171; 214
Lee DH 65
Leenders KL 131; 133; 141; 172; 193; 272
Left mind 307
Left-handedness
   Bush G, Clinton B, Perot R 190
Lehn JM (Nobel laureate,1987, chemistry) 340
Lehtmate J 217
Lem S 293; 386; 390
Lenat D 350
Leslie J 398
Lewontin RC 263
Liar 288
Libet B 326
Lie 287
   IQ test 265
   children 235
   information 319
   theory of 289
Lieberman P 231
Life
   carbon as carrier 58
   carrier, chemical force 52
   cosmic duration 52
   definition 58

emergence 59
energy and matter source 63
extraterrestrial, Mars 365
hydrogen as carrier 58
intelligence 56
nitrogen as carrier 59
oxygen as carrier 58
phosphorus as carrier 59
terrestriality 61
Life Satisfaction Scale 260
Life,
Lindahl T 147
Lindvall O. 272
Literacy 249
Living being
composition 43
definition 46
free energy 48
structural material 49
Lloyd S 17
Locke J 188; 287
Locus coeruleus
bonus-malus 109
psychoactive drug 175
Long-term potentiation 163
Lorenz EN 12
Lorenz K (Nobel laureate, 1973, physiology) 208
LSD, psychoactive agent 176; 291
LTP, long term potentiation 121; 140
Luria SE (Nobel laureate, 1969, physiology) 85; 207; 263
Lutz J 105

M

Magistretti PJ 145
Magnetoencephalography, MEG 136
Maimonides M 288
Malinov R 121
Malthus TR 386
Malus-bonus **Gls** 19
Manic-depressive illness 274
Mapping of brain **Gls** 131
Mars, meteorite and life traces 365
Martin KAC 111
Martin RD 73; 95
Martyrdom 280
Mathematical theorems 17
Matter, broad and narrow sense 7
Mayr E 389
McCarthy J 357
McGehee DS 171

McGrew WC 327
McLuhan HM 249
Medawar P (Nobel laureate, 1960, medicine) 278
Medium, message 249
Melzack R 317
Memory **Gls**
bonus-malus 169
emotional 171
explicit, implicit 163
forgetfulness 164
engram 167
evolution 163
information storage, memory 161
learning 163; 217
procedural 166
uncertainty 165
working, intuition 240
Memory loss, dementia 271
Mental **Gls**
abnormality, illness 272
faculties, loss 271
illnes 271
illness, 2020 year 272
object, 237
Mentalese, thinking 282
Metonymy 249
Michel Ch 135
Microtubule **Gls** 119
consciousness 339
Milky Way, distant future 376
Miller GA 246
Milner P 171
Mind **Gls**
excellent unification theory 399
immune system 144
theory of 354
Minsky M 80; 268; 276; 287; 303; 323; 350; 354; 358; 387
Mirenowicz J 142
Mishkin M 259
Mitochondrial Eve **Gls** 229
Mono-savants 268
Morality 221
intelligence 326
Moravec H 339; 346; 347; 364; 390
Morgan MJ 198
Morris JS 215; 250; 300
Mother tongue 234
Motors **Gls**
Mountcastle VB 16; 157; 277
Mousterian technology 91
MRI **Gls** 169

Multiplication, overabundant 18
Murray Ch 264
Myers DG 221
Myth 286
   self-consciousness, free will 287

N
NaDene people 96
Narcotic 140; 173
   bonus-malus 174
   culture 177
Natural selection **Gls** 93
Nature via nurture 280
Nature, or nurture 191; 278
Nature, via nurture 191
NC, network computer 353
NCC, neural correlate of consciousness 306
Neanderthals **Gls** 89
   lateralization 91
   speech 231
Nedergaard M 121
Neocortex, emergence 74
Neoteny **Gls** 85; 87; 232; 411
Nerve **Gls**
Nerve Growth Factor 125
Nervous system
   caenorhabditis elegans 68
   emergence 68
   disease 271
Network
   computer 353
   neuronal **Gls**
   plasticity 161
Neurogenetic determinism 279
Neuroglia 116
Neuron **Gls** 111
   action potential 121
   diversity 149
   life, death 124
   number, categories 114
   electrical properties 119
   energetics 134
   excitation, inhibition 117
   firing 121
   information processing 111
   microtubule 119
   non-spiking 113
   spiking 113
Neuronal messenger
   arachidonic acid 140
   nitric oxide, carbon monoxide 140
Neuronal networks 343
Neurosecretory cells 71

Neurotransmitter **Gls** 139
   acetylcholine, norepinephrine 140
   chemical weapon 141
   dopamine, GABA 140; 311
   dopamine, Parkinson's disease 272
   enkephalin 278
   hallucinogen 293
Nietzsche F 392
Nigel Th 350
Nobre AC 186
Non-self 276
Nottebohm F 194; 225
Nuclear war, future 386
Nucleus accumbens 105; 140; 175
Nurture, or nature 191; 278

O
Ockham's razor, intelligence 33
Olds J 171
Olson DR 248; 249; 289
Ontogeny **Gls** 103
Opiates, mimic by enkephalin 278
Optimism 222; 393
Orang-utan 217
Orgel LE 61
Out-of-Africa, origin 89
Overabundant production 18
Oxygen, origin 43

P
Pain - pleasure 19
Pain, representation 177
Pakkenberg B 196
Papert S 276
Paradox 283
Parapsychology 315
Parkinson's disease. 162; 272
Particle, superunified, elementary particles 38
Paul G 298
Peebles PJE 35
Pellerin L 145
Penrose R 9; 20; 35; 128; 155; 209; 220; 240; 282; 306; 308; 309; 339; 314; 359
Percept **Gls** 237
Personality **Gls** 274
   commissurotomy 277
   corpus callosum 277
   drugs, alcohol 294
   evolution stages 279
   fertilization 312
   lateralization 265

pathology 272
Perutz MF (Nobel laucrate, 1962, chemistry) 278; 387
Pessimism 393
PET **Gls** 179; 274
Peteresen SE 185
Petitto LA 235
Pfenninger A 207
Phantom seeing 317
  sounds 319
Phantomatics 387
Pheromone, communication 222
Phylogeny **Gls** 74
Physicality 3
Piaget J 275
Pich EM 291; 300
Pictures 'holy' 248
Pidgin, proto-language 224
Pinker S 213; 224; 231; 232; 244; 281; 283
Placebo **Gls** 177; 221; 280; 281
Planck M (Nobel laureate, 1918, physics) 262; 398
Planet
  extrasolar 365
  locations of intelligence 364; 365
Planning, multiple movements 188
Plasticity, brain 74; 150
Plautus 288; 401
Pleasure - pain 19; 21
Plomin R 278
PNA emergence of life 61
  peptide nucleic acid 61
Poincare H 297
Point of no return 385
  fifth kind 387
  first kind 383
  fourth kind 387
  second kind 384
  third kind 385
Pöppel E 155
Popper K 232; 264
Population bottleneck 95
Pornography 280
Posner MI 219
Preconcept 161
Prejudice 309
Premack D 216
Pre-writing 246
Prigogine I (Nobel laureate, 1977, chemistry) 305; 390
Prime number 285; 367
Prince A 354
Program, true, false 344

Prosopagnosia 169
Protein **Gls**
Proto-intelligence 239; 389
Proto-speech 89; 232
Prozac 177
Pseudoscience, science 288
Psilocybin 291
Psychedelic drugs, cave paintings 290
Psychedelics 173
Psychic disorders 271
Psychoactive agents, drugs **Gls** 175
Psychology, developmental 275
Psychons 305
Psychopathology 273
Psycho-physical laws 401
Psychosomatic effect 280
Pugh E 101
Pulvirenti L 294
Punishment - reward 19
Puzzle **Gls**
  real world 344
  jigsaw 296
  problems 344

Q

Qualia 240
Quantum computers 339
  mechanics **Gls** 35; 306
Question 241

R

Raff M.C. 126
Raichle ME 185; 186; 219
Rampino RM 95
Reading, writing 246
Real world, problems 344
Reality, virtual 345; 387
Reason **Gls** 205
Reeves H 385
Reflex **Gls** 75; 78
  artificial 346
  definition 78
  evolution 101; 210
  genetics 147
  occurrence 78; 412
  scheme 82
Regard M 192
Religion 287
REM sleep 177
Repertoire of performance 77
Representation, true **Gls** 23
Reptilian brain 128

Reward-punishment 19; 21
Riddle 296
Rifkin J 398
Right mind 307
RNA 61
Robin N 265
Robot **Gls**
   artificial reflex 346
   baby brain 351
   pain-pleasure 346
   common sense 351
Rolls ET 172; 325
Rose S 263; 279
Routtenberg A 142; 257
Rowe T 107
Royal Society (London) 288
Ruff CB 88; 89
Ruyter de RR 113

S

Sadato N 150
Saffran JR 219; 251
Sagan C 59
Salam A (Nobel laureate ,1979, physics) 34
Savage-Rumbaugh S 313
Schacter DL 312
Schank RC 209
Schiff SJ 136
Schillen TB 182
Schizophrenia **Gls** 150; 272; 273
   creative work 274
   diagnosis 273
   genetic impact 274
   memory 162
   pharmalogical treatment 273
Schlaug G 193; 270
School, intelligence 280
Schopf JW 61
Schramm DN 38
Schuler GD 152
Schulteis G 105
Schultz W 172; 334
Science 261
   future 398
   morality 327
   origin 262
   pseudoscience 288
   uncommon sense 262
Scott A 303
Scott SK 215
Search for extraterrestrial intelligence 368
Searle J 305; 324; 356

Seidenberg MS 251
Seitz RJ 107
Sejnowski TJ 114; 152; 177; 337; 354
Selective elimination 18
Self 276
   concept 302
Selfcognition, death own 331
Selfcognizability 3
Self-consciousness **Gls**
   free will 323
   definition 6; 302;303
   evolution of 314
   myths 287
Self-deception 332
   free will 303; 323
   illusion 180
Selfish gene 328
Selkoe DJ 109; 131
Semantics **Gls** 15
   information processing 345
   real world 78
   realworld 219
   understanding 219
Semaw S 88; 98
Sensor, detector **Gls** 13
Sergent J 270
SETI, see Search for extraterrestrial
   intelligence 368
Sex
   Barr-body 195
   brain differences 194
   brain lateralization 198
   brain mass 197
Sexual selection **Gls** 93; 94
Sexuality 280
Shannon CE 14; 16; 17; 20; 359
Shatz CJ 123
Shaywitz BA 197
Shirky C 338
Short-term memory
   synaptic connections 163
Sign language
   animal 226
Signal, definition **Gls** 12
Silbersweig DA 274
Simon HA Nobel laureate, 1978, economics
   258; 309; 346
Simpson GG 208
Singer S 85; 263
Singer W 149; 182; 184
Skills, artistic 270
Sleep **Gls** 177
   learning 178

REM 177
Smiling 275; 289
Smith EE 165
Smith SB 270
Smith-Churchland P 305
Snyder AZ 185
Snyder L 157
Snyder SH 291
Socialability 3
Software, human brain 102
Soul
 emergence 306
Speaker, and listener 241
Species **Gls** 66
Speech **Gls**
 area, handedness 198
 biological basis 228
 brain damage 243
 chimpanzee 216
 Cro-Magnon 96
 development at children 234
 genetics 229
 intelligence 222
 levels of 226
 Neanderthals 231
 origin of 229
 phonology, semantics 232
 private (children) 235
 vagueness 232
 writing 353
Sperry RW (Nobel laureate, 1981,medicine) 194; 305; 324; 327
Srinivasan MV 71
Star
 evolution 40
 mass and evolution 41
 neutronic 42
 supernova 41; 382
 three generations 42
Steinmetz H 105; 196
Steriade M 177
Sternberg EM 144
Sternberg RJ 207
Sternberg RJ 258
Stickgold R 317
Stonier T 10; 209
Stromswold K 235
Sub-intelligence 84; 214; 217; 239; 389
Sub-language, creole 224
Substantia nigra, Parkinson's disease 272
Suicide, Huntington's disease 272
Sun
 Black Dwarf 377
 distant future 378
 past, future 377
 Red Giant 379
 White Dwarf 377
Supercomputer, teraflops **Gls** 339
Superdiscent 84; 217
Superdrug 387; 386
Superintelligence **Gls**
 distant future 373
 science fiction 386
 supercomputer 390
 artificial 389; 393
Supernova, future catastrophe 382
Super-robot 395
Superunified force 36
Superunified theory 35; 393
Supplementary motor areas 188
Survival
 of the fittest 66
 of the luckiest 66
Symbol, speech 231
Symmetry
 beauty 296
 dual 35
Synapse **Gls** 111
 birth, death 127
 category, shape, number 116
 chemistry 139
 neuronal junction 117
 overabundant production 127
 selective selection 127
Synapsin 128
Synchronization **Gls** 184; 185
 nervous signals 184
Szent Gyorgyi A (Nobel laureate,1936, chemistry) 19

T

Taboo **Gls** 221; 280
Tanji J 188
Tarski A 285
Taube M 58; 367
Telepathy 315
Temporal lobe, pathology 271
Teracce H. 169
Terachips 341
Teraflops 339
Thalamus **Gls** 105
Theory of Everything 4; 6; 35; 401
Thieme H 88
Thinking **Gls**
 conscious 308

creative, sophisticated 266
   verbal 282
Thomas Aquinas St 287
tHooft G 16
Thorpe S 179
Thut G 172; 192
Time, internal representation 155
Tinbergen N (Nobel
   laureate,1973,physiology) 327
Tipler FJ 53
Tishkoff SA 89
Tobacco 174
Tomaso di E 293
Totality
   growth steadily 376
Truth, theory of 289
Tulving E 162
Turing AM 356
Turing Test 356
Turner G 386
Twins 280
   monozygotic, dizygotic 279

U

Uexküll von J 75
Ulam SM 285
Uncertainty, information processing 20
Unconsciousness **Gls**
Unhappiness 222
Unification theory **Gls** 399
Universal grammar 91; 191
Universe
   chemical composition 43
   closed 377
   distant future 376
   open 377

V

Vaadia E 183
Vague action 15
Vague signal 14; 15
Vagueness, information 28; 317
   information processing 20
Van Essen DC 117
Vandenberghe R 251
Vandermeersch B 89
Viaud G 81
Violent behaviour 331
Virtual reality **Gls** 345; 386
Visual illusion 180
   information, ambiguous 317
   information, vague 317

Vogel G 172
Vollmer G 268
Voltaire 287

W

Wald G 324
Wald G (Nobel laureate,1967,medicine) 305
War 329
Watanabe M 172
Weak Nuclear Force 39
Weaver W 17
Weinberg S (Nobel laureate ,1979, physics)
   35; 53
Weisberg R 287
Wernicke's area 110
   automatic speech 186
Wetware 339
Weyl H 297
White matter, brain **Gls** 116
Whittington MA 182
Wickelgreen I 162
Wiesel TN (Nobel laureate, 1981, physiology)
   110
Wigner EP (Nobel laureate, physics, 1960)
   260; 261
Wilczek F 8; 129
Wilson AC 95
Winson J 315; 317
Wisdom **Gls** 205; 258
Wise RA 142
Witten E 35
Wolpert L 391
Wood B 85; 98
Word
   processing 185
   read during life 246
   spoken during life 235
Working memory 168; 180; 309; 322
   language comprehension 16
   schizophrenia 274
World Health Organization, mental illnes 272
Wright FL 350
Writing
   Chinese, Japanese 247
   reading 246
   Sanskrit, Persian 246

Y

Yeh S-R 279
Young MP 169

Z

Zeki S 180
Zombies, consciousness 302

Zurek WH 15
Zwicky F 267